I0834697

# Ordinary and Partial Differential Equations

An Introduction to Dynamical Systems

John W. Cain, Ph.D. and Angela M. Reynolds, Ph.D.

Mathematics Textbook Series. Editor: Lon Mitchell

1. *Book of Proof* by Richard Hammack

2. *Linear Algebra* by Jim Hefferon

3. *Abstract Algebra: Theory and Applications* by Thomas Judson

4. *Ordinary and Partial Differential Equations* by John W. Cain and Angela M. Reynolds

Department of Mathematics & Applied Mathematics
Virginia Commonwealth University
Richmond, Virginia, 23284

Publication of this edition supported by the Center for Teaching Excellence at VCU

**Ordinary and Partial Differential Equations: An Introduction to Dynamical Systems**

Edition 1.0

© 2010 by John W. Cain and Angela Reynolds

This work is licensed under the Creative Commons Attribution-NonCommercial-No Derivative Works 3.0 License and is published with the express permission of the authors.

Typeset in 10pt Palladio L with Pazo Math fonts using PDFLaTeX

## Acknowledgements

John W. Cain expresses profound gratitude to his advisor, Dr. David G. Schaeffer, James B. Duke Professor of Mathematics at Duke University. The first five chapters are based in part upon Professor Schaeffer's introductory graduate course on ordinary differential equations. The material has been adapted to accommodate upper-level undergraduate students, essentially by omitting technical proofs of the major theorems and including additional examples. Other major influences on this book include the excellent texts of Perko [8], Strauss [10], and Strogatz [11]. In particular, the material presented in the last five chapters (including the ordering of the topics) is based heavily on Strauss' book. On the other hand, our exposition, examples, and exercises are more "user-friendly", making our text more accessible to readers with less background in mathematics.

Dr. Reynolds dedicates her portion of this textbook to her mother, father and sisters, she thanks them for all their support and love.

Finally, Dr. Cain dedicates his portion of this textbook to his parents Jeanette and Harry, who he loves more than words can express.

## Contents

# CHAPTER 1

# Introduction

The mathematical sub-discipline of *differential equations and dynamical systems* is foundational in the study of applied mathematics. Differential equations arise in a variety of contexts, some purely theoretical and some of practical interest. As you read this textbook, you will find that the qualitative and quantitative study of differential equations incorporates an elegant blend of linear algebra and advanced calculus. For this reason, it is expected that the reader has already completed courses in (i) linear algebra; (ii) multivariable calculus; and (iii) introductory differential equations. Familiarity with the following topics is especially desirable:

☞ From basic differential equations: separable differential equations and separation of variables; and solving linear, constant-coefficient differential equations using characteristic equations.

☞ From linear algebra: solving systems of $m$ algebraic equations with $n$ unknowns; matrix inversion; linear independence; and eigenvalues/eigenvectors.

☞ From multivariable calculus: parametrized curves; partial derivatives and gradients; and approximating a surface using a tangent plane.

Some of these topics will be reviewed as we encounter them later—in this chapter, we will recall a few basic notions from an introductory course in differential equations. Readers are encouraged to supplement this book with the excellent textbooks of Hubbard and West [5], Meiss [7], Perko [8], Strauss [10], and Strogatz [11].

**Question:** Why study differential equations?

**Answer:** When scientists attempt to mathematically model various natural phenomena, they often invoke physical "laws" or biological "principles" which govern the *rates of change* of certain quantities of interest. Hence, the equations in mathematical models tend to include derivatives. For example, suppose that a hot cup of coffee is placed in a room of constant ambient temperature $\alpha$. Newton's Law of Cooling states that the *rate of change* of the coffee temperature $T(t)$ is proportional to the difference between the coffee's temperature and the room temperature. Mathematically, this can be expressed as $\frac{dT}{dt} = k(T - \alpha)$, where $k$ is a proportionality constant.

Solution techniques for differential equations (DEs) depend in part upon how many independent variables and dependent variables the system has.

**Example 1.0.1.** One independent variable and one independent variable. In writing the equation

$$\frac{d^2y}{dx^2} + \cos(xy) = 3,$$

it is understood that $y$ is the dependent variable and $x$ is the independent variable.

When a differential equation involves a single independent variable, we refer to the equation as an *ordinary differential equation (ODE)*.

**Example 1.0.2.** If there are several dependent variables and a single independent variable, we might have equations such as

$$\frac{dy}{dx} = x^2y - xy^2 + z, \qquad \frac{dz}{dx} = z - y\cos x.$$

This is a *system* of two ODEs, and it is understood that $x$ is the independent variable.

**Example 1.0.3.** One dependent variable, several independent variables. Consider the DE

$$\frac{\partial u}{\partial t} = \frac{\partial^2 u}{\partial x^2} + \frac{\partial^2 u}{\partial y^2}.$$

This equation involves three independent variables ($x$, $y$, and $t$) and one dependent variable, $u$. This is an example of a *partial differential equation (PDE)*. If there are several independent variables and several dependent variables, one may have systems of PDEs.

Although these concepts are probably familiar to the reader, we give a more exact definition for what we mean by ODE. Suppose that $x$ and $y$ are independent and dependent variables, respectively, and let $y^{(k)}(x)$ denote the $k$th derivative of $y$ with respect to $x$. (If $k \leq 3$, we will use primes.)

**Definition 1.0.4.** Any equation of the form $F(x, y, y', y'', \ldots, y^{(n)}) = 0$ is called an *ordinary differential equation*. If $y^{(n)}$ is the highest derivative appearing in the equation, we say that the ODE is of *order n*.

**Example 1.0.5.**

$$\left(\frac{d^3y}{dx^3}\right)^2 - (\cos x)\frac{dy}{dx} = y\frac{d^2y}{dx^2}$$

can be written as $(y''')^2 - yy'' - (\cos x)y' = 0$, so using the notation in the above Definition, we would have $F(x, y, y', y'', y''') = (y''')^2 - yy'' - (\cos x)y'$. This is a third-order ODE.

**Definition 1.0.6.** A *solution* of the ODE $F(x, y, y', y'', \ldots, y^{(n)}) = 0$ on an interval $I$ is any function $y(x)$ which is $n$-times differentiable and satisfies the equation on $I$.

**Example 1.0.7.** For any choice of constant $A$, the function

$$y(x) = \frac{Ae^x}{1 + Ae^x}$$

is a solution of the first-order ODE $y' = y - y^2$ for all real $x$. To see why, we use the quotient rule to calculate

$$y' = \frac{Ae^x(1 + Ae^x) - (Ae^x)^2}{(1 + Ae^x)^2} = \frac{Ae^x}{(1 + Ae^x)^2}.$$

By comparison, we calculate that

$$y - y^2 = \frac{Ae^x}{(1 + Ae^x)} - \frac{(Ae^x)^2}{(1 + Ae^x)^2} = \frac{Ae^x}{(1 + Ae^x)^2}.$$

Therefore, $y' = y - y^2$, as claimed.

The definition of a solution of an ODE is easily extended to systems of ODEs (see below). In what follows, we will focus solely on systems of *first-order* ODEs. This may seem overly restrictive, until we make the following observation.

**Observation.** Any $n$th-order ODE can be written as a system of $n$ first-order ODEs. The process of doing so is straightforward, as illustrated in the following example:

**Example 1.0.8.** Consider the second-order ODE $y'' + (\cos x)y' + y^2 = e^x$. To avoid using second derivatives, we introduce a new dependent variable $z = y'$ so that $z' = y''$. Our ODE can be re-written as $z' + (\cos x)z + y^2 = e^x$. Thus, we have obtained a system of two first-order ODEs:

$$\frac{dy}{dx} = z, \qquad \frac{dz}{dx} = -(\cos x)z - y^2 + e^x.$$

A *solution* of the above system of ODEs on an open interval $I$ is any *vector* of differentiable functions $[y(x), z(x)]$ which simultaneously satisfy both ODEs when $x \in I$.

**Example 1.0.9.** Consider the system

$$\frac{dy}{dt} = z, \qquad \frac{dz}{dt} = -y.$$

We claim that for any choices of constants $C_1$ and $C_2$,

$$\begin{bmatrix} y(t) \\ z(t) \end{bmatrix} = \begin{bmatrix} C_1 \cos t + C_2 \sin t \\ -C_1 \sin t + C_2 \cos t \end{bmatrix}$$

is a solution of the system. To verify this, assume that $y$ and $z$ have this form. Differentiation reveals that $y' = -C_1 \sin t + C_2 \cos t$ and $z' = -C_1 \cos t - C_2 \sin t$. Thus, $y' = z$ and $z' = -y$, as required.

---

## 1.1. Initial and Boundary Value Problems

In the previous example, the solution of the system of ODEs contains arbitrary constants $C_1$ and $C_2$. Therefore, the system has infinitely many solutions. In practice, one often has additional information about the underlying system, allowing us to select a particular solution of practical interest. For example, suppose that a cup of coffee is cooling off and obeys Newton's Law of Cooling. In order to predict the coffee's temperature at future times, we would need to specify the temperature of the coffee at some reference time (usually considered to be the "initial" time). By specifying auxiliary conditions that solutions of an

ODE must satisfy, we may be able to single out a *particular* solution. There are two usual ways of specifying auxiliary conditions.

**Initial conditions.** Suppose $F(x, y, y', y'', \ldots, y^{(n)}) = 0$ is an $n$th order ODE which has a solution on an open interval $I$ containing $x = x_0$. Recall from your course on basic differential equations that, under reasonable assumptions, we would expect the general solution of this ODE to contain $n$ arbitrary constants. One way to eliminate these constants and single out one particular solution is to specify $n$ *initial conditions*. To do so, we may specify values for

$$y(x_0),\ y'(x_0),\ y''(x_0),\ \ldots\ y^{(n-1)}(x_0).$$

We regard $x_0$ as representing some "initial time". An ODE together with its initial conditions (ICs) forms an *initial value problem (IVP)*. Usually, initial conditions will be specified at $x_0 = 0$.

**Example 1.1.1.** Consider the second-order ODE $y''(x) + y(x) = 0$. You can check that the general solution is $y(x) = C_1 \cos x + C_2 \sin(x)$, where $C_1$ and $C_2$ are arbitrary constants. To single out a particular solution, we would need to specify two initial conditions. For example, if we require that $y(0) = 1$ and $y'(0) = 0$, we find that $C_1 = 1$ and $C_2 = 0$. Hence, we obtain a particular solution $y(x) = \cos x$.

If we have a system of $n$ first-order ODEs, we will specify one initial condition for each independent variable. If the dependent variables are

$$y_1(x), y_2(x), \ldots y_n(x),$$

we typically specify the values of

$$y_1(0), y_2(0), \ldots, y_n(0).$$

**Boundary conditions.** Instead of specifying requirements that $y$ and its derivatives must satisfy at *one* particular value of the independent variable $x$, we could instead impose requirements on $y$ and its derivatives at *different* $x$ values. The result is called a *boundary value problem (BVP)*.

**Example 1.1.2.** Consider the boundary value problem $y'' + y = 0$ with boundary conditions $y(0) = 1$ and $y(\pi/2) = 0$. The general solution of the ODE is $y(x) = C_1 \cos x + C_2 \sin x$. Using the first boundary condition, we find that

$C_1 = 1$. Since $y'(x) = -C_1 \sin x + C_2 \cos x$, the second boundary condition tells us that $-C_1 = 0$. Notice that the two boundary conditions produce conflicting requirements on $C_1$. Consequently, the BVP has no solutions.

As the previous example suggests, boundary value problems can be a tricky matter. In the ODE portion of this text, we consider only initial value problems.

---

## Exercises

1. Write the equation of the line that passes through the points $(-1, 2, 3)$ and $(4, 0, -1)$ in $\mathbb{R}^3$, three-dimensional Euclidean space.
2. Find the general solution of the differential equation

$$\frac{d^3y}{dx^3} + 2\frac{d^2y}{dx^2} + 5\frac{dy}{dx} = 0.$$

3. Find the general solution of the differential equation

$$\frac{d^2y}{dx^2} + 6\frac{dy}{dx} + 9y = 0.$$

4. Solve the IVP

$$y'' - 3y' + 2y = 0, \quad y(0) = 1, \quad y'(0) = 1.$$

5. Solve (if possible) the BVP

$$y'' - 3y' + 2y = 0, \quad y(0) = 0, \quad y(1) = e.$$

6. Solve the IVP

$$y^{(4)} - y'' = 0, \qquad \begin{array}{ll} y(0) = 1, & y'(0) = 0, \\ y''(0) = -1, & y'''(0) = 0. \end{array}$$

7. Solve the differential equation

$$\frac{dy}{dx} = (y+2)(y+1).$$

8. Solve the IVP

$$\frac{dy}{dx} = e^y \sin x, \qquad y(0) = 0.$$

9. Find the equations of the planes tangent to the surface

$$z = f(x,y) = x^2 - 2x + y^2 - 2y + 2$$

at the points $(x,y,z) = (1,1,0)$ and $(x,y,z) = (0,2,2)$.

10. Find the eigenvalues of the matrix

$$A = \begin{bmatrix} 1 & 4 \\ 4 & 1 \end{bmatrix}$$

and, for each eigenvalue, find a corresponding eigenvector.

11. Find the eigenvalues of the matrix

$$A = \begin{bmatrix} 1 & 3 & -1 \\ 0 & 3 & 0 \\ 0 & 1 & 2 \end{bmatrix}$$

and, for each eigenvalue, find a corresponding eigenvector.

12. Write the following differential equations as systems of first-order ODEs:

$$y'' - 5y' + 6y = 0$$
$$-y'' - 2y' = 7\cos(y')$$
$$y^{(4)} - y'' + 8y' + y^2 = e^x.$$

# CHAPTER 2

## Linear, Constant-Coefficient Systems

There are few classes of ODEs for which exact, analytical solutions can be obtained by hand. However, for many systems which cannot be solved explicitly, we may approximate the dynamics by using simpler systems of ODEs which can be solved exactly. This often allows us to extract valuable qualitative information about complicated dynamical systems. We now introduce techniques for systematically solving linear systems of first-order ODEs with constant coefficients.

**Notation.** Because we will be working with vectors of dependent variables, we should establish (or recall) some commonly used notation. We denote the set of real numbers by $\mathbb{R}$. We let $\mathbb{R}^n$ denote the set of all vectors with $n$ components, each of which is a real number. Usually, vectors will be denoted by bold letters such as $\mathbf{x}$, $\mathbf{y}$, and we will use capital letters such as $A$ to denote $n \times n$ matrices of real numbers. Generally, we shall not distinguish between row vectors and column vectors, as our intentions will usually be clear from the context. For example, if we write the product $\mathbf{x}A$, then $\mathbf{x}$ should be treated as a row vector, whereas if we write $A\mathbf{x}$, then $\mathbf{x}$ is understood to be a column vector. If we write $\mathbf{x}(t)$, we mean a vector of functions, each of which depends on a variable $t$. In such cases, the vector $\mathbf{x}(0)$ would be a constant vector in which each component function has been evaluated at $t = 0$. Moreover, the vector $\mathbf{x}'(t)$ is the vector consisting of the derivatives of the functions which form the components of $\mathbf{x}(t)$.

**Systems with constant coefficients.** Suppose that $y_1, y_2, \ldots y_n$ are variables which depend on a single variable $t$. The general form of a linear, constant-coefficient system of first-order ODEs is as follows:

$$\begin{aligned}\frac{dy_1}{dt} &= a_{11}y_1(t) + a_{12}y_2(t) + \cdots + a_{1n}y_n(t) + f_1(t)\\ \frac{dy_2}{dt} &= a_{21}y_1(t) + a_{22}y_2(t) + \cdots + a_{2n}y_n(t) + f_2(t)\\ &\vdots\\ \frac{dy_n}{dt} &= a_{n1}y_1(t) + a_{n2}y_2(t) + \cdots + a_{nn}y_n(t) + f_n(t).\end{aligned} \tag{2.1}$$

Here, each $a_{ij}$ is a constant $(1 \leq i,j \leq n)$, and $f_i(t)$ $(i = 1, 2, \ldots n)$ are functions of $t$ only.

**Example 2.0.3.** Soon, we will learn how to solve the linear, constant-coefficient system

$$\begin{aligned}\frac{dy_1}{dt} &= 3y_1 - 2y_2 + \cos t\\ \frac{dy_2}{dt} &= 10y_2 - t^2 + 6.\end{aligned} \tag{2.2}$$

The system (2.1) can be written more compactly if we introduce matrix/vector notation. Suppressing the dependence on $t$ for notational convenience, routine matrix multiplication will verify that

$$\begin{bmatrix} y_1' \\ y_2' \\ \vdots \\ y_n' \end{bmatrix} = \begin{bmatrix} a_{11} & a_{12} & \cdots & a_{1n} \\ a_{21} & a_{22} & \cdots & a_{2n} \\ \vdots & \vdots & \ddots & \vdots \\ a_{n1} & a_{n2} & \cdots & a_{nn} \end{bmatrix} \begin{bmatrix} y_1 \\ y_2 \\ \vdots \\ y_n \end{bmatrix} + \begin{bmatrix} f_1 \\ f_2 \\ \vdots \\ f_n \end{bmatrix} \tag{2.3}$$

is equivalent to the system (2.1). Furthermore, if we define

$$\mathbf{y} = \begin{bmatrix} y_1 \\ y_2 \\ \vdots \\ y_n \end{bmatrix} \qquad \mathbf{y}' = \begin{bmatrix} y_1' \\ y_2' \\ \vdots \\ y_n' \end{bmatrix} \qquad \mathbf{f} = \begin{bmatrix} f_1 \\ f_2 \\ \vdots \\ f_n \end{bmatrix} \tag{2.4}$$

and let $A$ denote the matrix of coefficients in Equation (2.3), then the entire system takes the form

$$\mathbf{y}' = A\mathbf{y} + \mathbf{f}. \tag{2.5}$$

**Definition 2.0.4.** The system of ODEs in Equation (2.5) is called *homogeneous* if the vector $\mathbf{f}$ is the zero vector. Otherwise, the system is *inhomogeneous*.

**Example 2.0.5.** The system

$$\frac{dy_1}{dt} = 5y_1 - y_2 \qquad \frac{dy_2}{dt} = -y_1 + 8y_2$$

is homogeneous, whereas the system (2.2) above is inhomogeneous. Notice that we may write (2.2) in matrix notation as

$$\begin{bmatrix} y_1' \\ y_2' \end{bmatrix} = \begin{bmatrix} 3 & -2 \\ 0 & 10 \end{bmatrix} \begin{bmatrix} y_1 \\ y_2 \end{bmatrix} + \begin{bmatrix} \cos t \\ -t^2 + 6 \end{bmatrix}.$$

**Road map of things to come.** In the next subsection, we will learn how to solve *homogeneous* constant-coefficient systems $\mathbf{y}' = A\mathbf{y}$. This will require lots of linear algebra! Next, we will learn how to solve *inhomogeneous* systems of the form (2.5). Finally, the next several chapters will be devoted to understanding how to qualitatively analyze solutions of *nonlinear* systems (which generally cannot be solved by hand). This will be accomplished by approximating nonlinear systems with linear ones of the form (2.5).

---

## 2.1. Homogeneous Systems

In order to motivate the techniques we will use to solve homogeneous systems $\mathbf{y}' = A\mathbf{y}$, we draw an analogy with a simple, first-order initial value problem.

**Motivating example.** Consider the ODE $\frac{dy}{dt} = ay$ where $a$ is a constant, and suppose we have an initial condition $y(0) = y_0$. This initial value problem is easy to solve using techniques from your first course in differential equations. The solution is $y(t) = e^{at}y_0$. Notice that the solution involves an exponential function.

**Question:** Can we extend this example to homogeneous, constant-coefficient systems of equations?

More specifically, consider the system $\mathbf{y}' = A\mathbf{y}$, where $\mathbf{y} = \mathbf{y}(t)$ is a vector of length $n$ and $A$ is an $n \times n$ constant matrix. Suppose we form an initial value problem by specifying a vector of initial conditions; i.e., we assume that

$$\mathbf{y}_0 = \begin{bmatrix} y_1(0) \\ y_2(0) \\ \vdots \\ y_n(0) \end{bmatrix}$$

is given. The resulting IVP has the form $\mathbf{y}' = A\mathbf{y}$, $\mathbf{y}(0) = \mathbf{y}_0$, which resembles the form of the IVP in the motivating example. By analogy with the above system, can we say that the solution of our new IVP is given by $\mathbf{y}(t) = \mathrm{e}^{tA}\mathbf{y}_0$? If so, we would need some way of assigning meaning to the object $\mathrm{e}^{tA}$ where $A$ is a *matrix*. Certainly $\mathrm{e}^{tA}$ would need to be an $n \times n$ matrix as well, because both $\mathbf{y}(t)$ and $\mathbf{y}_0$ are vectors in $\mathbb{R}^n$. Fortunately, for square matrices such as $A$, there is a natural way to assign meaning to $\mathrm{e}^A$.

**Definition 2.1.1.** Suppose $A$ is an $n \times n$ constant matrix. The *matrix exponential* $\mathrm{e}^A$ is defined in terms of the Maclaurin series expansion of the usual exponential function. That is,

$$\mathrm{e}^A = \sum_{k=0}^{\infty} \frac{1}{k!} A^k, \tag{2.6}$$

which is a sum involving powers of the matrix $A$.

We make several remarks about this definition:

☞ For positive integers $k$, recall that $k!$ is read "$k$ factorial" and is defined to be the product of the first $k$ natural numbers: $k(k-1)(k-2)\cdots(3)(2)(1)$. Additionally, we define $0! = 1$.

☞ If $A$ is a matrix, then $A^0 = I$, the identity matrix. (This is analogous to the fact that $a^0 = 1$ for scalars $a$.)

☞ If $t$ is a scalar, then $tA$ is matrix, so

$$\mathrm{e}^{tA} = \sum_{k=0}^{\infty} \frac{1}{k!}(tA)^k = \sum_{k=0}^{\infty} \frac{t^k}{k!} A^k.$$

**Example 2.1.2.** Suppose $A$ is the $2 \times 2$ matrix consisting entirely of zeros. Then $A^0 = I$, and for each $k \geq 1$ we have

$$A^k = \begin{bmatrix} 0 & 0 \\ 0 & 0 \end{bmatrix}.$$

Therefore, $e^{tA} = I$ because the only non-zero term in the series expansion (2.6) is the one corresponding to $k = 0$. Notice the parallel between the fact that $e^0 = 1$ for scalars and the fact that the exponential of the zero matrix is the identity matrix.

As we shall soon see, every constant, square matrix $A$ has a matrix exponential $e^A$. Moreover, we will find that the solution of the initial value problem $\mathbf{y}' = A\mathbf{y}$, $\mathbf{y}(0) = \mathbf{y}_0$ really *is* $\mathbf{y}(t) = e^{tA}\mathbf{y}_0$, just as we would hope. Hence, our main immediate challenge is to devise a procedure for calculating the matrix exponential $e^{tA}$ for an arbitrary square matrix $A$. Because $e^{tA}$ is defined by its series representation, we need a method for computing powers of $A$, which will require us to review some facts from linear algebra.

**2.1.1 Diagonalizable Matrices.** Before we recall what it means for a matrix to be *diagonalizable*, we consider a very simple class of matrices: the diagonal matrices. A matrix is *diagonal* if all of the entries off of the main diagonal are 0. In other words, $a_{ij} = 0$ whenever $i \neq j$. When writing diagonal matrices, we typically do not write the entries which lie off the main diagonal, as it is understood that all of these entries are 0.

**Diagonal matrices.** Suppose $D$ is an $n \times n$ diagonal matrix

$$D = \begin{bmatrix} d_1 & & & \\ & d_2 & & \\ & & \ddots & \\ & & & d_n \end{bmatrix},$$

which we will sometimes denote by $D = \text{diag}\{d_1, d_2, \ldots d_n\}$. Straightforward induction verifies that powers of $D$ are given by

$$D^k = \begin{bmatrix} d_1^k & & & \\ & d_2^k & & \\ & & \ddots & \\ & & & d_n^k \end{bmatrix} = \text{diag}\{d_1^k, d_2^k, \ldots d_n^k\}.$$

According to the series representation for the matrix exponential, we have

$$e^{tD} = \sum_{k=0}^{\infty} \frac{t^k}{k!} D^k = \begin{bmatrix} \sum_{k=0}^{\infty} \frac{t^k}{k!} d_1^k & & & \\ & \sum_{k=0}^{\infty} \frac{t^k}{k!} d_2^k & & \\ & & \ddots & \\ & & & \sum_{k=0}^{\infty} \frac{t^k}{k!} d_n^k \end{bmatrix}.$$

The entries in this matrix are simply the Maclaurin series representations for the functions $e^{d_1 t}, e^{d_2 t}, \ldots e^{d_n t}$. Therefore, we have shown that for the diagonal matrix $D$,

$$e^{tD} = \begin{bmatrix} e^{d_1 t} & & & \\ & e^{d_2 t} & & \\ & & \ddots & \\ & & & e^{d_n t} \end{bmatrix}.$$

**Example 2.1.3.** Consider the initial value problem $\frac{dx}{dt} = 6x$, $\frac{dy}{dt} = -3y$, $x(0) = 2$, and $y(0) = 1$. These ODEs are *uncoupled* in the sense that they are effectively independent of one another—changes in $x(t)$ have no impact on $y(t)$ and vice-versa. In matrix notation, this system takes the form

$$\begin{bmatrix} x' \\ y' \end{bmatrix} = \begin{bmatrix} 6 & 0 \\ 0 & -3 \end{bmatrix} \begin{bmatrix} x \\ y \end{bmatrix}, \qquad \begin{bmatrix} x(0) \\ y(0) \end{bmatrix} = \begin{bmatrix} 2 \\ 1 \end{bmatrix}.$$

Since the coefficient matrix $D$ is diagonal, we immediately have

$$e^{tD} = \begin{bmatrix} e^{6t} & 0 \\ 0 & e^{-3t} \end{bmatrix}.$$

Multiplying the matrix exponential by the vector of initial conditions should give the solution of the initial value problem:

$$\begin{bmatrix} x(t) \\ y(t) \end{bmatrix} = e^{tD} \begin{bmatrix} x(0) \\ y(0) \end{bmatrix} = \begin{bmatrix} e^{6t} & 0 \\ 0 & e^{-3t} \end{bmatrix} \begin{bmatrix} 2 \\ 1 \end{bmatrix} = \begin{bmatrix} 2e^{6t} \\ e^{-3t} \end{bmatrix}.$$

You can verify that $x(t) = 2e^{6t}$, $y(t) = e^{-3t}$ is, indeed, a solution of this IVP.

Obviously the diagonal matrices form a very narrow class of matrices, and most of the systems of ODEs we wish to consider will be coupled. After the diagonal matrices, the next easiest class of matrices to exponentiate are the *diagonalizable* matrices. Before defining what this means, we recall one other definition.

**Definition 2.1.4.** Two $n \times n$ matrices $A$ and $B$ are called *similar* if there exists an invertible matrix $P$ such that $A = PBP^{-1}$.

Note that every matrix $A$ is similar to itself (just take $P = I$, the identity matrix). It is also easy to show that if $A$ is similar to $B$, then $B$ is similar to $A$. Finally, if $A$ is similar to $B$ and $B$ is similar to $C$, then $A$ is similar to $C$. In other words, similarity of matrices forms an equivalence relation, and matrices can be partitioned into equivalence classes according to similarity.

**Definition 2.1.5.** A square matrix $A$ is called *diagonalizable* if it is similar to a diagonal matrix.

Above, we implied that diagonalizable matrices are "nice" because it is straightforward to exponentiate such matrices. To see why, suppose that $A$ is diagonalizable. Then there exists a diagonal matrix $D$ and an invertible matrix $P$ such that $A = PDP^{-1}$. It follows that $A^2 = (PDP^{-1})(PDP^{-1})$. Since matrix multiplication is associative (although NOT commutative), we may regroup the terms in this product as long as we preserve their ordering:

$$A^2 = PD(P^{-1}P)DP^{-1} = PDIDP^{-1} = PD^2P^{-1}.$$

The same idea can be extended to higher powers of $A$:

$$A^n = \underbrace{(PDP^{-1})(PDP^{-1})\cdots(PDP^{-1})}_{\text{n times}} = PD(P^{-1}P)D(P^{-1}P)\cdots(P^{-1}P)DP^{-1},$$

which simplifies to $PD^nP^{-1}$. Recalling the series expansion for $e^{tA}$, we have

$$e^{tA} = \sum_{k=0}^{\infty} \frac{t^k}{k!}A^k = \lim_{n\to\infty} \sum_{k=0}^{n} \frac{t^k}{k!}A^k.$$

Replacing $A^k$ with $PD^kP^{-1}$, the series becomes

$$e^{tA} = \lim_{n\to\infty} \sum_{k=0}^{n} \frac{t^k}{k!} PD^kP^{-1}.$$

Since each term in the (finite) sum is pre-multiplied by $P$ and post-multiplied by $P^{-1}$, we may use associativity of matrix multiplication to write

$$e^{tA} = P\left(\lim_{n\to\infty} \sum_{k=0}^{n} \frac{t^k}{k!} D^k\right) P^{-1} = P\left(\sum_{k=0}^{\infty} \frac{t^k}{k!} D^k\right) P^{-1} = Pe^{tD}P^{-1}. \qquad (2.7)$$

**Nice consequence:** Formula (2.7) suggests a convenient procedure for exponentiating a diagonalizable matrix $A$. First, find a diagonal matrix $D$ and an invertible matrix $P$ such that $A = PDP^{-1}$. After decomposing $A$ in this way, it follows that $e^{tA} = Pe^{tD}P^{-1}$, and $e^{tD}$ is easy to compute. Of course, the challenging part is finding the matrices $D$ and $P$, a task that we shall undertake shortly.

Before giving a general procedure for exponentiating diagonalizable matrices, we address a more basic question, namely, "Which matrices are diagonalizable?" This question, which we answer in Theorem 2.1.7 below, will require us to recall a definition from basic linear algebra:

**Definition 2.1.6.** Suppose $A$ is an $n \times n$ matrix. A scalar $\lambda$ is called an *eigenvalue* for $A$ if there exists a *non-zero* vector $\mathbf{v}$ such that $A\mathbf{v} = \lambda\mathbf{v}$. Any non-zero vector $\mathbf{v}$ satisfying this equality is called an *eigenvector* corresponding to the eigenvalue $\lambda$.

**Theorem 2.1.7.** An $n \times n$ matrix $A$ is diagonalizable if and only if there exists a set of $n$ linearly independent eigenvectors for $A$.

*Proof.* See your linear algebra textbook. One direction of the proof is fairly straightforward: If there are $n$ linearly independent eigenvectors, form the matrix $P$ by letting the columns of $P$ be the eigenvectors. Then, show that $A = PDP^{-1}$, where $D$ is a diagonal matrix whose diagonal entries are the eigenvalues. □

It is important to note that eigenvalues need not be real numbers, and that eigenvectors may contain complex entries. In what follows, we focus on matrices which can be diagonalized over the *real* number field. That is, whenever we write $A = PDP^{-1}$ where $D$ is diagonal, we will insist that both $D$ and $P$ have

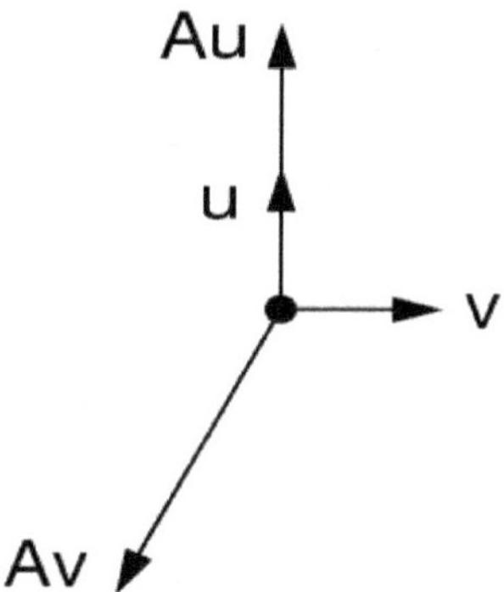


**Figure 2.1.** The vector $\mathbf{u}$ is an eigenvector for $A$ because the vectors $\mathbf{u}$ and $A\mathbf{u}$ span the same line. However, $\mathbf{v}$ is not an eigenvector because the lines spanned by $\mathbf{v}$ and $A\mathbf{v}$ are different.

*real* entries. Later, when we study matrices with complex eigenvalues, we will adopt a very different approach to diagonalization.

Geometrically, non-zero real eigenvalues $\lambda$ and their corresponding eigenvectors are easily understood. If $\lambda > 0$ and $\mathbf{v}$ is a corresponding eigenvector, then the product $A\mathbf{v} = \lambda\mathbf{v}$ effectively stretches $\mathbf{v}$ without altering its direction. Likewise, if $\lambda < 0$, then $A\mathbf{v} = \lambda\mathbf{v}$ stretches the vector $\mathbf{v}$ by a factor $|\lambda|$ and reverses its orientation. Importantly, the lines in $\mathbb{R}^n$ spanned by eigenvectors are *unaffected* by multiplication by the matrix $A$. This is what makes eigenvectors very special: whereas most vectors are both stretched *and* rotated via matrix-vector multiplication, eigenvectors have their directions preserved.

**Example 2.1.8.** If

$$A = \begin{bmatrix} -1 & 0 \\ -2 & 2 \end{bmatrix}, \qquad \mathbf{u} = \begin{bmatrix} 0 \\ 1 \end{bmatrix}, \quad \text{and} \quad \mathbf{v} = \begin{bmatrix} 1 \\ 0 \end{bmatrix},$$

then

$$A\mathbf{u} = \begin{bmatrix} 0 \\ 2 \end{bmatrix} \quad \text{and} \quad A\mathbf{v} = \begin{bmatrix} -1 \\ -2 \end{bmatrix}.$$

Observe that $A\mathbf{u} = 2\mathbf{u}$, which means that $\mathbf{u}$ is an eigenvector and $\lambda = 2$ is its corresponding eigenvalue. Multiplication by $A$ stretches $\mathbf{u}$ by a factor of 2 and preserves its orientation. On the other hand, $\mathbf{v}$ is not an eigenvector, because $A\mathbf{v}$ is not simply a scalar multiple of $\mathbf{v}$ (see Figure 2.1).

Calculation of eigenvalues is typically done as follows. If we seek a non-zero vector $\mathbf{v}$ satisfying a relationship $A\mathbf{v} = \lambda\mathbf{v}$ for some scalar $\lambda$, then $A\mathbf{v} - \lambda I\mathbf{v} = 0$. Equivalently, $(A - \lambda I)\mathbf{v} = 0$. Since $\mathbf{v} \neq 0$, it must be the case that the matrix $A - \lambda I$ is singular which, in turn, implies that its determinant is 0. Therefore, the eigenvalues must satisfy the equation $\det(A - \lambda I) = 0$. Once we have the eigenvalues, finding eigenvectors is straightforward, and then the proof of Theorem 2.1.7 suggests a procedure for diagonalizing $A$ (see below).

**Definition 2.1.9.** Given a square matrix $A$, the equation $\det(A - \lambda I) = 0$ is called the *characteristic equation* of $A$. The expression $\det(A - \lambda I)$ is a polynomial in the variable $\lambda$, and is called the *characteristic polynomial* of $A$.

**How to diagonalize a (diagonalizable) matrix.** We will illustrate this method via an example, using

$$A = \begin{bmatrix} 1 & -1 \\ 2 & 4 \end{bmatrix}.$$

**Step 1:** Find the eigenvalues of $A$ by solving the characteristic equation. Since

$$A - \lambda I = \begin{bmatrix} 1 & -1 \\ 2 & 4 \end{bmatrix} - \begin{bmatrix} \lambda & 0 \\ 0 & \lambda \end{bmatrix} = \begin{bmatrix} 1-\lambda & -1 \\ 2 & 4-\lambda \end{bmatrix},$$

the characteristic polynomial $p(\lambda)$ is given by

$$p(\lambda) \;=\; \det(A - \lambda I) \;=\; (1-\lambda)(4-\lambda) + 2 \;=\; (\lambda - 2)(\lambda - 3).$$

The roots of this polynomial, $\lambda = 2$ and $\lambda = 3$, are the eigenvalues of $A$.

**Step 2:** For each eigenvalue, find a maximal linearly independent set of eigenvectors. For a given eigenvalue $\lambda$, if you convert the matrix $(A - \lambda I)$ to reduced row-echelon form, the number of all-zero rows tells you how many linearly independent eigenvectors you must produce. Let's start with $\lambda = 2$. Eigenvectors satisfy the equation $(A - \lambda I)\mathbf{v} = 0$ which, in this case, means

$$\begin{bmatrix} -1 & -1 \\ 2 & 2 \end{bmatrix} \begin{bmatrix} v_1 \\ v_2 \end{bmatrix} = \begin{bmatrix} 0 \\ 0 \end{bmatrix}.$$

Row-reducing this linear, homogeneous system yields

$$\begin{bmatrix} 1 & 1 \\ 0 & 0 \end{bmatrix} \begin{bmatrix} v_1 \\ v_2 \end{bmatrix} = \begin{bmatrix} 0 \\ 0 \end{bmatrix}.$$

Remark: You should *always* obtain at least one row of zeros when row-reducing $A - \lambda I$. After all, we knew in advance that $\det(A - \lambda I) = 0$. Since our reduced matrix has one row of zeros, we have one free variable: let us choose $v_2$. Expanding the system, we have $v_1 + v_2 = 0$, which means $v_1 = -v_2$. The set of solutions is therefore

$$\begin{bmatrix} v_1 \\ v_2 \end{bmatrix} = \begin{bmatrix} -v_2 \\ v_2 \end{bmatrix} = v_2 \begin{bmatrix} -1 \\ 1 \end{bmatrix},$$

where $v_2$ is any real constant. It follows that

$$\begin{bmatrix} -1 \\ 1 \end{bmatrix}$$

is an eigenvector corresponding to $\lambda = 2$ (as is any non-zero multiple of this vector). For $\lambda = 3$, the same procedure produces another eigenvector. In this case, we must solve $(A - 3I)\mathbf{v} = 0$, which gives

$$\begin{bmatrix} -2 & -1 \\ 2 & 1 \end{bmatrix} \begin{bmatrix} v_1 \\ v_2 \end{bmatrix} = \begin{bmatrix} 0 \\ 0 \end{bmatrix}.$$

Row-reducing as before,

$$\begin{bmatrix} 1 & \frac{1}{2} \\ 0 & 0 \end{bmatrix} \begin{bmatrix} v_1 \\ v_2 \end{bmatrix} = \begin{bmatrix} 0 \\ 0 \end{bmatrix}.$$

Expanding the last equation reveals that $v_1 + \frac{1}{2}v_2 = 0$. Treating $v_2$ as the free variable, we write $v_1 = -\frac{1}{2}v_2$ so that

$$\begin{bmatrix} v_1 \\ v_2 \end{bmatrix} = \begin{bmatrix} -\frac{1}{2}v_2 \\ v_2 \end{bmatrix} = v_2 \begin{bmatrix} -\frac{1}{2} \\ 1 \end{bmatrix}.$$

Thus, any non-zero scalar multiple of

$$\begin{bmatrix} -\frac{1}{2} \\ 1 \end{bmatrix}$$

is an eigenvector corresponding to $\lambda = 3$. To facilitate computation by hand, let us choose an eigenvector with integer entries:

$$\begin{bmatrix} -1 \\ 2 \end{bmatrix}.$$

**Step 3:** Work with a basis of eigenvectors instead of the standard basis vectors. To do so, we arrange the eigenvectors as columns in a matrix $P$. (We usually order the eigenvectors according to which eigenvalue they correspond to, starting with the lowest eigenvalues—feel free to choose whatever order you wish.) Specifically, let

$$P = \begin{bmatrix} -1 & -1 \\ 1 & 2 \end{bmatrix},$$

the first column being the eigenvector corresponding to $\lambda = 2$ and the second column being the eigenvector corresponding to $\lambda = 3$. Next, form the diagonal matrix $D$ whose entries are the eigenvalues:

$$D = \begin{bmatrix} 2 & 0 \\ 0 & 3 \end{bmatrix}.$$

Note that the ordering of the eigenvalues in the matrix $D$ should correspond to the ordering of the columns in $P$. Next, the reader can verify that

$$P^{-1} = \begin{bmatrix} -2 & -1 \\ 1 & 1 \end{bmatrix}.$$

Moreover, straightforward matrix multiplication reveals that

$$A = PDP^{-1}. \tag{2.8}$$

The process of writing $A$ in the form (2.8) is called *diagonalization*. After diagonalizing $A$, the process of calculating the matrix exponential $e^{tA}$ is straightforward.

Recalling that $e^{tA} = Pe^{tD}P^{-1}$, we may now calculate

$$e^{tA} = \begin{bmatrix} -1 & -1 \\ 1 & 2 \end{bmatrix} \begin{bmatrix} e^{2t} & 0 \\ 0 & e^{3t} \end{bmatrix} \begin{bmatrix} -2 & -1 \\ 1 & 1 \end{bmatrix}$$
$$= \begin{bmatrix} 2e^{2t} - e^{3t} & e^{2t} - e^{3t} \\ -2e^{2t} + 2e^{3t} & -e^{2t} + 2e^{3t} \end{bmatrix}.$$

Now that we have substantially widened the class of matrices that we know how to exponentiate, we formally state a theorem regarding how the matrix exponential can be used to solve homogeneous systems of constant-coefficient ODEs.

**Theorem 2.1.10.** Suppose that $\mathbf{y}' = A\mathbf{y}$ is a system of constant coefficient ODEs and let $\mathbf{y}_0 = \mathbf{y}(0)$ be a vector of initial conditions. Then this initial value problem has exactly one solution, which is given by $\mathbf{y}(t) = e^{tA}\mathbf{y}(0)$. Moreover, the solution exists for all time $t$.

*Proof.* See Sections 1.3-1.4 of Perko [8]. The proof is not difficult, but is best done by introducing the concept of a matrix norm (which we shall not work with here). □

In the absence of a vector of initial conditions, the *general* solution of a constant-coefficient system $\mathbf{y}' = A\mathbf{y}$ is $\mathbf{y}(t) = e^{tA}\mathbf{u}$, where $\mathbf{u}$ is a vector of arbitrary constants.

**Example 2.1.11.** Solve the initial value problem

$$\frac{dx}{dt} = x - y \qquad x(0) = C_1$$
$$\frac{dy}{dt} = 2x + 4y \qquad y(0) = C_2.$$

To do so, we write the system in matrix form

$$\begin{bmatrix} x' \\ y' \end{bmatrix} = \begin{bmatrix} 1 & -1 \\ 2 & 4 \end{bmatrix} \begin{bmatrix} x \\ y \end{bmatrix}.$$

The coefficient matrix is the same matrix $A$ used in the previous example, so we have already computed its matrix exponential. According to Theorem 2.1.10, the solution is

$$\begin{bmatrix} x(t) \\ y(t) \end{bmatrix} = e^{tA} \begin{bmatrix} x(0) \\ y(0) \end{bmatrix} = \begin{bmatrix} 2e^{2t} - e^{3t} & e^{2t} - e^{3t} \\ -2e^{2t} + 2e^{3t} & -e^{2t} + 2e^{3t} \end{bmatrix} \begin{bmatrix} C_1 \\ C_2 \end{bmatrix}$$
$$= \begin{bmatrix} (2C_1 + C_2)e^{2t} - (C_1 + C_2)e^{3t} \\ (-2C_1 - C_2)e^{2t} + (2C_1 + 2C_2)e^{3t} \end{bmatrix}.$$

This represents the general solution of the system of ODEs. For the specific choice of initial conditions $x(0) = 8$ and $y(0) = 1$, the solution would be

$$\begin{bmatrix} x(t) \\ y(t) \end{bmatrix} = \begin{bmatrix} 17e^{2t} - 9e^{3t} \\ -17e^{2t} + 18e^{3t} \end{bmatrix}.$$

**2.1.2 Algebraic and Geometric Multiplicities of Eigenvalues.** The process of computing matrix exponentials can be tedious, especially for matrices which are not diagonalizable over $\mathbb{R}$. In preparation for things to come, we introduce some notions from linear algebra which may not be familiar to all readers.

**Definition 2.1.12.** (Algebraic multiplicity.) Suppose $\lambda^*$ is an eigenvalue of $A$ and, more specifically, that the factor $(\lambda - \lambda^*)$ is repeated $m$ times in the factorization of the characteristic polynomial. We say that $\lambda^*$ is an eigenvalue of *algebraic multiplicity $m$*.

**Example 2.1.13.** If we find that $A$ has characteristic equation $\lambda(\lambda - 3)^4(\lambda^2 + 1) = 0$, then $\lambda = 0$ has algebraic multiplicity 1 and $\lambda = 3$ has algebraic multiplicity 4. There are actually complex eigenvalues $\lambda = \pm i$, both of which have algebraic multiplicity 1. We will deal with complex eigenvalues later.

**Definition 2.1.14.** (Eigenspace and geometric multiplicity.) Suppose $\lambda$ is an eigenvalue of $A$. The *eigenspace* of $\lambda$ is the span of all eigenvectors associated with $\lambda$. The dimension of the eigenspace of $\lambda$ is called the *geometric multiplicity* of $\lambda$.

**Example 2.1.15.** Consider the matrices

$$A = \begin{bmatrix} 1 & 1 & 0 \\ 0 & 1 & 0 \\ 0 & 0 & 2 \end{bmatrix} \qquad B = \begin{bmatrix} 1 & 0 & 1 \\ 0 & 1 & 0 \\ 0 & 0 & 2 \end{bmatrix}.$$

You can check that both $A$ and $B$ have the same characteristic polynomial, namely $p(\lambda) = (\lambda - 1)^2(\lambda - 2)$. Therefore, $\lambda = 1$ is an eigenvalue with algebraic multiplicity 2, and $\lambda = 2$ is an eigenvalue with algebraic multiplicity 1. To calculate the geometric multiplicities, we start with the matrix $A$ and the eigenvalue $\lambda = 1$. In this case,

$$A - \lambda I = \begin{bmatrix} 0 & 1 & 0 \\ 0 & 0 & 0 \\ 0 & 0 & 1 \end{bmatrix},$$

which is row equivalent to

$$\begin{bmatrix} 0 & 1 & 0 \\ 0 & 0 & 1 \\ 0 & 0 & 0 \end{bmatrix}.$$

To find eigenvectors, observe that solutions of $(A - \lambda I)\mathbf{v} = 0$ must satisfy

$$\begin{bmatrix} 0 & 1 & 0 \\ 0 & 0 & 1 \\ 0 & 0 & 0 \end{bmatrix} \begin{bmatrix} v_1 \\ v_2 \\ v_3 \end{bmatrix} = \begin{bmatrix} 0 \\ 0 \\ 0 \end{bmatrix}.$$

Expanding reveals that $v_2 = v_3 = 0$, while $v_1$ is a free variable. The solutions of the above system are

$$\begin{bmatrix} v_1 \\ v_2 \\ v_3 \end{bmatrix} = \begin{bmatrix} v_1 \\ 0 \\ 0 \end{bmatrix} = v_1 \begin{bmatrix} 1 \\ 0 \\ 0 \end{bmatrix},$$

from which we conclude that

$$\begin{bmatrix} 1 \\ 0 \\ 0 \end{bmatrix}$$

is an eigenvector corresponding to $\lambda = 1$. The span of this eigenvector is a one-dimensional subspace of $\mathbb{R}^3$, which means the eigenspace is one-dimensional. Equivalently, $\lambda = 1$ has geometric multiplicity 1. A similar calculation shows that, for the matrix $A$, the geometric multiplicity of the eigenvalue $\lambda = 2$ is also 1.

Now for the matrix $B$, the eigenvalue $\lambda = 1$ gives rise to the matrix

$$B - \lambda I = \begin{bmatrix} 0 & 0 & 1 \\ 0 & 0 & 0 \\ 0 & 0 & 1 \end{bmatrix}.$$

This time, row reduction leads to a matrix which has *two* rows consisting entirely of zeros. Eigenvectors must satisfy

$$\begin{bmatrix} 0 & 0 & 1 \\ 0 & 0 & 0 \\ 0 & 0 & 0 \end{bmatrix} \begin{bmatrix} v_1 \\ v_2 \\ v_3 \end{bmatrix} = \begin{bmatrix} 0 \\ 0 \\ 0 \end{bmatrix},$$

which implies that $v_3 = 0$ but that *both* $v_1$ and $v_2$ are free variables. Solutions of $(B - \lambda I)\mathbf{v} = 0$ are of the form

$$\begin{bmatrix} v_1 \\ v_2 \\ v_3 \end{bmatrix} = \begin{bmatrix} v_1 \\ v_2 \\ 0 \end{bmatrix} = v_1 \begin{bmatrix} 1 \\ 0 \\ 0 \end{bmatrix} + v_2 \begin{bmatrix} 0 \\ 1 \\ 0 \end{bmatrix}.$$

Therefore, we have obtained a set of two linearly independent eigenvectors

$$\begin{bmatrix} 1 \\ 0 \\ 0 \end{bmatrix} \quad \text{and} \quad \begin{bmatrix} 0 \\ 1 \\ 0 \end{bmatrix},$$

which means that the eigenspace for $\lambda = 1$ is two-dimensional. In other words, for the matrix $B$, the geometric multiplicity of $\lambda = 1$ is 2. The reader can verify that the geometric multiplicity of $\lambda = 2$ is 1.

The above example gives rise to several important observations, namely

☞ The sum of the *algebraic* multiplicities of the eigenvalues always equals $n$, the dimension of the underlying space. (In the above example, $n = 3$.)

☞ In the example, the sum of the *geometric* multiplicities of the eigenvalues of $A$ is 2, whereas for the matrix $B$, the sum of the geometric multiplicities is 3.

☞ In order to calculate the geometric multiplicities, it was actually not necessary to compute the eigenvectors. The geometric multiplicity of an eigenvalue $\lambda$ of a matrix $A$ is the same as the number of zero rows in the reduced row echelon form of the matrix $(A - \lambda I)$. Since $\det(A - \lambda I) = 0$, the matrix $(A - \lambda I)$ is singular, implying that there will be at least one row of zeros in the reduced row-echelon form. Thus, every eigenvalue has a geometric multiplicity *at least 1*. The following Lemma is useful in proving Theorem 2.1.17, which gives another criterion for determining whether a matrix $M$ is diagonalizable.

**Lemma 2.1.16.** Eigenvectors corresponding to different eigenvalues are linearly independent.

*Proof.* We prove this statement for a set of 2 eigenvectors; the reader can extend the proof to the general case. Let $v_1$ and $v_2$ be eigenvectors of a matrix $A$ corresponding to *different* eigenvalues $\lambda_1$ and $\lambda_2$. Suppose indirectly that these two eigenvectors are linearly *dependent*. Then there exists a constant $c$ such that $v_2 = cv_1$. Moreover, since eigenvectors are non-zero, it must be the case that $c \neq 0$. Multiplying both sides of the equation by $A$, we have $Av_2 = cAv_1$. Equivalently, $\lambda_2 v_2 = c\lambda_1 v_1$. Replacing $v_2$ with $cv_1$, this implies that $\lambda_2 cv_1 = \lambda_1 cv_1$. But since $v_1 \neq 0$ and $c \neq 0$, this would mean that $\lambda_2 = \lambda_1$, contradicting our assumption that these were *different* eigenvalues. □

**Theorem 2.1.17.** Suppose $M$ is an $n \times n$ matrix with real eigenvalues. Then $M$ is diagonalizable if and only if the sum of the geometric multiplicities of the eigenvalues is equal to $n$. (Note: If an eigenvalue has algebraic multiplicity larger than 1, we count its geometric multiplicity only once when forming the sum.)

Notice that, according to this Theorem, the matrix $A$ in the previous example is *not* diagonalizable, whereas the matrix $B$ is diagonalizable because we can produce a basis for $\mathbb{R}^3$ consisting entirely of eigenvectors. By contrast, the matrix $A$ fails to be diagonalizable because it is "deficient" in the sense that its eigenvectors can only span a 2-dimensional subspace of $\mathbb{R}^3$. In general, the sum of the geometric multiplicities of the distinct eigenvalues can never exceed $n$, but this sum can certainly be less than $n$ (c.f., the matrix $A$ in the previous example). One nice consequence of Lemma 2.1.16 and Theorem 2.1.17 is the following

**Corollary 2.1.18.** If an $n \times n$ matrix $M$ has $n$ *distinct* real eigenvalues, then $M$ is diagonalizable.

**Example 2.1.19** (Triangular Matrices). If a square matrix $M$ has no non-zero entries above its main diagonal, then $M$ is called *lower triangular*, and if it has no non-zero entries below its main diagonal, then $M$ is called *upper triangular*. More rigorously, $M$ is lower triangular if $M_{ij} = 0$ whenever $i < j$, and $M$ is upper triangular if $M_{ij} = 0$ whenever $i > j$. Writing down the characteristic equation for a triangular matrix is easy. For example, suppose that $A$ is upper-triangular of the form

$$A = \begin{bmatrix} a_{11} & a_{12} & \cdots & a_{1n} \\ 0 & a_{22} & \cdots & a_{2n} \\ \vdots & \vdots & \ddots & \vdots \\ 0 & 0 & \cdots & a_{nn} \end{bmatrix}.$$

Then the characteristic equation $\det(A - \lambda I) = 0$ is simply

$$(\lambda - a_{11})(\lambda - a_{22}) \cdots (\lambda - a_{nn}) = 0,$$

which means that the *eigenvalues of a triangular matrix are the entries on the main diagonal.* In light of Corollary 2.1.18, we immediately see that the triangular matrix

$$\begin{bmatrix} 1 & 0 & 0 \\ 6 & 8 & 0 \\ -6 & 3 & -4 \end{bmatrix}$$

is diagonalizable because its eigenvalues 1, 8, and $-4$ are real and distinct.

We now work through an example which connects much of the material in this subsection.

**Example 2.1.20.** Solve the system

$$\begin{aligned} \frac{dx_1}{dt} &= x_1 + x_3 & x_1(0) &= 1 \\ \frac{dx_2}{dt} &= x_2 + 2x_3 & x_2(0) &= 1 \\ \frac{dx_3}{dt} &= 3x_3 & x_3(0) &= 1. \end{aligned}$$

**Solution:** In matrix notation, this system takes the form

$$\begin{bmatrix} x_1' \\ x_2' \\ x_3' \end{bmatrix} = \begin{bmatrix} 1 & 0 & 1 \\ 0 & 1 & 2 \\ 0 & 0 & 3 \end{bmatrix} \begin{bmatrix} x_1 \\ x_2 \\ x_3 \end{bmatrix}.$$

Letting $\mathbf{x}$ denote the vector of unknowns and $A$ the coefficient matrix, the solution of the system is $\mathbf{x}(t) = e^{tA}\mathbf{x}(0)$. To calculate the matrix exponential, first observe that since $A$ is triangular, we may immediately conclude that the eigenvalues are the entries on the main diagonal. The eigenvalue $\lambda = 1$ has algebraic multiplicity 2. To find eigenvectors associated with this eigenvalue, we row-reduce the matrix $(A - \lambda I) = A - I$:

$$A - I = \begin{bmatrix} 0 & 0 & 1 \\ 0 & 0 & 2 \\ 0 & 0 & 2 \end{bmatrix} \rightarrow \begin{bmatrix} 0 & 0 & 1 \\ 0 & 0 & 0 \\ 0 & 0 & 0 \end{bmatrix}.$$

We immediately see that $\lambda = 1$ has geometric multiplicity 2, and we expect to find a set of two linearly independent eigenvectors for this eigenvalue. Solving the system $(A - \lambda I)\mathbf{v} = 0$ as usual, we find that $v_3 = 0$ while $v_1$ and $v_2$ are free. Thus,

$$\begin{bmatrix} v_1 \\ v_2 \\ v_3 \end{bmatrix} = \begin{bmatrix} v_1 \\ v_2 \\ 0 \end{bmatrix} = v_1 \begin{bmatrix} 1 \\ 0 \\ 0 \end{bmatrix} + v_2 \begin{bmatrix} 0 \\ 1 \\ 0 \end{bmatrix}$$

is the set of solutions of $(A - I)\mathbf{v} = 0$, and we have obtained eigenvectors

$$\begin{bmatrix} 1 \\ 0 \\ 0 \end{bmatrix} \quad \text{and} \quad \begin{bmatrix} 0 \\ 1 \\ 0 \end{bmatrix}.$$

The reader can show that the eigenvalue $\lambda = 3$ has geometric multiplicity 1 and gives rise to an eigenvector

$$\begin{bmatrix} \frac{1}{2} \\ 1 \\ 1 \end{bmatrix}.$$

By Lemma 2.1.16, we know that the three eigenvectors we have produced are linearly independent and form a basis for our space $\mathbb{R}^3$. This means that $A$ is diagonalizable, so form the matrix $P$ by arranging the eigenvectors as columns:

$$P = \begin{bmatrix} 1 & 0 & \frac{1}{2} \\ 0 & 1 & 1 \\ 0 & 0 & 1 \end{bmatrix}.$$

Then $A = PDP^{-1}$, where $D$ is the diagonal matrix

$$D = \begin{bmatrix} 1 & 0 & 0 \\ 0 & 1 & 0 \\ 0 & 0 & 3 \end{bmatrix}$$

and

$$P^{-1} = \begin{bmatrix} 1 & 0 & -\frac{1}{2} \\ 0 & 1 & -1 \\ 0 & 0 & 1 \end{bmatrix}.$$

Notice that $D$ has the eigenvalues on its main diagonal, arranged in the same order that the corresponding eigenvectors were arranged in $P$. The matrix exponential $e^{tA}$ is given by

$$e^{tA} = Pe^{tD}P^{-1} = \begin{bmatrix} 1 & 0 & \frac{1}{2} \\ 0 & 1 & 1 \\ 0 & 0 & 1 \end{bmatrix} \begin{bmatrix} e^t & 0 & 0 \\ 0 & e^t & 0 \\ 0 & 0 & e^{3t} \end{bmatrix} \begin{bmatrix} 1 & 0 & -\frac{1}{2} \\ 0 & 1 & -1 \\ 0 & 0 & 1 \end{bmatrix}$$

which, after some tedious matrix multiplication, simplifies to

$$e^{tA} = \begin{bmatrix} e^t & 0 & -\frac{1}{2}e^t + \frac{1}{2}e^{3t} \\ 0 & e^t & -e^t + e^{3t} \\ 0 & 0 & e^{3t} \end{bmatrix}.$$

Finally, we multiply $e^{tA}$ by our vector of initial conditions to obtain the solution of the initial value problem:

$$\begin{bmatrix} x_1(t) \\ x_2(t) \\ x_3(t) \end{bmatrix} = \begin{bmatrix} e^t & 0 & -\frac{1}{2}e^t + \frac{1}{2}e^{3t} \\ 0 & e^t & -e^t + e^{3t} \\ 0 & 0 & e^{3t} \end{bmatrix} \begin{bmatrix} 1 \\ 1 \\ 1 \end{bmatrix} = \begin{bmatrix} \frac{1}{2}e^t + \frac{1}{2}e^{3t} \\ e^{3t} \\ e^{3t} \end{bmatrix}.$$

**Observation.** In the above example, notice that the exponential functions involved in the solution contain the eigenvalues in their exponents. Since $\lambda = 1$ and $\lambda = 3$ are the eigenvalues, the functions $e^t$ and $e^{3t}$ appear in the solutions.

**Observation.** The previous Observation suggests that, if eigenvalues are positive, then the exponential functions in the solutions would increase without bound as $t \to \infty$. Negative eigenvalues would give rise to solutions which decay to 0 as

$t \to \infty$. For example, if $\lambda = -4$ is an eigenvalue, we would expect the decaying exponential function $e^{-4t}$ to appear within the general solution.

Now that we know to expect exponential functions to serve as the "building blocks" of our solutions, there is a much faster way of solving certain systems $\mathbf{y}' = A\mathbf{y}$ without exponentiating the matrix $A$.

**Proposition 2.1.21.** Suppose that $\mathbf{y}' = A\mathbf{y}$ is a system of ODEs, where $A$ is an $n \times n$ matrix which is diagonalizable (i.e., we can find a set of $n$ linearly independent eigenvectors for $A$). Let $\lambda_1, \lambda_2, \ldots \lambda_n$ denote the (possibly repeated) eigenvalues, and let $\mathbf{v}_1, \mathbf{v}_2, \ldots \mathbf{v}_n$ denote their corresponding eigenvectors. Then the general solution of the system $\mathbf{y}' = A\mathbf{y}$ is given by

$$\mathbf{y}(t) = c_1 e^{\lambda_1 t}\mathbf{v}_1 + c_2 e^{\lambda_2 t}\mathbf{v}_2 + \cdots + c_n e^{\lambda_n t}\mathbf{v}_n. \tag{2.9}$$

*Proof.* Suppose that $\mathbf{y}(t)$ is as in Equation (2.9). We must show that $\mathbf{y}(t)$ satisfies $\mathbf{y}' = A\mathbf{y}$. Differentiating with respect to $t$, we calculate

$$\begin{aligned}
\mathbf{y}'(t) &= c_1\lambda_1 e^{\lambda_1 t}\mathbf{v}_1 + c_2\lambda_2 e^{\lambda_2 t}\mathbf{v}_2 + \cdots + c_n\lambda_n e^{\lambda_n t}\mathbf{v}_n \\
&= c_1 e^{\lambda_1 t}(\lambda_1\mathbf{v}_1) + c_2 e^{\lambda_2 t}(\lambda_2\mathbf{v}_2) + \cdots + c_n e^{\lambda_n t}(\lambda_n\mathbf{v}_n) \\
&= c_1 e^{\lambda_1 t}A\mathbf{v}_1 + c_2 e^{\lambda_2 t}A\mathbf{v}_2 + \cdots + c_n e^{\lambda_n t}A\mathbf{v}_n \\
&= A\left(c_1 e^{\lambda_1 t}\mathbf{v}_1 + c_2 e^{\lambda_2 t}\mathbf{v}_2 + \cdots + c_n e^{\lambda_n t}\mathbf{v}_n\right) \;=\; A\mathbf{y}(t).
\end{aligned}$$

□

**Example 2.1.22.** Solve the initial value problem

$$\begin{aligned}
x_1' &= -3x_1 + 5x_2 \qquad & x_1(0) &= 2 \\
x_2' &= x_1 + x_2 & x_2(0) &= -1.
\end{aligned}$$

**Solution:** In matrix notation, the system has the form $\mathbf{x}' = A\mathbf{x}$, where the coefficient matrix $A$ is given by

$$A = \begin{bmatrix} -3 & 5 \\ 1 & 1 \end{bmatrix}.$$

The characteristic polynomial is

$$\det(A - \lambda I) = (-3 - \lambda)(1 - \lambda) - 5 \;=\; \lambda^2 + 2\lambda - 8 \;=\; (\lambda + 4)(\lambda - 2),$$

which means that the eigenvalues are $\lambda = -4$ and $\lambda = 2$. Because we have distinct, real eigenvalues, Corollary 2.1.18 tells us that $A$ is diagonalizable. The reader can check that

$$\begin{bmatrix} 1 \\ 1 \end{bmatrix} \quad \text{and} \quad \begin{bmatrix} -5 \\ 1 \end{bmatrix}$$

are eigenvectors corresponding to $\lambda = 2$ and $\lambda = -4$, respectively. Proposition 2.1.21 tells us that the general solution of our system is given by

$$\begin{bmatrix} x_1(t) \\ x_2(t) \end{bmatrix} = c_1 e^{2t} \begin{bmatrix} 1 \\ 1 \end{bmatrix} + c_2 e^{-4t} \begin{bmatrix} -5 \\ 1 \end{bmatrix},$$

where $c_1$ and $c_2$ are arbitrary constants. We determine the values of these constants by using the initial conditions. Setting $t = 0$, the formula for the general solution becomes

$$\begin{bmatrix} 2 \\ -1 \end{bmatrix} = c_1 \begin{bmatrix} 1 \\ 1 \end{bmatrix} + c_2 \begin{bmatrix} -5 \\ 1 \end{bmatrix} = \begin{bmatrix} 1 & -5 \\ 1 & 1 \end{bmatrix} \begin{bmatrix} c_1 \\ c_2 \end{bmatrix}.$$

This is an inhomogeneous system of two equations in two unknowns, and basic linear algebra techniques will lead you to the solution $c_1 = -1/2$ and $c_2 = -1/2$. Finally, we see that the solution of the initial value problem is

$$\begin{bmatrix} x_1(t) \\ x_2(t) \end{bmatrix} = \begin{bmatrix} -\frac{1}{2}e^{2t} + \frac{5}{2}e^{-4t} \\ -\frac{1}{2}e^{2t} - \frac{1}{2}e^{-4t} \end{bmatrix}.$$

**2.1.3 Complex Eigenvalues.** Unfortunately, not all matrices are diagonalizable, and not all matrices have real eigenvalues. We will now learn how to exponentiate matrices with complex conjugate eigenvalues. Recall that the imaginary unit i is defined according to the rule $i^2 = -1$. All complex numbers $z$ can be written in the form $z = \alpha + \beta i$, where $\alpha$ and $\beta$ are real numbers. Here, $\alpha$ is called the *real part* of $z$, and $\beta$ is called the *imaginary part* of $z$. Finally, we remark that the complex numbers $\alpha + \beta i$ and $\alpha - \beta i$ are called *complex conjugates* of each other.

**Example 2.1.23.** The matrix

$$A = \begin{bmatrix} 1 & -2 \\ 2 & 1 \end{bmatrix}$$

has characteristic polynomial $\det(A - \lambda I) = (1-\lambda)(1-\lambda)+4 = \lambda^2 - 2\lambda + 5$. The roots of this polynomial are $\lambda = 1 \pm 2\mathrm{i}$, and hence our eigenvalues are complex conjugates.

We still wish to solve constant-coefficient systems of ODEs for which the coefficient matrix has complex conjugate eigenvalues. How can we exponentiate such matrices? As a first step towards answering this question, we tackle a special case. Suppose

$$M = \begin{bmatrix} 0 & -\beta \\ \beta & 0 \end{bmatrix}, \tag{2.10}$$

where $\beta$ is a non-zero real number. Notice that the characteristic equation for $M$ is given by $\lambda^2 + \beta^2 = 0$. The roots of this equation are $\lambda = \pm|\beta|\mathrm{i}$, which are pure imaginary numbers (i.e., the real part is zero).

**Lemma 2.1.24.** If $M$ is the matrix in Equation (2.10), then

$$\mathrm{e}^{tM} = \begin{bmatrix} \cos\beta t & -\sin\beta t \\ \sin\beta t & \cos\beta t \end{bmatrix}. \tag{2.11}$$

*Proof.* The series representation for the matrix exponential

$$\mathrm{e}^{tM} = \sum_{k=0}^{\infty} \frac{t^k}{k!} M^k$$

requires us to calculate powers of $M$. Fortunately, since

$$M^2 = \begin{bmatrix} -\beta^2 & 0 \\ 0 & -\beta^2 \end{bmatrix} = -\beta^2 I,$$

computing higher powers of $M$ is very straightforward. We focus our attention on the $(1,1)$ entries of the matrices in the above summation. The upper-left entry in the matrix $\mathrm{e}^{tM}$ would be

$$\frac{t^0}{0!} - \frac{\beta^2 t^2}{2!} + \frac{\beta^4 t^4}{4!} \cdots = \sum_{k=0}^{\infty} (-1)^k \frac{\beta^{2k} t^{2k}}{(2k)!}.$$

This is precisely the Maclaurin series for the function $\cos\beta t$. A similar calculation works for the other entries in the matrix. □

Of course, we shall wish to exponentiate matrices with complex conjugate eigenvalues which are not pure imaginary. In order to extend Lemma 2.1.24 to handle more general matrices, we will need another Lemma:

**Lemma 2.1.25.** Suppose $A$ and $B$ are $n \times n$ matrices. If the matrices commute ($AB = BA$) then $e^{A+B} = e^A e^B$.

*Proof.* See Section 1.3 of Perko [8]. □

**Warning:** If matrices $A$ and $B$ do *not* commute, then we cannot conclude that $e^{A+B} = e^A e^B$. The reader is encouraged to test this, using the matrices

$$A = \begin{bmatrix} 1 & 0 \\ 0 & 0 \end{bmatrix} \quad \text{and} \quad B = \begin{bmatrix} 0 & 1 \\ 0 & 0 \end{bmatrix}.$$

Lemmas 2.1.24 and 2.1.25 can be combined to prove

**Proposition 2.1.26.** Suppose

$$A = \begin{bmatrix} \alpha & -\beta \\ \beta & \alpha \end{bmatrix},$$

where $\alpha$ and $\beta$ are real numbers. Then

$$e^{tA} = \begin{bmatrix} e^{\alpha t}\cos\beta t & -e^{\alpha t}\sin\beta t \\ e^{\alpha t}\sin\beta t & e^{\alpha t}\cos\beta t \end{bmatrix} = e^{\alpha t}\begin{bmatrix} \cos\beta t & -\sin\beta t \\ \sin\beta t & \cos\beta t \end{bmatrix}.$$

*Proof.* We write $A$ as a sum of two matrices:

$$A = \begin{bmatrix} \alpha & 0 \\ 0 & \alpha \end{bmatrix} + \begin{bmatrix} 0 & -\beta \\ \beta & 0 \end{bmatrix} = B + M,$$

where $M$ is the matrix from Lemma 2.1.24. Clearly $B$ and $M$ commute because $B = \alpha I$ is simply a scalar multiple of the identity matrix. Therefore, Lemma 2.1.25 tells us that

$$e^{tA} = e^{t(B+M)} = e^{tB+tM} = e^{tB}e^{tM}.$$

We know how to compute both of these matrix exponentials:

$$e^{tB} = \begin{bmatrix} e^{\alpha t} & 0 \\ 0 & e^{\alpha t} \end{bmatrix} = e^{\alpha t} I \quad \text{and} \quad e^{tM} = \begin{bmatrix} \cos\beta t & -\sin\beta t \\ \sin\beta t & \cos\beta t \end{bmatrix}.$$

Therefore,

$$e^{tA} = e^{tB}e^{tM} = e^{\alpha t}\begin{bmatrix} \cos\beta t & -\sin\beta t \\ \sin\beta t & \cos\beta t \end{bmatrix},$$

as claimed. □

**Remark.** In the above Proposition, you may check that the characteristic equation for $A$ is given by

$$\lambda^2 - 2\alpha\lambda + (\alpha^2 + \beta^2) = 0,$$

and has roots $\lambda = \alpha \pm \beta i$.

We now show that every $2 \times 2$ matrix with complex conjugate eigenvalues $\alpha \pm \beta i$ is similar to a matrix of the form in Proposition 2.1.26. We will exploit this fact to learn how to exponentiate any matrix with complex eigenvalues.

**Definition 2.1.27.** If $A$ is a matrix with complex conjugate eigenvalues $\alpha \pm \beta i$, then the matrix

$$M = \begin{bmatrix} \alpha & -\beta \\ \beta & \alpha \end{bmatrix}$$

is called the *real canonical form* for $A$.

The next Theorem states that every $2 \times 2$ matrix with complex conjugate eigenvalues is similar to its real canonical form. Since we know how to exponentiate matrices in real canonical form, we should be able to solve systems of constant coefficient ODEs in cases where the coefficient matrix has complex conjugate eigenvalues. Notice that we do not attempt to diagonalize such matrices, because we would prefer to work with matrices containing real entries (to facilitate finding real-valued solutions of ODEs).

**Theorem 2.1.28.** Suppose $A$ is a $2 \times 2$ matrix with eigenvalues $\alpha \pm \beta i$, and suppose $\mathbf{w}$ is a complex eigenvector corresponding to the eigenvalue $\alpha + \beta i$. Write the eigenvector as $\mathbf{w} = \mathbf{u} + i\mathbf{v}$, where $\mathbf{u}$ and $\mathbf{v}$ are real vectors, and form the matrix $P = [\mathbf{v}|\mathbf{u}]$. Then

$$A = P\begin{bmatrix} \alpha & -\beta \\ \beta & \alpha \end{bmatrix}P^{-1}.$$

In other words, $A$ is similar to a real canonical form.

Theorem 2.1.28 tells us exactly how to put a matrix in real canonical form, as we illustrate in the following example.

**Example 2.1.29.** Solve the system

$$\frac{dx_1}{dt} = x_1 - 12x_2 \qquad \frac{dx_2}{dt} = 3x_1 + x_2.$$

**Solution:** As usual, we write the system in matrix form

$$\begin{bmatrix} x_1' \\ x_2' \end{bmatrix} = \begin{bmatrix} 1 & -12 \\ 3 & 1 \end{bmatrix} \begin{bmatrix} x_1 \\ x_2 \end{bmatrix}.$$

Letting $A$ denote the coefficient matrix, the characteristic equation is $\lambda^2 - 2\lambda + 37 = 0$, which has roots $\lambda = 1 \pm 6i$. According to Theorem 2.1.28, we should find a complex eigenvector for the eigenvalue with positive imaginary part: $\lambda = 1 + 6i$. We calculate

$$A - \lambda I = \begin{bmatrix} 1-(1+6i) & -12 \\ 3 & 1-(1+6i) \end{bmatrix} = \begin{bmatrix} -6i & -12 \\ 3 & -6i \end{bmatrix}.$$

When computing the reduced row-echelon form of this $2 \times 2$ matrix, we may immediately replace one of the rows with zeros—we know the matrix is singular because $\lambda$ was an eigenvalue. This saves us a bit of time with elementary row operations, because it is not necessary that we notice that Row 1 is $-2i$ times Row 2. The reduced row-echelon form is easy to calculate:

$$\begin{bmatrix} -6i & -12 \\ 3 & -6i \end{bmatrix} \rightarrow \begin{bmatrix} 0 & 0 \\ 3 & -6i \end{bmatrix} \rightarrow \begin{bmatrix} 3 & -6i \\ 0 & 0 \end{bmatrix} \rightarrow \begin{bmatrix} 1 & -2i \\ 0 & 0 \end{bmatrix}.$$

We seek a complex eigenvector $\mathbf{w}$ which satisfies $(A - \lambda I)\mathbf{w} = 0$, so set

$$\begin{bmatrix} 1 & -2i \\ 0 & 0 \end{bmatrix} \begin{bmatrix} w_1 \\ w_2 \end{bmatrix} = \begin{bmatrix} 0 \\ 0 \end{bmatrix}.$$

We find that $w_1 - 2iw_2 = 0$, so $w_1 = 2iw_2$ and we treat $w_2$ as a free variable. Thus,

$$\begin{bmatrix} w_1 \\ w_2 \end{bmatrix} = \begin{bmatrix} 2iw_2 \\ w_2 \end{bmatrix} = w_2 \begin{bmatrix} 2i \\ 1 \end{bmatrix}.$$

Next, the theorem says we should write this (complex) eigenvector in terms of its real and imaginary parts:

$$\begin{bmatrix} 2i \\ 1 \end{bmatrix} = \mathbf{w} = \mathbf{u} + i\mathbf{v} = \begin{bmatrix} 0 \\ 1 \end{bmatrix} + i \begin{bmatrix} 2 \\ 0 \end{bmatrix}.$$

Form the matrix $P = [\mathbf{v}|\mathbf{u}]$ by placing the *imaginary* part of the eigenvector $\mathbf{w}$ in the first column, and the *real* part of $\mathbf{w}$ in the second column:

$$P = \begin{bmatrix} 2 & 0 \\ 0 & 1 \end{bmatrix}.$$

This matrix is particularly easy to invert:

$$P^{-1} = \begin{bmatrix} \frac{1}{2} & 0 \\ 0 & 1 \end{bmatrix}.$$

Since our eigenvalue was $\lambda = 1 + 6i$, the notation used in the statement of Theorem 2.1.28 says that $\alpha = 1$ and $\beta = 6$. The reader can verify that $A = PMP^{-1}$, where

$$M = \begin{bmatrix} 1 & -6 \\ 6 & 1 \end{bmatrix}$$

is in real canonical form. It follows that $e^{tA} = Pe^{tM}P^{-1}$, and by Proposition 2.1.26, we have

$$e^{tA} = \begin{bmatrix} 2 & 0 \\ 0 & 1 \end{bmatrix} e^t \begin{bmatrix} \cos 6t & -\sin 6t \\ \sin 6t & \cos 6t \end{bmatrix} \begin{bmatrix} \frac{1}{2} & 0 \\ 0 & 1 \end{bmatrix} = e^t \begin{bmatrix} \cos 6t & -2\sin 6t \\ \frac{1}{2}\sin 6t & \cos 6t \end{bmatrix}.$$

Finally, since no initial conditions were specified, we obtain the general solution by multiplying $e^{tA}$ by a vector of arbitrary constants:

$$\begin{bmatrix} x_1(t) \\ x_2(t) \end{bmatrix} = e^t \begin{bmatrix} \cos 6t & -2\sin 6t \\ \frac{1}{2}\sin 6t & \cos 6t \end{bmatrix} \begin{bmatrix} c_1 \\ c_2 \end{bmatrix} = \begin{bmatrix} c_1 e^t \cos 6t - 2c_2 e^t \sin 6t \\ \frac{1}{2}c_1 e^t \sin 6t + c_2 e^t \cos 6t \end{bmatrix}.$$

**Observation.** Complex conjugate eigenvalues give rise to *oscillatory* solutions.

**Example 2.1.30.** Solve the system $\mathbf{x}' = A\mathbf{x}$ where

$$A = \begin{bmatrix} 3 & -2 & 0 \\ 1 & 1 & 0 \\ 0 & 0 & 4 \end{bmatrix} \quad \text{and} \quad \mathbf{x}(0) = \begin{bmatrix} 1 \\ 2 \\ 3 \end{bmatrix}.$$

**Cool observation:** The matrix $A$ has a *block diagonal structure*

$$A = \left[\begin{array}{cc|c} 3 & -2 & 0 \\ 1 & 1 & 0 \\ \hline 0 & 0 & 4 \end{array}\right].$$

The upper-left corner is a $2 \times 2$ block, the bottom-right corner is a $1 \times 1$ block, and all other entries are zero. The reader is encouraged to prove that the exponential of a block-diagonal matrix is obtained by exponentiating each diagonal block separately. In the present case, the only part that will require any work is the $2 \times 2$ block. To find the eigenvalues of A, we calculate

$$\det(A - \lambda I) = \begin{vmatrix} 3-\lambda & -2 & 0 \\ 1 & 1-\lambda & 0 \\ 0 & 0 & 4-\lambda \end{vmatrix}.$$

Performing a co-factor expansion using the last row, this determinant simplifies to

$$\det(A - \lambda I) = (4-\lambda) \begin{vmatrix} 3-\lambda & -2 \\ 1 & 1-\lambda \end{vmatrix} = (4-\lambda)(\lambda^2 - 4\lambda + 5).$$

Thus, $\lambda = 4$ is one eigenvalue, and the other two are $\lambda = 2 \pm \mathrm{i}$. You can show that

$$\begin{bmatrix} 0 \\ 0 \\ 1 \end{bmatrix}$$

is an eigenvector corresponding to $\lambda = 4$. As in the previous example, we must find a complex eigenvector for the eigenvalue with positive imaginary part ($\lambda = 2 + \mathrm{i}$):

$$A - \lambda I = \begin{bmatrix} 1-\mathrm{i} & -2 & 0 \\ 1 & -1-\mathrm{i} & 0 \\ 0 & 0 & 2-\mathrm{i} \end{bmatrix}.$$

Clearly the third row is independent of the others. However, since $\lambda = 2 + \mathrm{i}$ is an eigenvalue, we know that these rows must be linearly dependent. This means that rows 1 and 2 must form a linearly dependent set and, since there are only two rows, must be scalar multiples of each other. Row reduction will annihilate one of these two rows—we will replace the first row with zeros since the second row looks more convenient to work with. The reduced row-echelon form for $(A - \lambda I)$ is

$$\begin{bmatrix} 1 & -1-\mathrm{i} & 0 \\ 0 & 0 & 1 \\ 0 & 0 & 0 \end{bmatrix}.$$

Eigenvectors $\mathbf{w}$ must satisfy

$$\begin{bmatrix} 1 & -1-\mathrm{i} & 0 \\ 0 & 0 & 1 \\ 0 & 0 & 0 \end{bmatrix} \begin{bmatrix} w_1 \\ w_2 \\ w_3 \end{bmatrix} = \begin{bmatrix} 0 \\ 0 \\ 0 \end{bmatrix},$$

from which we conclude that $w_1 + (-1 - \mathrm{i})w_2 = 0$ and $w_3 = 0$. Treating $w_2$ as the free variable, we find that $w_1 = (1 + \mathrm{i})w_2$. To find our complex eigenvector, we write

$$\begin{bmatrix} w_1 \\ w_2 \\ w_3 \end{bmatrix} = \begin{bmatrix} (1+\mathrm{i})w_2 \\ w_2 \\ 0 \end{bmatrix} = w_2 \begin{bmatrix} 1+\mathrm{i} \\ 1 \\ 0 \end{bmatrix}.$$

As before, we follow the instructions of Theorem 2.1.28 by splitting $\mathbf{w}$ into its real and imaginary parts:

$$\begin{bmatrix} 1+\mathrm{i} \\ 1 \\ 0 \end{bmatrix} = \mathbf{w} = \mathbf{u} + \mathrm{i}\mathbf{v} = \begin{bmatrix} 1 \\ 1 \\ 0 \end{bmatrix} + \mathrm{i} \begin{bmatrix} 1 \\ 0 \\ 0 \end{bmatrix}.$$

Next, we form the matrix $P$ by arranging the vectors as suggested by Theorem 2.1.28. Column 1 should be the vector $\mathbf{v}$, Column 2 should be $\mathbf{u}$, and Column 3 should be the eigenvector corresponding to the real eigenvalue $\lambda = 4$. In summary,

$$P = \begin{bmatrix} 1 & 1 & 0 \\ 0 & 1 & 0 \\ 0 & 0 & 1 \end{bmatrix}.$$

It is easily checked that

$$P^{-1} = \begin{bmatrix} 1 & -1 & 0 \\ 0 & 1 & 0 \\ 0 & 0 & 1 \end{bmatrix}$$

and that $A = PMP^{-1}$ where

$$M = \begin{bmatrix} 2 & -1 & 0 \\ 1 & 2 & 0 \\ 0 & 0 & 4 \end{bmatrix}.$$

Notice that $M$ has the same block-diagonal form as the original matrix $A$. The upper-left block of $M$ contains the real canonical form of the upper-left block of $A$.

Our last step is to calculate $e^{tA}$ and to multiply this matrix by the vector of initial conditions. Using Proposition 2.1.26, we compute

$$e^{tA} = Pe^{tM}P^{-1} = P \begin{bmatrix} e^{2t}\cos t & -e^{2t}\sin t & 0 \\ e^{2t}\sin t & e^{2t}\cos t & 0 \\ 0 & 0 & e^{4t} \end{bmatrix} P^{-1},$$

where we have exponentiated the two diagonal blocks separately. After evaluating the above matrix product (which is tedious), multiply the result by the vector of initial conditions to obtain

$$\begin{bmatrix} x_1(t) \\ x_2(t) \\ x_3(t) \end{bmatrix} = e^{tA} \begin{bmatrix} 1 \\ 2 \\ 3 \end{bmatrix} = \begin{bmatrix} e^{2t}(\cos t - 3\sin t) \\ e^{2t}(2\cos t - \sin t) \\ 3e^{4t} \end{bmatrix}.$$

**2.1.4 Repeated Eigenvalues and Non-Diagonalizable Matrices.** We now turn our attention to devising a method for exponentiating matrices that are not diagonalizable. In order for a matrix to be diagonalizable, we need the geometric multiplicities of the different eigenvalues to sum to $n$, the dimension of the underlying space. Sometimes, eigenvalues with high algebraic multiplicity may have low geometric multiplicity.

**Example 2.1.31.** Compute $e^{tA}$, where

$$A = \begin{bmatrix} 0 & 1 & 0 \\ 0 & 0 & 1 \\ 0 & 0 & 0 \end{bmatrix}.$$

Since $A$ is upper-triangular, we immediately see that $\lambda = 0$ is an eigenvalue of algebraic multiplicity 3. To find eigenvectors, we must solve $(A - \lambda I)\mathbf{v} = 0$ which, in this case, simplifies to $A\mathbf{v} = 0$. The matrix $A$ is already in reduced row-echelon form, and the vectors $\mathbf{v}$ which satisfy $A\mathbf{v} = 0$ are of the form $v_2 = v_3 = 0$, with $v_1$ free. Thus,

$$\begin{bmatrix} v_1 \\ v_2 \\ v_3 \end{bmatrix} = \begin{bmatrix} v_1 \\ 0 \\ 0 \end{bmatrix} = v_1 \begin{bmatrix} 1 \\ 0 \\ 0 \end{bmatrix},$$

and we see that the eigenspace is only one-dimensional. Since our only eigenvalue $\lambda = 0$ has geometric multiplicity 1, Theorem 2.1.17 tells us that $A$ is *not* diagonalizable.

In order to compute the matrix exponential, we resort to a direct use of the definition

$$e^{tA} = \sum_{k=0}^{\infty} \frac{t^k}{k!} A^k = I + tA + \frac{t^2}{2!} A^2 + \cdots$$

Fortunately, for this particular matrix, the powers are not so difficult to calculate. You can show that

$$A^2 = \begin{bmatrix} 0 & 0 & 1 \\ 0 & 0 & 0 \\ 0 & 0 & 0 \end{bmatrix}$$

and that $A^3$ is the zero matrix. It follows that the matrix $A^k = 0$ if $k \geq 3$, which means that the matrix exponential is actually a *finite* sum

$$e^{tA} = I + tA + \frac{t^2}{2!} A^2 = \begin{bmatrix} 1 & t & \frac{1}{2}t^2 \\ 0 & 1 & t \\ 0 & 0 & 1 \end{bmatrix}.$$

This example motivates the following definition.

**Definition 2.1.32.** Let $k$ be a positive integer. An $n \times n$ matrix $N$ is called *nilpotent of order k* if $N^{k-1} \neq 0$ but $N^k = 0$.

In the above example, $A$ is nilpotent of order 3. The fact that a non-zero matrix can be nilpotent is quite different from what the reader may be accustomed to. Certainly the concept of nilpotency does not apply to real numbers—if $\alpha$ is a real number and $k \geq 1$, it is impossible for $\alpha^k = 0$ unless $\alpha = 0$.

The matrix exponential of a nilpotent matrix is represented by a *finite* sum, a fact that we alluded to in the previous example and now state formally.

**Lemma 2.1.33.** If $N$ is nilpotent of order $k$, then

$$e^{tN} = I + tN + \frac{t^2}{2!}N^2 + \cdots + \frac{t^{k-1}}{(k-1)!}N^{k-1} = \sum_{j=0}^{k-1} \frac{t^j}{j!}N^j.$$

*Proof.* Since $N$ is nilpotent of order $k$, we know that $N^m = 0$ whenever $m \geq k$. This means that we can "drop" every term of index $k$ or higher in the series representation of $e^{tN}$. □

Before we present a method for exponentiating non-diagonalizable matrices, we need one last definition from linear algebra.

**Definition 2.1.34.** Suppose $A$ is an $n \times n$ matrix and that $\lambda$ is an eigenvalue of algebraic multiplicity $m \leq n$. Then for each $k = 1, 2, \ldots m$, any non-zero solution $\mathbf{v}$ of the equation $(A - \lambda I)^k \mathbf{v} = 0$ is called a *generalized eigenvector of A*.

The following Theorem tells us how to decompose a non-diagonalizable matrix as a sum of two types of matrices that we know how to exponentiate.

**Theorem 2.1.35.** (Simple Jordan Decompositions): Suppose that $A$ is an $n \times n$ matrix with real eigenvalues $\lambda_1, \lambda_2, \ldots \lambda_n$, repeated according to algebraic multiplicity. Then there exists a family of $n$ generalized eigenvectors $\mathbf{v}_1, \mathbf{v}_2, \ldots \mathbf{v}_n$ such that

☞ The matrix $P = [\mathbf{v}_1|\mathbf{v}_2|\cdots|\mathbf{v}_n]$ is invertible.

☞ $A = S + N$ where $N$ is nilpotent and $S$ is diagonalizable. Specifically, if we define $D = \text{diag}\{\lambda_1, \lambda_2, \ldots \lambda_n\}$, then $S = PDP^{-1}$.

☞ $S$ and $N$ commute: $SN = NS$.

We illustrate the usefulness of this Theorem via an example.

**Example 2.1.36.** Compute $e^{tA}$ where

$$A = \begin{bmatrix} 1 & 0 & 0 \\ -1 & 2 & 0 \\ 1 & 1 & 2 \end{bmatrix}.$$

**Solution:** Since $A$ is lower-triangular, the eigenvalues are the entries on the main diagonal. You can check that

$$\begin{bmatrix} 1 \\ 1 \\ -2 \end{bmatrix}$$

is an eigenvector for the eigenvalue $\lambda = 1$. The other eigenvalue $\lambda = 2$ has algebraic multiplicity 2. To compute corresponding eigenvectors, we row-reduce $(A - \lambda I)$ as

$$A - 2I = \begin{bmatrix} -1 & 0 & 0 \\ -1 & 0 & 0 \\ 1 & 1 & 0 \end{bmatrix} \rightarrow \begin{bmatrix} 1 & 0 & 0 \\ 0 & 1 & 0 \\ 0 & 0 & 0 \end{bmatrix}.$$

Hence, solutions of $(A - 2I)\mathbf{v} = 0$ have the form $v_1 = v_2 = 0$, with $v_3$ free. It follows that

$$\begin{bmatrix} 0 \\ 0 \\ 1 \end{bmatrix}$$

is an eigenvector for $\lambda = 2$. Unfortunately, since the geometric multiplicity of this eigenvalue is only 1, we have failed to produce a set of 3 linearly independent eigenvectors for the matrix $A$, which means that $A$ is not diagonalizable. Theorem 2.1.35 suggests that we compute a generalized eigenvector for $\lambda = 2$. We must solve $(A - 2I)^2\mathbf{v} = 0$, or equivalently

$$\begin{bmatrix} 1 & 0 & 0 \\ 1 & 0 & 0 \\ -2 & 0 & 0 \end{bmatrix} \begin{bmatrix} v_1 \\ v_2 \\ v_3 \end{bmatrix} = \begin{bmatrix} 0 \\ 0 \\ 0 \end{bmatrix}.$$

Row-reduction gives

$$\begin{bmatrix} 1 & 0 & 0 \\ 0 & 0 & 0 \\ 0 & 0 & 0 \end{bmatrix} \begin{bmatrix} v_1 \\ v_2 \\ v_3 \end{bmatrix} = \begin{bmatrix} 0 \\ 0 \\ 0 \end{bmatrix},$$

which has solution $v_1 = 0$ with both $v_2$ and $v_3$ free. Thus, solutions of $(A - 2I)\mathbf{v} = 0$ are

$$\begin{bmatrix} v_1 \\ v_2 \\ v_3 \end{bmatrix} = \begin{bmatrix} 0 \\ v_2 \\ v_3 \end{bmatrix} = v_2 \begin{bmatrix} 0 \\ 1 \\ 0 \end{bmatrix} + v_3 \begin{bmatrix} 0 \\ 0 \\ 1 \end{bmatrix}.$$

We recognize

$$\begin{bmatrix} 0 \\ 0 \\ 1 \end{bmatrix}$$

as the eigenvector that we had already produced, which means that we can take

$$\begin{bmatrix} 0 \\ 1 \\ 0 \end{bmatrix}$$

as our *generalized* eigenvector. Next, we form the matrix

$$P = \begin{bmatrix} 1 & 0 & 0 \\ 1 & 0 & 1 \\ -2 & 1 & 0 \end{bmatrix}.$$

Notice that Column 1 contains the eigenvector for the eigenvalue $\lambda = 1$, Column 2 contains the eigenvector for $\lambda = 2$, and Column 3 contains the generalized eigenvector for $\lambda = 2$. Inverting $P$, we find that

$$P^{-1} = \begin{bmatrix} 1 & 0 & 0 \\ 2 & 0 & 1 \\ -1 & 1 & 0 \end{bmatrix}.$$

Let $D$ denote the diagonal matrix $D = \operatorname{diag}\{1, 2, 2\}$ consisting of the eigenvalues, arranged in an order corresponding to the order we used for the columns of $P$. The matrix $S$ in the statement of Theorem 2.1.35 is given by

$$S = PDP^{-1} = \begin{bmatrix} 1 & 0 & 0 \\ -1 & 2 & 0 \\ 2 & 0 & 2 \end{bmatrix},$$

from which we define

$$N = A - S = \begin{bmatrix} 0 & 0 & 0 \\ 0 & 0 & 0 \\ -1 & 1 & 0 \end{bmatrix}.$$

The Theorem says that $N$ should be a nilpotent matrix—indeed, the reader can check that $N^2 = 0$. Importantly, the Theorem also guarantees that $S$ and $N$ *commute*, which means that $e^{tA} = e^{t(S+N)} = e^{tS}e^{tN}$. Since $S$ is diagonalizable and $N$ is nilpotent, we know how to exponentiate both of these matrices. According to Lemma 2.1.33, since $N$ is nilpotent of order 2, we have

$$e^{tN} = I + tN = \begin{bmatrix} 1 & 0 & 0 \\ 0 & 1 & 0 \\ -t & t & 1 \end{bmatrix}.$$

Moreover,

$$e^{tS} = Pe^{tD}P^{-1} = P \begin{bmatrix} e^t & 0 & 0 \\ 0 & e^{2t} & 0 \\ 0 & 0 & e^{2t} \end{bmatrix} P^{-1}.$$

Finally, since $e^{tA} = e^{tS}e^{tN}$, after tedious matrix multiplication we find that

$$e^{tA} = \begin{bmatrix} e^t & 0 & 0 \\ e^t - e^{2t} & e^{2t} & 0 \\ -2e^t + (2-t)e^{2t} & te^{2t} & e^{2t} \end{bmatrix}.$$

**Observation.** Repeated real eigenvalues can cause the matrix exponential to contain products of polynomials with exponentials; e.g., $(2-t)e^{2t}$ or $t^7e^{-5t}$.

**Example 2.1.37.** Solve the initial value problem

$$\begin{aligned} x_1' &= x_2 & x_1(0) &= 2 \\ x_2' &= -9x_1 - 6x_2 & x_2(0) &= 1. \end{aligned}$$

**Solution:** If we write the system in matrix form, the coefficient matrix is

$$A = \begin{bmatrix} 0 & 1 \\ -9 & -6 \end{bmatrix}.$$

The characteristic equation is $\lambda^2 + 6\lambda + 9 = 0$, which factors as $(\lambda + 3)^2 = 0$. Thus, $\lambda = -3$ is an eigenvalue of algebraic multiplicity 2. The matrix $A - \lambda I$ row reduces to

$$\begin{bmatrix} 3 & 1 \\ 0 & 0 \end{bmatrix},$$

which means that eigenvectors $\mathbf{v}$ must satisfy the equation

$$\begin{bmatrix} 3 & 1 \\ 0 & 0 \end{bmatrix} \begin{bmatrix} v_1 \\ v_2 \end{bmatrix} = \begin{bmatrix} 0 \\ 0 \end{bmatrix}.$$

This implies that $3v_1 + v_2 = 0$ and, treating $v_2$ as the free variable, we may write $v_1 = -\frac{1}{3}v_2$. Solutions of $(A - \lambda I)\mathbf{v} = 0$ have the form

$$\begin{bmatrix} v_1 \\ v_2 \end{bmatrix} = v_2 \begin{bmatrix} -\frac{1}{3} \\ 1 \end{bmatrix},$$

which means that our eigenspace is only one-dimensional. Theorem 2.1.35 suggests that we find a generalized eigenvector by solving $(A - \lambda I)^2\mathbf{v} = 0$. Interestingly,

$$(A - \lambda I)^2 = 0 \qquad \longleftarrow \qquad \text{Remember this later!!!}$$

(Used properly, this observation will allow for a *much* quicker solution to our problem. We will see how after going through the tedious motions of Theorem 2.1.35.) Consequently, if we attempt to solve $(A - \lambda I)^2\mathbf{v} = 0$, then both $v_1$ and $v_2$ are free variables (i.e., any vector will satisfy this equation). For our generalized eigenvector, we may select *any* vector which is independent of the eigenvector we already computed. It is convenient to choose

$$\begin{bmatrix} 0 \\ 1 \end{bmatrix}$$

as our generalized eigenvector, and form the matrix

$$P = \begin{bmatrix} -\frac{1}{3} & 0 \\ 1 & 1 \end{bmatrix}.$$

Again, notice that Column 1 contains our eigenvector and Column 2 contains our generalized eigenvector. You can check that

$$P^{-1} = \begin{bmatrix} -3 & 0 \\ 3 & 1 \end{bmatrix}.$$

Letting $D = \text{diag}\{-3, -3\}$ denote the diagonal matrix with our repeated eigenvalue as its entries, we compute the matrix $S$ in the statement of Theorem 2.1.35:

$$S = PDP^{-1} = \begin{bmatrix} -\frac{1}{3} & 0 \\ 1 & 1 \end{bmatrix} \begin{bmatrix} -3 & 0 \\ 0 & -3 \end{bmatrix} \begin{bmatrix} -3 & 0 \\ 3 & 1 \end{bmatrix} = \begin{bmatrix} -3 & 0 \\ 0 & -3 \end{bmatrix}.$$

(Again, we make the curious observation that $S$ is not simply diagonalizable—it's diagonal!). Theorem 2.1.35 tells us that

$$N = A - S = \begin{bmatrix} 3 & 1 \\ -9 & -3 \end{bmatrix}$$

should be nilpotent, and you can check that $N^2 = 0$. We have $A = S + N$ where $S$ and $N$ commute, which means that $e^{tA} = e^{tN}e^{tS}$. Since $N$ is nilpotent of order 2, Lemma 2.1.33 tells us that $e^{tN} = I + tN$. Thus,

$$\begin{aligned} e^{tA} = [I + tN]e^{tS} &= \begin{bmatrix} 1+3t & t \\ -9t & 1-3t \end{bmatrix} \begin{bmatrix} e^{-3t} & 0 \\ 0 & e^{-3t} \end{bmatrix} \\ &= \begin{bmatrix} (1+3t)e^{-3t} & te^{-3t} \\ -9te^{-3t} & (1-3t)e^{-3t} \end{bmatrix}. \end{aligned}$$

Finally, we multiply $e^{tA}$ by our vector of initial conditions to get the overall solution

$$\begin{bmatrix} x_1(t) \\ x_2(t) \end{bmatrix} = \begin{bmatrix} (1+3t)e^{-3t} & te^{-3t} \\ -9te^{-3t} & (1-3t)e^{-3t} \end{bmatrix} \begin{bmatrix} 2 \\ 1 \end{bmatrix} = \begin{bmatrix} (2+7t)e^{-3t} \\ (1-21t)e^{-3t} \end{bmatrix}.$$

**Remark.** As soon as we noticed that $(A - \lambda I)^2 = 0$ we could have saved *lots* of time. The whole purpose of Theorem 2.1.35 is to decompose a matrix $A$ as a sum of a diagonalizable matrix $S$ and a nilpotent matrix $N$. So upon observing that $A - \lambda I$ is nilpotent (of order 2), we could immediately define $A - \lambda I = N$. According to the notation of the Theorem, this would mean $S = \lambda I$. We have written $A = S + N$ where $S$ is diagonal, $N$ is nilpotent, and $SN = NS$. Hence, we could have immediately computed the matrix exponential for $A$ without any further effort!

The techniques we have learned so far can be combined to handle cases we have not considered (such as repeated complex conjugate eigenvalues). In principle, it is possible to solve any linear constant-coefficient system. We have not attempted to state the most general possible theorems regarding solutions of constant-coefficient systems. Instead, we will move ahead to a topic that will help us visualize the solutions.

---

## 2.2. Phase Portraits and Planar Systems

We now introduce the concept of the *phase plane*, a useful way of visualizing solutions of planar systems (two dependent variables) of ODEs. We begin by considering the system

$$\begin{bmatrix} x_1' \\ x_2' \end{bmatrix} = \begin{bmatrix} 0 & -1 \\ 1 & 0 \end{bmatrix} \begin{bmatrix} x_1 \\ x_2 \end{bmatrix}.$$

Observe that the coefficient matrix $A$ is already in real canonical form, and

$$e^{tA} = \begin{bmatrix} \cos t & -\sin t \\ \sin t & \cos t \end{bmatrix}.$$

If we were interested in the solution satisfying initial conditions $x_1(0) = 1$ and $x_2(0) = 0$, we would multiply $e^{tA}$ by the vector $[1 \;\; 0]$ to obtain $x_1(t) = \cos t$

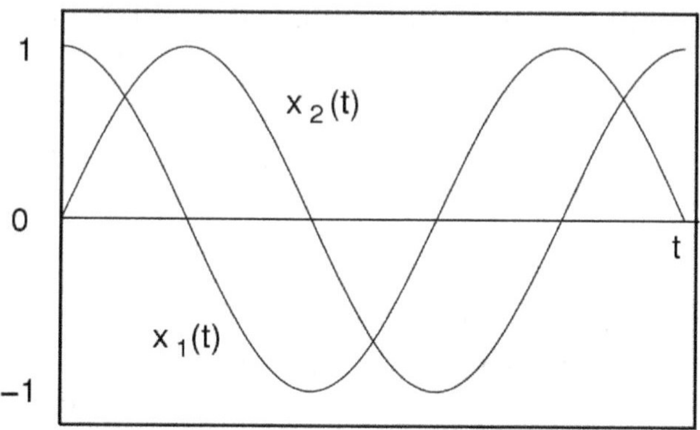


**Figure 2.2.** Graphs of $x_1(t) = \cos t$ and $x_2(t) = \sin t$.

and $x_2(t) = \sin t$. The most natural way to visualize these solutions is to simply graph both $x_1$ and $x_2$ as a function of $t$ as in Figure 2.2. An alternate way to visualize the solution is to plot $x_2(t)$ versus $x_1(t)$ as a curve parametrized by time $t$. Such a plot is given in Figure 2.3, which illustrates the parametrized curve $(\cos t, \sin t)$ for $t \in [0, 2\pi]$. In this case, the parametrized curve is a circle of radius 1 because $x_1(t)$ and $x_2(t)$ satisfy the relationship

$$x_1^2 + x_2^2 \;=\; \cos^2(t) + \sin^2(t) \;=\; 1$$

for all $t$. Notice that as $t$ increases, the curve is traversed in the counterclockwise direction.

Other initial conditions would give rise to other parametric curves, giving a visual representation of the "flow" of the system. A parametric plot of $x_2(t)$ versus $x_1(t)$ for various choices of initial conditions is called a *phase portrait* or *phase plane diagram* for the system.

We now discuss three canonical examples of phase portraits for planar systems, and then explain how these special cases can be used to sketch phase portraits for more general planar systems. From the previous subsections, we know that every $2 \times 2$ matrix $A$ can be written in the form $A = PMP^{-1}$, where $M$ has one of three forms:

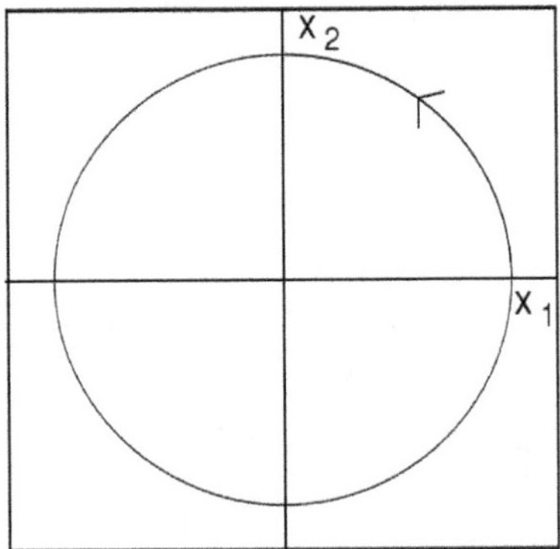


**Figure 2.3.** Parametric plot of $x_2(t) = \sin t$ versus $x_1(t) = \cos t$ for $t \in [0, 2\pi]$.

☞ If $A$ is diagonalizable, then

$$M = \begin{bmatrix} \alpha & 0 \\ 0 & \beta \end{bmatrix}.$$

☞ If $A$ is non-diagonalizable and has a real, repeated eigenvalue $\alpha$, then

$$M = \begin{bmatrix} \alpha & 1 \\ 0 & \alpha \end{bmatrix}.$$

☞ If $A$ has complex-conjugate eigenvalues $\alpha \pm \beta i$, then

$$M = \begin{bmatrix} \alpha & -\beta \\ \beta & \alpha \end{bmatrix}.$$

**Case 1:** Suppose $\mathbf{x}' = M\mathbf{x}$ where $M = \text{diag}\{\alpha, \beta\}$. The general solution is $\mathbf{x}(t) = e^{tM}\mathbf{x}(0)$; i.e.,

$$\begin{bmatrix} x_1(t) \\ x_2(t) \end{bmatrix} = \begin{bmatrix} e^{\alpha t} & 0 \\ 0 & e^{\beta t} \end{bmatrix} \begin{bmatrix} c_1 \\ c_2 \end{bmatrix} = \begin{bmatrix} c_1 e^{\alpha t} \\ c_2 e^{\beta t} \end{bmatrix}.$$

The exponential solutions $x_1 = c_1 e^{\alpha t}$ and $x_2 = c_2 e^{\beta t}$ either grow or decay depending upon the signs of $\alpha$ and $\beta$. Suppose that both $\alpha$ and $\beta$ are positive. Then both $e^{\alpha t}$ and $e^{\beta t}$ will increase without bound as $t \to \infty$. We can sketch

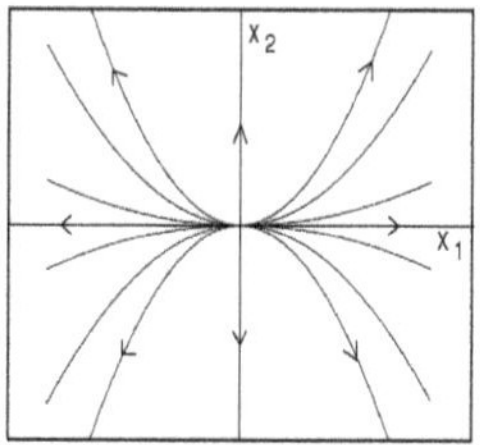

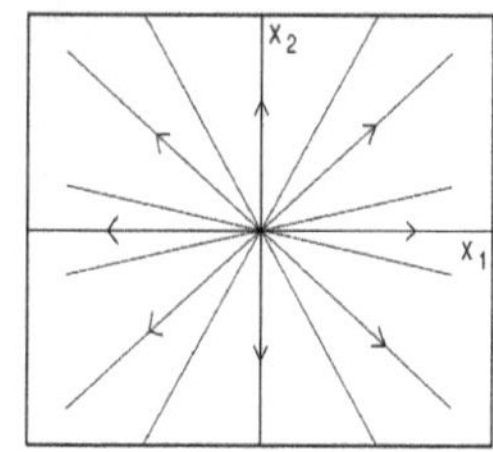

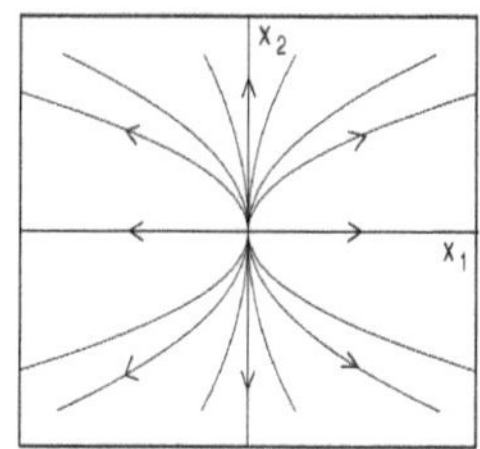


**Figure 2.4.** Phase portraits corresponding to Case 1 in which both $\alpha$ and $\beta$ are positive. Left panel: $0 < \alpha < \beta$. Middle panel: $0 < \alpha = \beta$. Right panel: $0 < \beta < \alpha$.

the phase portrait by eliminating the parameter $t$. Notice that $x_1^\beta = c_1^\beta e^{\alpha\beta t}$ and $x_2^\alpha = c_2^\alpha e^{\alpha\beta t}$. Assuming that $c_1 \neq 0$, taking the ratio gives

$$\frac{x_2^\alpha}{x_1^\beta} = \frac{c_2^\alpha}{c_1^\beta} = c,$$

where $c$ is a constant. By algebra, we have $x_2 = c x_1^{\beta/\alpha}$, and graphing such power functions is easy. Figure 2.4 shows the possible phase portraits in cases where both $\alpha$ and $\beta$ are positive. Notice that all trajectories point outward from the origin. In particular, if $\beta = \alpha$ (and thus $\beta/\alpha = 1$), the trajectories in the phase plane are straight lines directed outward from the origin.

If both $\alpha$ and $\beta$ are negative, then both $x_1$ and $x_2$ will decay to zero exponentially fast as $t \to \infty$. As indicated in Figure 2.5, the phase portraits are similar to the previous ones, except that all trajectories are directed inward towards the origin.

A more interesting situation occurs when $\alpha$ and $\beta$ have different signs. In this case, we have exponential growth for one of our variables and exponential decay for the other variable. The possible phase portraits are shown in Figure 2.6.

The various phase portraits in Figures 2.4–2.6 all have one thing in common: the special solution corresponding to $c_1 = c_2 = 0$ is a constant solution in which we stay "stuck" at the origin in the phase plane for all time $t$. Constant solutions of ODEs will be of particular interest to us, and we give them a special name.

**Definition 2.2.1.** An *equilibrium* of a system of ODEs is a solution that is constant. That is, all dependent variables are constant for all time $t$.

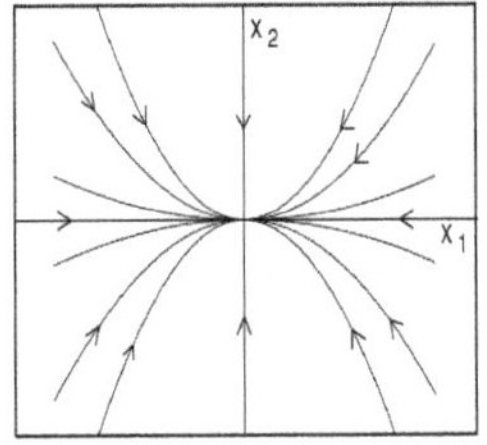

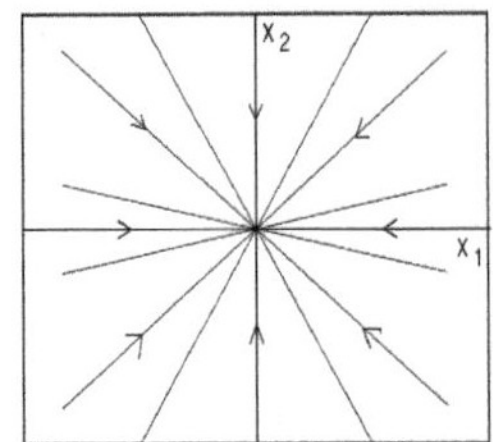

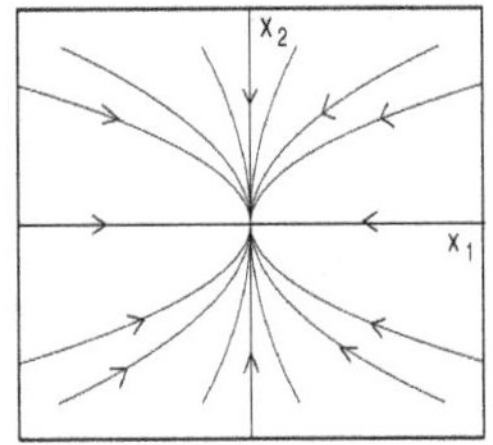


**Figure 2.5.** Phase portraits corresponding to Case 1 in which both $\alpha$ and $\beta$ are negative. Left panel: $\beta < \alpha < 0$. Middle panel: $\alpha = \beta < 0$. Right panel: $\alpha < \beta < 0$.

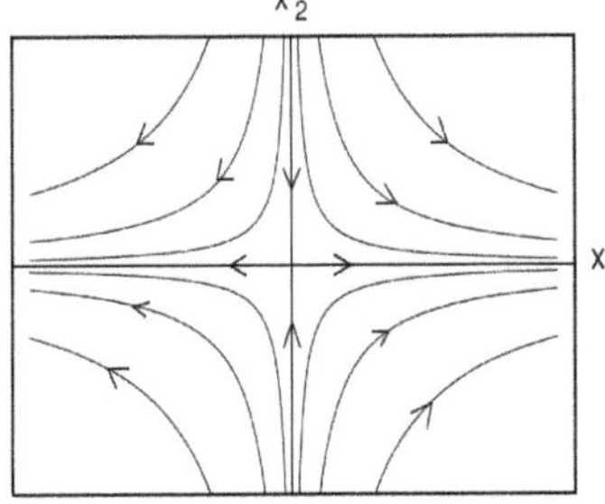

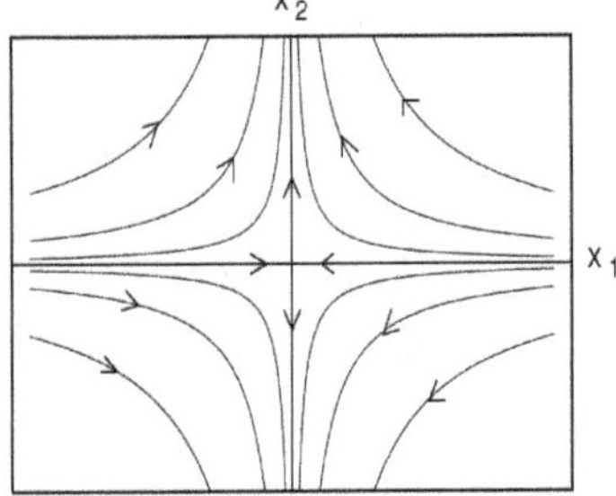


**Figure 2.6.** Phase portraits corresponding to Case 1 in which both $\alpha$ and $\beta$ have different signs. Left panel: $\beta < 0 < \alpha$. Right panel: $\alpha < 0 < \beta$.

**Observation.** For linear, constant-coefficient systems $\mathbf{x}' = A\mathbf{x}$ of ODEs, $\mathbf{x} = 0$ is *always* an equilibrium solution. In terms of the phase portrait, this means that the origin always corresponds to an equilibrium solution.

One way of further classifying equilibria is provided by the phase portraits discussed above. Namely,

**Definition 2.2.2.** In Figure 2.4, the origin is called an *unstable node*—all trajectories point outward from the origin. In Figure 2.5, the origin is called a *stable node*—all trajectories point inward to the origin. In Figure 2.6, the origin is called a *saddle*.

Later, we will actually re-state this Definition in a more precise way. We remark that saddle equilibria are different from stable/unstable nodes in that only certain, special trajectories actually touch the equilibrium point. In Figure 2.6, these are the trajectories which lie along the $x_1$ and $x_2$ axes. The four trajectories which approach the origin as $t \to \pm\infty$ are called *separatrices*. (Again, we will make this definition more precise later.)

**Case 2:** Suppose $\mathbf{x}' = M\mathbf{x}$ where $M$ is the non-diagonalizable matrix

$$M = \begin{bmatrix} \alpha & 1 \\ 0 & \alpha \end{bmatrix}$$

with $\alpha$ as a repeated, real eigenvalue. The phase portrait is more difficult to plot without computer assistance. To solve the system of ODEs, we decompose $M$ as the sum of a diagonal matrix and a nilpotent matrix:

$$M = D + N = \begin{bmatrix} \alpha & 0 \\ 0 & \alpha \end{bmatrix} + \begin{bmatrix} 0 & 1 \\ 0 & 0 \end{bmatrix}.$$

Clearly $D$ and $N$ commute since $D$ is a scalar multiple of the identity matrix, and you can also verify that $N$ is nilpotent of order 2. It follows that

$$e^{tM} = e^{tD}e^{tN} = e^{tD}[I + tN] = \begin{bmatrix} e^{\alpha t} & 0 \\ 0 & e^{\alpha t} \end{bmatrix}\begin{bmatrix} 1 & t \\ 0 & 1 \end{bmatrix} = \begin{bmatrix} e^{\alpha t} & te^{\alpha t} \\ 0 & e^{\alpha t} \end{bmatrix}.$$

The general solution of the system is therefore $\mathbf{x}(t) = e^{tM}\mathbf{x}(0)$, or

$$\begin{bmatrix} x_1(t) \\ x_2(t) \end{bmatrix} = \begin{bmatrix} e^{\alpha t} & te^{\alpha t} \\ 0 & e^{\alpha t} \end{bmatrix}\begin{bmatrix} c_1 \\ c_2 \end{bmatrix} = \begin{bmatrix} c_1 e^{\alpha t} + c_2 te^{\alpha t} \\ c_2 e^{\alpha t} \end{bmatrix}.$$

These solutions give a parametrization for the solution curves in the phase portrait, which appears in Figure 2.7. If $\alpha < 0$, the trajectories approach the origin as $t \to \infty$, and if $\alpha > 0$, the trajectories are oriented outward from the origin. Notice that, unlike the case in which the origin is a saddle equilibrium, there is only one separatrix in this case (the $x_1$-axis). In the left panel of Figure 2.7, the origin is a stable node, and in the right panel the origin is an unstable node.

We remark that the characteristic equation for $M$ is $(\lambda - \alpha)^2 = 0$, which is a "critical" case in the sense that we are just on the verge of having complex conjugate roots. Indeed, quadratic equations with negative discriminants have complex conjugate solutions, and repeated real roots occur when the discriminant is 0.

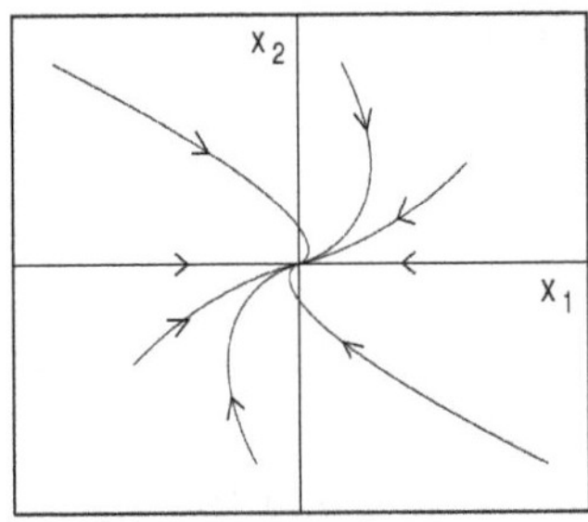

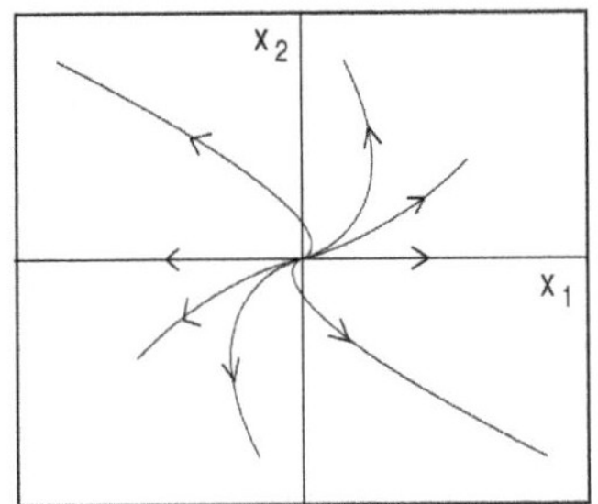


**Figure 2.7.** Phase portraits corresponding to Case 2, in which $M$ has a repeated eigenvalue. Left panel: $\alpha = -1$. Right panel: $\alpha = 1$.

**Case 3:** Suppose $\mathbf{x}' = M\mathbf{x}$ where $M$ is in real canonical form

$$M = \begin{bmatrix} \alpha & -\beta \\ \beta & \alpha \end{bmatrix}.$$

We know that

$$e^{tM} = e^{\alpha t} \begin{bmatrix} \cos \beta t & -\sin \beta t \\ \sin \beta t & \cos \beta t \end{bmatrix},$$

which means that the general solution of the system is

$$\begin{bmatrix} x_1(t) \\ x_2(t) \end{bmatrix} = e^{tM} \begin{bmatrix} c_1 \\ c_2 \end{bmatrix} = \begin{bmatrix} c_1 e^{\alpha t} \cos \beta t - c_2 e^{\alpha t} \sin \beta t \\ c_1 e^{\alpha t} \sin \beta t + c_2 e^{\alpha t} \cos \beta t \end{bmatrix}.$$

Solutions involve combinations of exponential functions $e^{\alpha t}$ with periodic functions $\sin \beta t$ and $\cos \beta t$. If $\alpha < 0$, the exponential factors will decay as $t \to \infty$, while the other factors simply oscillate. If we graph some solution curves in the phase portrait, we will see that trajectories spiral inward to the origin. Likewise, if $\alpha > 0$, trajectories will spiral outward from the origin. Finally, if $\alpha = 0$, the exponential factor is constant and the solutions are purely oscillatory. In this case, the phase portrait consists of concentric circular trajectories.

For matrices $M$ in real canonical form[1], the orientation of the trajectories can be determined from the sign of $\beta$: clockwise if $\beta < 0$ and counter-clockwise if $\beta > 0$. These various possibilities are summarized in Figure 2.8. The equilibrium at the

[1]If a matrix with complex conjugate eigenvalues is *not* in real canonical form, one simple way to determine the orientation of the trajectories is to plot a few "slope field vectors" as illustrated in the next example.

origin is called a *stable focus* if trajectories spiral inward ($\alpha < 0$), an *unstable focus* if trajectories spiral outward ($\alpha > 0$), and a *center* if trajectories form concentric, closed curves ($\alpha = 0$).

**Remark.** Above, we did not consider the possibility that the matrix $M$ in the equation $\mathbf{x}' = M\mathbf{x}$ has $\lambda = 0$ as an eigenvalue. In such cases, the origin is called a *degenerate* equilibrium. Notice that if $\lambda = 0$ is an eigenvalue, then $\det M = 0$, which means that $M$ is a singular matrix. It follows that the solutions of the equation $M\mathbf{x} = 0$ form a subspace of dimension *at least 1*, and any vector $\mathbf{x}$ in this subspace would be an equilibrium for our system of ODEs. In the remainder of the course, we will typically work only with systems which have *isolated* equilibrium points (defined later), as opposed to systems with infinitely many equilibrium points.

It is also worth noting that for the autonomous systems of ODEs that we have considered, *solution trajectories in the phase portrait cannot intersect each other*. Can you explain why?

Suppose a planar system of ODEs has a coefficient matrix $A$ which is not in one of the three canonical forms we discussed above. To sketch the phase portrait, we need to determine which canonical form $A$ is similar to, and this is accomplished by finding the eigenvalues and eigenvectors. If $A$ is any $2 \times 2$ matrix with non-zero eigenvalues, then the associated phase portrait is always a "skewed" version of one of the portraits in the above figures. We illustrate this via two examples.

**Example 2.2.3.** Sketch the phase portrait for $\mathbf{x}' = A\mathbf{x}$, where

$$A = \begin{bmatrix} 0 & -4 \\ 1 & 0 \end{bmatrix}.$$

**Solution:** The characteristic equation is $\lambda^2 + 4 = 0$, from which we infer that the eigenvalues are complex conjugate: $\lambda = \pm 2i$. We will put $A$ in real canonical form by finding an eigenvector for $\lambda = 2i$, the eigenvalue with positive imaginary part. Row-reducing the matrix $(A - \lambda I)$ gives

$$A - \lambda I = \begin{bmatrix} -2i & -4 \\ 1 & -2i \end{bmatrix} \longrightarrow \begin{bmatrix} 1 & -2i \\ 0 & 0 \end{bmatrix}.$$

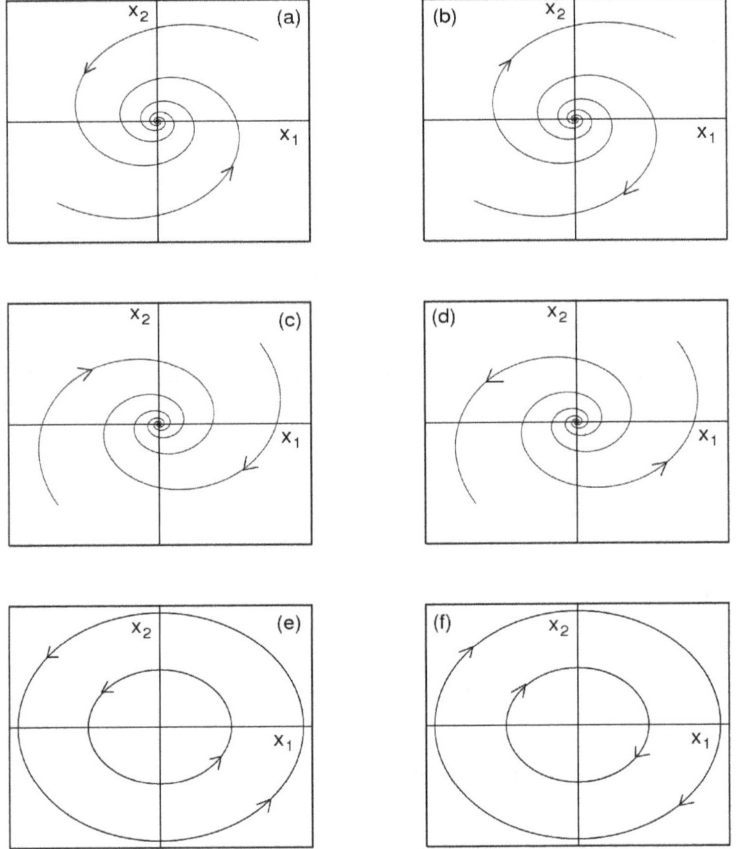


**Figure 2.8.** Phase portraits corresponding to Case 3, in which $M$ has a complex conjugate eigenvalues.

(a) Origin is a stable focus ($\alpha < 0$) with trajectories oriented counter-clockwise ($\beta > 0$).

(b) Origin is an unstable focus ($\alpha > 0$) with trajectories oriented clockwise ($\beta < 0$).

(c) Origin is a stable focus ($\alpha < 0$) with trajectories oriented clockwise ($\beta < 0$).

(d) Origin is an unstable focus ($\alpha > 0$) with trajectories oriented counter-clockwise ($\beta > 0$).

(e) Origin is a center ($\alpha = 0$) with trajectories oriented counter-clockwise ($\beta > 0$).

(f) Origin is a center ($\alpha = 0$) with trajectories oriented clockwise ($\beta < 0$).

Eigenvectors $\mathbf{w}$ must satisfy $w_1 - 2iw_2 = 0$, so we set $w_1 = 2iw_2$ and treat $w_2$ as a free variable. Next, express the eigenvector as a sum of its real and imaginary parts:

$$\begin{bmatrix} w_1 \\ w_2 \end{bmatrix} = \begin{bmatrix} 2iw_2 \\ w_2 \end{bmatrix} = w_2 \begin{bmatrix} 2i \\ 1 \end{bmatrix} = w_2 \left( \begin{bmatrix} 0 \\ 1 \end{bmatrix} + i \begin{bmatrix} 2 \\ 0 \end{bmatrix} \right).$$

Form the matrix

$$P = \begin{bmatrix} 2 & 0 \\ 0 & 1 \end{bmatrix},$$

which contains the imaginary part of $\mathbf{w}$ in its first column and the real part of $\mathbf{w}$ in its second column. Clearly

$$P^{-1} = \begin{bmatrix} \frac{1}{2} & 0 \\ 0 & 1 \end{bmatrix},$$

and the real canonical form for $A$ is given by

$$M = P^{-1}AP = \begin{bmatrix} 0 & -2 \\ 2 & 0 \end{bmatrix}.$$

Routine calculation yields the matrix exponential

$$e^{tA} = Pe^{tM}P^{-1} = P \begin{bmatrix} \cos 2t & -\sin 2t \\ \sin 2t & \cos 2t \end{bmatrix} P^{-1} = \begin{bmatrix} \cos 2t & -2\sin 2t \\ \frac{1}{2}\sin 2t & \cos 2t \end{bmatrix},$$

from which it follows that the general solution of our system is

$$\begin{bmatrix} x_1(t) \\ x_2(t) \end{bmatrix} = e^{tA} \begin{bmatrix} c_1 \\ c_2 \end{bmatrix} = \begin{bmatrix} c_1 \cos 2t - 2c_2 \sin 2t \\ \frac{1}{2}c_1 \sin 2t + c_2 \cos 2t \end{bmatrix}.$$

The parameter $t$ can be eliminated by algebra, leading to the relationship

$$x_1^2 + 4x_2^2 = c_1^2 + 4c_2^2,$$

where the right hand side is an arbitrary non-negative constant. You should recognize this as the equation of an ellipse. The phase portrait is sketched in Figure 2.9, and we see that the origin is a *center*. One way to determine the orientation of the trajectories is to pick a few convenient points in the phase

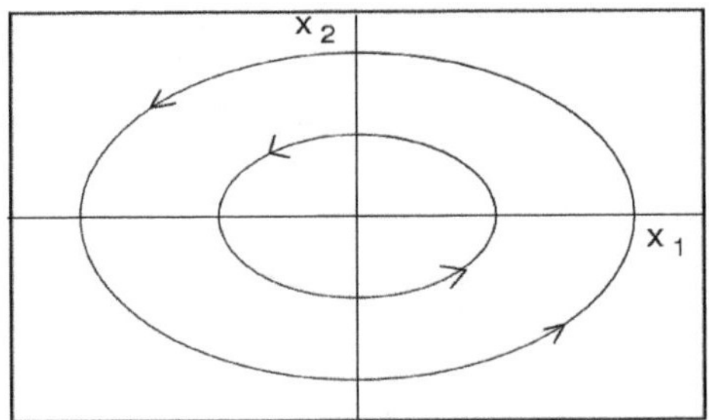


**Figure 2.9.** Phase portrait showing the elliptic trajectories $x_1^2 + 4x_2^2 =$ constant.

plane and sketch the associated "slope field" vectors. For example, suppose we start from the point $(x, y) = (1, 0)$ on the $x$-axis in the phase plane. To determine the direction of motion from that point, we multiply the coefficient matrix $A$ by $(1,0)$, obtaining $(0,1)$. The vector $(0,1)$, which points straight upward, is tangent to the solution trajectory passing through $(1,0)$. This indicates that trajectories are oriented counter-clockwise, as shown in Figure 2.9. Notice that our phase portrait is simply a "skewed" version of our canonical example of a center equilibrium (which had circular trajectories).

**Example 2.2.4.** Sketch the phase portrait for $\mathbf{x}' = A\mathbf{x}$, where

$$A = \begin{bmatrix} 2 & -3 \\ 1 & -2 \end{bmatrix}.$$

The characteristic equation is $\lambda^2 - 1 = 0$, so we have real eigenvalues $\lambda = \pm 1$ with different sign. We immediately conclude that the origin in our phase portrait will be a *saddle*. You should show that the eigenvalues $\lambda = 1$ and $\lambda = -1$ give rise to eigenvectors

$$\begin{bmatrix} 3 \\ 1 \end{bmatrix} \quad \text{and} \quad \begin{bmatrix} 1 \\ 1 \end{bmatrix},$$

respectively. By Proposition 2.1.21, the general solution is

$$\begin{bmatrix} x_1 \\ x_2 \end{bmatrix} = c_1 e^t \begin{bmatrix} 3 \\ 1 \end{bmatrix} + c_2 e^{-t} \begin{bmatrix} 1 \\ 1 \end{bmatrix}.$$

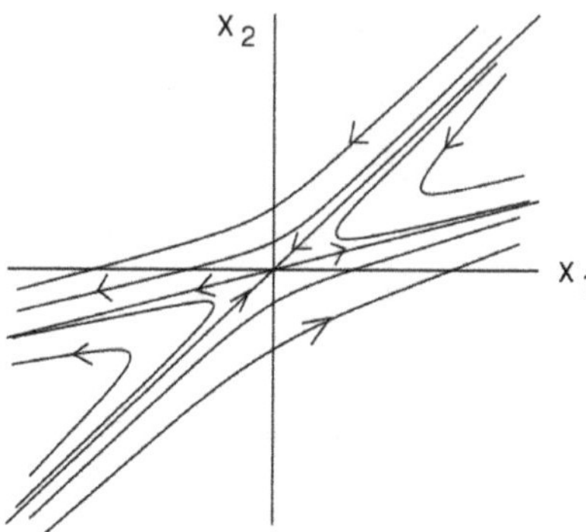


**Figure 2.10.** Phase portrait showing the skewed saddle The separatrices are the lines spanned by the eigenvectors $[1,1]$ and $[3,1]$.

**Consequence:** If we set $c_1 = 0$ and plot the trajectories

$$c_2 e^{-t} \begin{bmatrix} 1 \\ 1 \end{bmatrix}$$

in the phase plane, we find that this is a parametrization of a *line of slope 1* as $t$ varies from $-\infty$ to $\infty$. The value of $c_2$ merely selects our "starting point" as we traverse the line. Similarly, if we had set $c_2 = 0$, we would have obtained the parametrization of a line of slope $\frac{1}{3}$. In other words *the lines spanned by the two eigenvectors of A form the separatrices in our phase portrait.* This tells us precisely how to "skew" our canonical example of a saddle equilibrium to obtain a sketch of the phase portrait, which is shown in Figure 2.10. Notice that the motion along the separatrix corresponding to the negative eigenvalue is directed inward to the origin, while the motion along the separatrix corresponding to the positive eigenvalue is directed outward from the origin.

Before continuing our investigation of the "geometry" associated with planar, constant-coefficient systems, we follow up our earlier remark that the origin is always an equilibrium of the linear, homogeneous constant-coefficient system $\mathbf{x}' = A\mathbf{x}$. This admittedly narrow class of systems is not quite as exclusive as it may appear. Having an equilibrium at the origin is not a severe restriction at all, because equilibria can always be relocated via a change of variables. For example, consider the constant-coefficient system $\mathbf{x}' = A\mathbf{x} - \mathbf{b}$, where $\mathbf{b}$ is a constant vector. Assuming that $A$ is an invertible matrix (i.e., $\lambda = 0$ is not an eigenvalue), we may solve for equilibrium solutions by setting $\mathbf{x}' = 0$. Equivalently, any vector $\mathbf{x}$ satisfying $A\mathbf{x} = \mathbf{b}$ is an equilibrium. Let $\mathbf{x}^* = A^{-1}\mathbf{b}$ denote the equilibrium solution. Then our original system of ODEs can be written as $\mathbf{x}' = A\mathbf{x} - A\mathbf{x}^* = A(\mathbf{x} - \mathbf{x}^*)$. Making the substitution $\mathbf{y} = \mathbf{x} - \mathbf{x}^*$, we obtain a new system $\mathbf{y}' = A\mathbf{y}$. The new constant-coefficient system has its equilibrium at the origin, and we can solve it using the techniques discussed in previous sections. We may then recover the solution of our original system by writing $\mathbf{x} = \mathbf{y} + \mathbf{x}^*$.

The system $\mathbf{x}' = A\mathbf{x} - \mathbf{b}$ is actually a special case of the inhomogeneous systems we will learn how to solve soon.

---

## 2.3. Stable, Unstable, and Center Subspaces

In the previous subsection, we used the words 'stable' and 'unstable' without giving a careful definition of what those terms mean. We know that for homogeneous constant-coefficient systems $\mathbf{x}' = A\mathbf{x}$, the stability of the equilibrium at the origin is somehow determined by the eigenvalues of $A$. The eigenvectors of $A$ determine how the phase portrait is "skewed" from one of the three canonical phase portraits.

We will give a rigorous definition of stability of equilibria in the next chapter; for now, a loose definition will suffice. An equilibrium solution $\mathbf{x}^*$ of a system of ODEs is *stable* if, whenever we start from initial conditions that are appropriately "close" to $\mathbf{x}^*$, the resulting solution trajectory never strays too far from $\mathbf{x}^*$. A stable equilibrium is called *asymptotically stable* if the solution trajectory actually approaches $\mathbf{x}^*$ as $t \to \infty$. If an equilibrium is not stable, it is called *unstable*. This means that there exist initial conditions arbitrarily "close" to $\mathbf{x}^*$ for which the solution trajectory is repelled from $\mathbf{x}^*$.

For example, if the origin is a center, then we would say the origin is a stable equilibrium but is not asymptotically stable. Saddle equilibria are unstable, because it is always possible to choose initial conditions arbitrarily close to the equilibrium for which the direction of motion in the phase portrait is away from the equilibrium as $t$ increases. (Just choose any initial conditions *not* lying on the separatrix that is oriented towards the equilibrium.)

Saddles are interesting in that there are "special" trajectories in the phase portrait on which the direction of motion is directed *toward* the unstable equilibrium. In our canonical example of a saddle, one separatrix was oriented toward the origin, and the other separatrix was oriented away from the origin. The notion that our underlying space can be decomposed into stable and unstable "directions" is the subject of our discussion below. First, we recall a familiar definition from linear algebra.

**Definition 2.3.1.** Let $\mathbf{v}_1, \mathbf{v}_2, \ldots, \mathbf{v}_k$ be vectors in $\mathbb{R}^n$. The *span* of these vectors is the set of all linear combinations

$$c_1\mathbf{v}_1 + c_2\mathbf{v}_2 + \cdots + c_k\mathbf{v}_k,$$

where $c_1, c_2, \ldots c_k$ are real numbers.

Notice that if these vectors are linearly independent, then their span forms a $k$-dimensional subspace of $\mathbb{R}^n$.

**Example 2.3.2.** The span of the vectors

$$\begin{bmatrix} 1 \\ 0 \\ 0 \end{bmatrix} \quad \text{and} \quad \begin{bmatrix} 1 \\ 1 \\ 0 \end{bmatrix}$$

in $\mathbb{R}^3$ is the $xy$-plane.

Now consider the homogeneous system $\mathbf{x}' = A\mathbf{x}$ of ODEs, where $A$ is an $n \times n$ matrix. Let $\lambda_1, \lambda_2, \ldots \lambda_n$ denote the eigenvalues of $A$, repeated according to algebraic multiplicity. Each eigenvalue can be written in the form $\lambda_j = \alpha_j + i\beta_j$, where $\alpha_j$ and $\beta_j$ are real. (Of course, $\beta_j = 0$ if the eigenvalue $\lambda_j$ is real.) Associated with each eigenvalue is a set of eigenvectors (and possibly generalized eigenvectors).

**Definition 2.3.3.** The *stable subspace* of the system $\mathbf{x}' = A\mathbf{x}$ is the span of all eigenvectors and generalized eigenvectors associated with eigenvalues having *negative real part* ($\alpha_j < 0$). The *unstable subspace* of the system $\mathbf{x}' = A\mathbf{x}$ is the span of all eigenvectors and generalized eigenvectors associated with eigenvalues having *positive real part* ($\alpha_j > 0$). The *center subspace* of the system $\mathbf{x}' = A\mathbf{x}$ is the span of all eigenvectors and generalized eigenvectors associated with eigenvalues having *zero real part* ($\alpha_j = 0$).

**Notation:** The stable, unstable, and center subspaces are denoted by $E^s$, $E^u$, and $E^c$, respectively.

**Example 2.3.4.** Consider the system

$$\mathbf{x}' = \begin{bmatrix} 1 & 2 \\ 5 & 4 \end{bmatrix} \mathbf{x}.$$

Letting $A$ denote the coefficient matrix, the characteristic equation is given by

$$\lambda^2 - 5\lambda - 6 = 0.$$

Factoring the characteristic equation as $(\lambda - 6)(\lambda + 1) = 0$, we obtain eigenvalues $\lambda = -1$ and $\lambda = 6$. The roots are real and have opposite sign, indicating that the origin is a saddle. You can verify that

$$\begin{bmatrix} -1 \\ 1 \end{bmatrix} \quad \text{and} \quad \begin{bmatrix} \frac{2}{5} \\ 1 \end{bmatrix}$$

are eigenvectors corresponding to $\lambda = -1$ and $\lambda = 6$, respectively. By the above definition, the stable subspace is given by

$$E^s = \text{span} \left\{ \begin{bmatrix} -1 \\ 1 \end{bmatrix} \right\}$$

and the unstable subspace is given by

$$E^u = \text{span} \left\{ \begin{bmatrix} \frac{2}{5} \\ 1 \end{bmatrix} \right\}.$$

The center subspace $E^c$ consists only of the zero vector. Notice that $E^s$ and $E^u$ really *are* subspaces of $\mathbb{R}^2$, the underlying space. They are straight lines through the origin and, in this example of a saddle, they correspond to the separatrices (see Figure 2.11).

We remark that the general solution of this system is given by

$$\begin{bmatrix} x_1(t) \\ x_2(t) \end{bmatrix} = c_1 e^{-t} \begin{bmatrix} -1 \\ 1 \end{bmatrix} + c_2 e^{6t} \begin{bmatrix} \frac{2}{5} \\ 1 \end{bmatrix}.$$

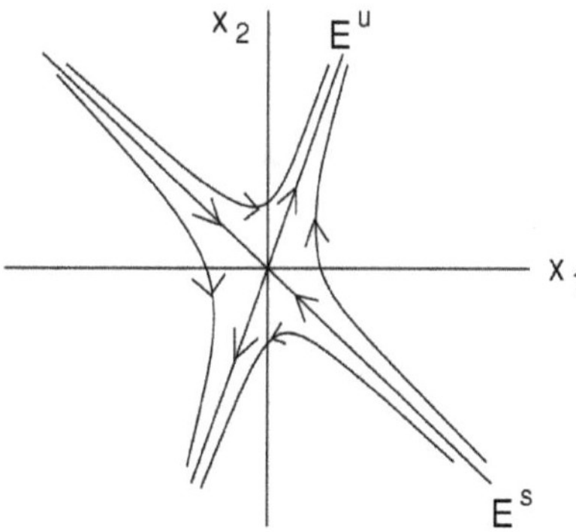


**Figure 2.11.** Phase portrait showing the stable and unstable subspaces $E^s$ and $E^u$.

If we start from (non-zero) initial conditions inside the stable subspace ($c_1 \neq 0$ and $c_2 = 0$), then our solution trajectory will remain in the stable subspace for all time $t$, and we will approach the origin as $t \to \infty$. Likewise, if we start from initial conditions inside the unstable subspace, ($c_1 = 0$ and $c_2 \neq 0$), then our solution trajectory remains in the unstable subspace for all time $t$ but we always move away from the origin.

**Example 2.3.5.** The system

$$\mathbf{x}' = \begin{bmatrix} 1 & -8 \\ 8 & 1 \end{bmatrix} \mathbf{x}$$

has a coefficient matrix that is already in real canonical form, and its eigenvalues are $1 \pm 8i$. Since both eigenvalues have positive real part, $E^u = \mathbb{R}^2$ whereas both $E^s$ and $E^c$ consist only of the zero vector. Likewise, the system

$$\mathbf{x}' = \begin{bmatrix} 0 & -8 \\ 8 & 0 \end{bmatrix} \mathbf{x}$$

has a coefficient matrix with eigenvalues $\pm 8i$, both of which have zero real part. In this case, $E^c = \mathbb{R}^2$ while the stable and unstable subspaces consist only of the zero vector.

The notions of stable, unstable, and center subspaces are not restricted to planar systems, as we illustrate in the following example.

**Example 2.3.6.** Consider the system

$$\mathbf{x}' = \begin{bmatrix} -2 & -1 & 0 \\ 1 & -2 & 0 \\ 0 & 0 & 8 \end{bmatrix} \mathbf{x},$$

whose coefficient matrix has a convenient block-diagonal structure. The eigenvalues are $-2 \pm \mathrm{i}$ and 8. The eigenvalue $-2 + \mathrm{i}$ gives rise to a complex eigenvector

$$\begin{bmatrix} 0 \\ 1 \\ 0 \end{bmatrix} + \mathrm{i} \begin{bmatrix} 1 \\ 0 \\ 0 \end{bmatrix},$$

and the real eigenvalue has

$$\begin{bmatrix} 0 \\ 0 \\ 1 \end{bmatrix}$$

as an eigenvector. Since $-2 + \mathrm{i}$ has negative real part, we conclude that the stable subspace is

$$E^s = \text{span}\left\{ \begin{bmatrix} 0 \\ 1 \\ 0 \end{bmatrix}, \begin{bmatrix} 1 \\ 0 \\ 0 \end{bmatrix} \right\}.$$

Graphically, this is the $xy$-plane in $\mathbb{R}^3$. The unstable subspace

$$E^u = \text{span}\left\{ \begin{bmatrix} 0 \\ 0 \\ 1 \end{bmatrix} \right\}$$

is one-dimensional and corresponds to the $z$-axis in $\mathbb{R}^3$. The center subspace $E^c$ consists only of the zero vector.

Several observations will help us sketch the phase portrait. Any trajectory starting from non-zero initial conditions in $E^s$ (the $xy$-plane) will spiral inward toward the origin while always remaining within $E^s$. Any non-zero trajectory starting in $E^u$ (the $z$-axis) will remain within $E^u$ for all time $t$ and will be oriented outward from the origin. Finally, any trajectory starting outside $E^s$ and $E^u$ will spiral away from the $xy$-plane but will draw closer to the $z$-axis as $t$ advances. A

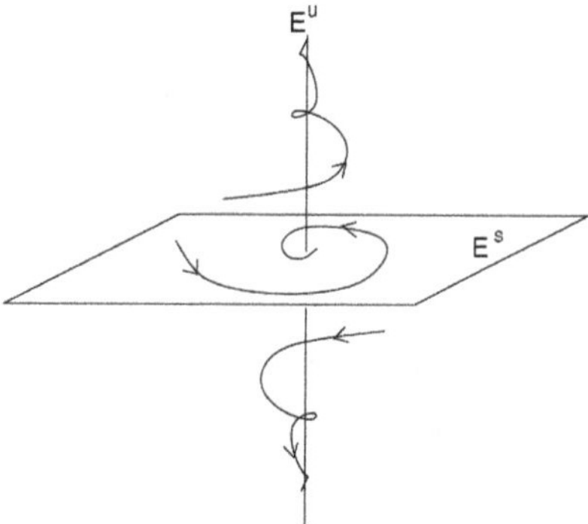


**Figure 2.12.** Phase portrait of a 3-D system with a two-dimensional stable subspace $E^s$ and a one-dimensional unstable subspace $E^u$.

sketch of the phase portrait appears in Figure 2.12. As another illustration, the system

$$\mathbf{x}' = \begin{bmatrix} 0 & -1 & 0 \\ 1 & 0 & 0 \\ 0 & 0 & 8 \end{bmatrix} \mathbf{x},$$

has a coefficient matrix with eigenvalues $\pm$i and 8. The corresponding eigenvectors are exactly as in the previous example. The only difference is that the $xy$-plane is now the *center* subspace because the eigenvectors that span that plane correspond to an eigenvalue with *zero* real part. The $z$-axis is the unstable subspace. Any trajectory starting from non-zero initial conditions in $E^c$ (the $xy$-plane) will circle around the origin, remaining in $E^c$ for all time $t$ without being attracted or repelled by the origin. Any non-zero trajectory starting in $E^u$ (the $z$-axis) will remain within $E^u$ for all time $t$ and will be oriented outward from the origin. Finally, any trajectory starting outside $E^c$ and $E^u$ will spiral away from the $xy$-plane but will always maintain a constant distance from the $z$-axis as $t$ advances. In other words, such trajectories are confined to "infinite cylinders." A sketch of this phase portrait appears in Figure 2.13.

Before moving on, we introduce some important terminology related to our above discussion. As usual, consider the system $\mathbf{x}' = A\mathbf{x}$ where $A$ is an $n \times n$ constant matrix. Given an initial condition $\mathbf{x}_0 = \mathbf{x}(0)$, we know that the unique solution of this initial value problem is given by $\mathbf{x}(t) = e^{tA}\mathbf{x}_0$. In terms of the phase portrait, once we pick the point $\mathbf{x}_0$, pre-multiplying by the matrix

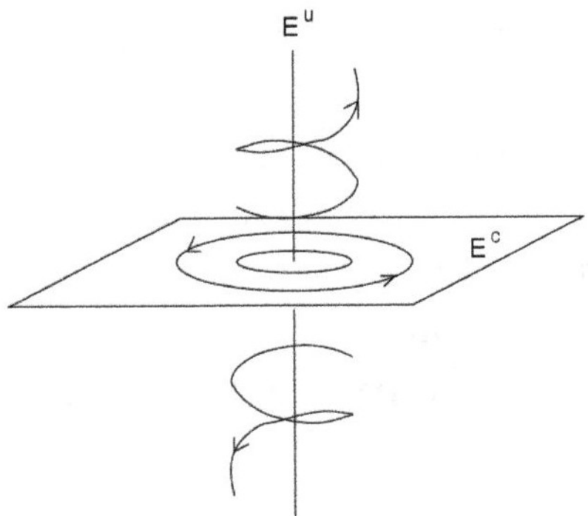


**Figure 2.13.** Phase portrait of a 3-D system with a two-dimensional center subspace $E^c$ and one-dimensional unstable subspace $E^u$.

$e^{tA}$ and letting $t$ increase will trace out a curve in $\mathbb{R}^n$. Let $\phi_t$ be the function which associates each different initial condition with its solution trajectory: $\phi_t(\mathbf{x}_0) = e^{tA}\mathbf{x}_0$.

**Definition 2.3.7.** The set of functions $\phi_t = e^{tA}$ is called the *flow* of the system $\mathbf{x}' = A\mathbf{x}$ of ODEs.

The reason for using the name "flow" is that $\phi_t$ describes the motion along trajectories in the phase space starting from various choices of initial conditions $\mathbf{x}_0$.

**Definition 2.3.8.** If *all* eigenvalues of $A$ have non-zero real part, the flow is called a *hyperbolic flow* the system $\mathbf{x}' = A\mathbf{x}$ is called a *hyperbolic system*, and the origin is called a *hyperbolic equilibrium* point.

**Example 2.3.9.** The system $\mathbf{x}' = A\mathbf{x}$ where

$$A = \begin{bmatrix} 0 & 1 \\ -9 & 0 \end{bmatrix},$$

is non-hyperbolic because the eigenvalues of $A$ are $\lambda = \pm 3i$, both of which have zero real part. The origin, a center in this case, is a non-hyperbolic equilibrium. Note that, if an equilibrium is hyperbolic, then the dimension of its center subspace is zero.

Our next comments concern properties of the stable, unstable, and center subspaces associated with the system $\mathbf{x}' = A\mathbf{x}$. In the examples we gave, there are several common themes. First, the pairwise intersections of $E^s$, $E^u$, and $E^c$ consist only of the zero vector. Second, the sum of the dimensions of these three subspaces is always equal to the dimension of the underlying space $\mathbb{R}^n$. Finally, if we start from an initial condition $\mathbf{x}_0$ which lies inside one of these three subspaces, then $\phi_t(\mathbf{x}_0)$ remains within that subspace for all real $t$. These observations are now stated formally.

Recall that if $S_1$ and $S_2$ are subspaces of a vector space $V$, then the *sum* of the subspaces is defined as

$$S_1 + S_2 = \{\mathbf{x} + \mathbf{y} : \mathbf{x} \in S_1 \text{ and } \mathbf{y} \in S_2\}.$$

The sum $S_1 + S_2$ is itself a subspace of $V$, and the concept of sums of vector spaces is easily extended to larger finite sums. In the special case where $S_1 + S_2 = V$ and $S_1 \cap S_2$ consists only of the zero vector, we refer to the sum as a *direct sum* and write $S_1 \oplus S_2 = V$.

**Theorem 2.3.10.** Consider the system $\mathbf{x}' = A\mathbf{x}$, where $A$ is an $n \times n$ constant matrix, and let $E^s$, $E^u$, and $E^c$ denote the stable, unstable, and center subspaces associated with the equilibrium at the origin. Then $\mathbb{R}^n = E^s \oplus E^u \oplus E^c$.

**Theorem 2.3.11.** The subspaces $E^s$, $E^u$ and $E^c$ are *invariant with respect to the flow* $\phi_t = e^{tA}$ in the following sense: If $\mathbf{x}_0$ is any initial condition in $E^s$, then $e^{tA}\mathbf{x}_0$ is in $E^s$ for all $t$. Similar statements hold for $E^c$ and $E^u$.

In the phase portrait, if we start from within one of these three invariant subspaces, our solution trajectory will never escape from that subspace.

The final theorem in this section is one of the most important qualitative results for the constant-coefficient systems we have studied up to now. It tells us what sort of behavior we can expect if we know the eigenvalues of the coefficient matrix $A$.

**Theorem 2.3.12.** Consider the linear system $\mathbf{x}' = A\mathbf{x}$, where $A$ is an $n \times n$ constant matrix. Then the following statements are equivalent:

☞ Given any $\mathbf{x}_0 \in \mathbb{R}^n$, we have

$$\lim_{t \to \infty} e^{tA}\mathbf{x}_0 = 0,$$

and for any non-zero $\mathbf{x}_0 \in \mathbb{R}^n$, the distance from $e^{tA}\mathbf{x}_0$ to the origin tends to $\infty$ as $t \to -\infty$.

☞ All eigenvalues of $A$ have negative real part.

☞ The stable subspace $E^s$ at the origin is the entire space $\mathbb{R}^n$.

In other words, if all eigenvalues have negative real part, then the flow directs us towards the origin as $t \to \infty$ and we are repelled from the origin as $t \to -\infty$. The proofs of the three preceding theorems are not difficult, but are omitted. For linear, homogeneous, constant-coefficient systems, it is easy to classify the stability of the origin. Namely, the origin is

☞ Asymptotically stable if *all* eigenvalues of $A$ have negative real part.

☞ Stable if none of the eigenvalues has positive real part.

☞ Unstable if *any* of the eigenvalues has positive real part.

**Example 2.3.13.** If

$$A = \begin{bmatrix} 1 & 0 \\ 0 & -1 \end{bmatrix},$$

then the origin is unstable because one of the eigenvalues is positive. If

$$A = \begin{bmatrix} 0 & 1 \\ -1 & 0 \end{bmatrix},$$

then the origin is stable but not asymptotically stable. If

$$A = \begin{bmatrix} -1 & 0 \\ 0 & -6 \end{bmatrix},$$

then the origin is asymptotically stable.

---

## 2.4. Trace and Determinant

A more compact way of classifying phase portraits of planar systems can be stated in terms of the trace and determinant of the coefficient matrix. Recall the following

**Definition 2.4.1.** The *trace* of a square matrix $A$ is the sum of the entries on its main diagonal and is denoted by $\mathrm{tr}A$.

The eigenvalues of a $2 \times 2$ matrix $A$ can be expressed in terms of tr$A$ and det $A$. Suppose

$$A = \begin{bmatrix} a & b \\ c & d \end{bmatrix}.$$

Then the matrix

$$A - \lambda I = \begin{bmatrix} a - \lambda & b \\ c & d - \lambda \end{bmatrix}$$

has determinant $(a - \lambda)(d - \lambda) - bc$. Equivalently, the characteristic equation is

$$\lambda^2 - (a + d)\lambda + (ad - bc) = 0.$$

Since $\text{tr}A = a + d$ and $\det A = ad - bc$, the characteristic equation can also be written as

$$\lambda^2 - (\text{tr}A)\lambda + \det A = 0. \tag{2.12}$$

The roots of (2.12) are

$$\lambda = \frac{\text{tr}A \pm \sqrt{(\text{tr}A)^2 - 4\det A}}{2}. \tag{2.13}$$

Note that the sum of these eigenvalues is tr$A$. This is true of all square matrices, not just $2 \times 2$ matrices (see exercises). We also know that det $A$ is the product of the eigenvalues. For $2 \times 2$ matrices, we have either two real eigenvalues or a complex conjugate pair of eigenvalues, $\alpha \pm \beta i$. Thus, for planar systems $\mathbf{x}' = A\mathbf{x}$, we can use tr$A$ and det $A$ to classify the origin as a saddle, unstable node, stable node, unstable focus, stable focus or center:

**Case 1:** If $\det A < 0$, we claim that the origin is a saddle. To see why, we must show that the eigenvalues of $A$ are real and have opposite sign. Suppose indirectly that $A$ has complex conjugate eigenvalues $\alpha \pm \beta i$. Then the product of the eigenvalues (which equals det $A$) would be positive, contradicting our assumption that $\det A < 0$. It follows that the eigenvalues must be real, and they must have opposite sign in order for $\det A < 0$. Therefore, the origin is a saddle, as claimed.

**Case 2:** Next, suppose that $\det A > 0$ and $(\text{tr}A)^2 - 4\det A \geq 0$. From formula (2.13), we know that the eigenvalues are real because the discriminant is positive. Since $\det A > 0$, the eigenvalues have the same sign, and it follows that the origin is a node. Whether the origin is stable or unstable depends upon tr$A$ (the sum of the eigenvalues):

☞ If tr$A > 0$, then both eigenvalues are positive and the node is unstable.

☞ If tr$A < 0$, then both eigenvalues are negative and the node is stable.

**Case 3:** Finally, suppose that $\det A > 0$ and $(\text{tr}A)^2 - 4\det A < 0$. The discriminant in formula (2.13) is negative, implying that the eigenvalues are complex conjugate. The origin is either a focus or a center depending upon the trace of $A$. The sum of the eigenvalues $\alpha \pm \beta i$ is $2\alpha$, or equivalently tr$A = 2\alpha$.

☞ If tr$A > 0$, the real part of the eigenvalues is positive and the origin is an unstable focus.

☞ If tr$A < 0$, the real part of the eigenvalues is negative and the origin is a stable focus.

☞ If tr$A = 0$, the real part of the eigenvalues is zero and the origin is a center.

The above observations are useful in classifying equilibria of planar systems.

**Example 2.4.2.** Consider the system $\mathbf{x}' = A\mathbf{x}$ where

$$A = \begin{bmatrix} -2 & 10 \\ -3 & 1 \end{bmatrix}.$$

Since $\det A = 28 > 0$, the origin is not a saddle. Since $(\text{tr}A)^2 - 4\det A = -111 < 0$, the eigenvalues are complex conjugate, and since tr$A = -1 < 0$, the eigenvalues have negative real part. Therefore, the origin is a *stable focus*.

Admittedly, for planar systems it is relatively easy to actually *compute* the eigenvalues of the coefficient matrix, avoiding the need for the trace-determinant classification. However, the above formalism can be very useful for systems in which the coefficient matrix $A$ contains unspecified parameters. The inequalities appearing in *Cases 1–3* can be used to determine the ranges of parameter values for which various types of dynamical behavior will be observed. Such issues will be explored in our chapter on bifurcations.

---

## 2.5. Inhomogeneous Systems

We now extend our earlier work to inhomogeneous linear systems of the form

$$\mathbf{x}' = A\mathbf{x} + \mathbf{b}(t),$$

where $A$ is an $n \times n$ constant matrix and $\mathbf{b}(t)$ is a vector of continuous functions. Note that $\mathbf{b}(t)$ is allowed to involve the independent variable $t$ but none of the dependent variables. To solve such systems, we will extend the variation of parameters method covered in introductory differential equations courses.

**Variation of parameters.** Let us briefly review how to solve first-order, linear, homogeneous ODEs with only one dependent variable. Specifically, consider the ODE $x'(t) = a(t)x(t) + b(t)$, where $x(t)$ is scalar-valued, not vector-valued[2]. The first step is to write the ODE in the form

$$x'(t) - a(t)x(t) = b(t). \tag{2.14}$$

We cannot simply integrate both sides with respect to $t$, because this would lead to an integral equation

$$x(t) - \int a(t)x(t)\,\mathrm{d}t \;=\; \int b(t)\,\mathrm{d}t,$$

which does not help us solve for $x(t)$. Instead, the trick is to multiply Equation (2.14) by an *integrating factor*

$$\mathrm{e}^{-\int a(t)\,\mathrm{d}t} \tag{2.15}$$

to obtain

$$\mathrm{e}^{-\int a(t)\,\mathrm{d}t}\left[x' - a(t)x\right] \;=\; \mathrm{e}^{-\int a(t)\,\mathrm{d}t}\, b(t).$$

Equivalently,

$$\frac{\mathrm{d}}{\mathrm{d}t}\left\{\mathrm{e}^{-\int a(t)\,\mathrm{d}t}\, x(t)\right\} \;=\; \mathrm{e}^{-\int a(t)\,\mathrm{d}t}\, b(t).$$

Now we are in a position to integrate both sides with respect to $t$, after which we may (hopefully) solve for $x(t)$.

**Example 2.5.1.** Solve the initial value problem

$$\frac{\mathrm{d}x}{\mathrm{d}t} \;=\; -\frac{2x}{t} + \ln t \qquad x(1) = 1.$$

[2]For this scalar ODE, the coefficient function $a(t)$ is actually allowed to depend upon the independent variable $t$. Later, when we solve $\mathbf{x}' = A\mathbf{x} + \mathbf{b}(t)$, we must insist that $A$ be a constant matrix.

**Solution:** In our above notation, $a(t) = -2/t$, from which we calculate

$$\int \frac{2}{t}\,\mathrm{d}t = 2\ln|t|.$$

(We need not include the integration constant here. After reading ahead a few lines, make sure you can explain why.) Since our initial condition is given at $t = 1$, we may drop the absolute value bars. The integrating factor is $e^{2\ln t} = e^{\ln t^2} = t^2$. Multiplying both sides of our ODE by the integrating factor gives

$$t^2\frac{\mathrm{d}x}{\mathrm{d}t} + 2tx = t^2\ln t.$$

Equivalently,

$$\frac{\mathrm{d}}{\mathrm{d}t}\left\{t^2x\right\} = t^2\ln t,$$

and integrating both sides yields

$$t^2x = \int t^2\ln t\,\mathrm{d}t.$$

Integrating by parts,

$$t^2x = \frac{t^3}{3}\ln t - \frac{t^3}{9} + C,$$

where $C$ is a constant of integration. The initial condition $x(1) = 1$ can be used to calculate $C = 10/9$, which means that the solution of our initial value problem is

$$x(t) = \frac{t}{3}\ln t - \frac{t}{9} + \frac{10}{9t^2}.$$

We can extend the variation of parameters technique to inhomogeneous systems of ODEs of the form $\mathbf{x}' = A\mathbf{x} + \mathbf{b}(t)$, where $A$ is a constant matrix. To facilitate this process, we first state a lemma which tells us how to write our usual initial value problem as an integral equation. Reformulating ODEs as integral equations will help us frequently in subsequent chapters.

**Lemma 2.5.2.** Consider the initial value problem $x' = f(x,t)$ with $x(0) = x_0$. If $f$ is continuous, then this initial value problem can be written as an integral equation

$$x(t) = x_0 + \int_0^t f(x(s),s)\,\mathrm{d}s. \tag{2.16}$$

Here, $s$ is a "dummy variable" of integration.

*Proof.* First, observe that if we set $t = 0$ in Equation (2.16), then the equation reduces to $x(0) = x_0$, which means the initial condition is satisfied. If we differentiate both sides of (2.16) using the Fundamental Theorem of Calculus, then we obtain $x'(t) = f(x(t), t)$, and we see that the ODE is also satisfied. □

Now consider the initial value problem

$$\mathbf{x}' = A\mathbf{x} + \mathbf{b}(t) \qquad \mathbf{x}(0) = \mathbf{x}_0, \tag{2.17}$$

where $A$ is an $n \times n$ constant matrix. By analogy with the one-variable problem above, we re-write the equation as $\mathbf{x}' - A\mathbf{x} = \mathbf{b}(t)$. If this were a one-variable problem with $A$ constant, we would use $e^{-tA}$ as the integrating factor. We claim that for the initial value problem (2.17), we can use the *matrix* $e^{-tA}$ as an integrating factor. Multiplying both sides of our ODE system by $e^{-tA}$ yields

$$\overbrace{e^{-tA}}^{\text{matrix}} \underbrace{\{\mathbf{x}' - A\mathbf{x}\}}_{\text{vector}} = \overbrace{e^{-tA}}^{\text{matrix}} \underbrace{\mathbf{b}(t)}_{\text{vector}}.$$

Equivalently,

$$\frac{d}{dt}\left\{e^{-tA}\mathbf{x}\right\} = e^{-tA}\mathbf{b}(t),$$

and by Lemma 2.5.2 we have

$$e^{-tA}\mathbf{x}(t) = e^{-tA}\mathbf{x}(t)\Big|_{t=0} + \int_0^t e^{-sA}\mathbf{b}(s)\,ds = I\mathbf{x}(0) + \int_0^t e^{-sA}\mathbf{b}(s)\,ds,$$

where $I$ is the identity matrix. Multiplying through by $e^{tA}$, we have established

**Theorem 2.5.3.** The solution of the inhomogeneous initial value problem (2.17) is given by

$$\mathbf{x}(t) = e^{tA}\mathbf{x}(0) + e^{tA}\int_0^t e^{-sA}\mathbf{b}(s)\,ds. \tag{2.18}$$

When applying formula (2.18), our main challenge is to actually evaluate the integral of the vector-valued function $e^{-sA}\mathbf{b}(s)$.

**Example 2.5.4.** Solve the initial value problem

$$x_1' = -x_1 + x_2 + e^{-t} \qquad x_1(0) = 1$$
$$x_2' = -x_1 - x_2 + 2e^{-t} \qquad x_2(0) = 1.$$

**Solution:** In matrix form, the system becomes

$$\underbrace{\begin{bmatrix} x_1' \\ x_2' \end{bmatrix}}_{\mathbf{x}'} = \underbrace{\begin{bmatrix} -1 & 1 \\ -1 & -1 \end{bmatrix}}_{A} \underbrace{\begin{bmatrix} x_1 \\ x_2 \end{bmatrix}}_{\mathbf{x}} + \underbrace{\begin{bmatrix} e^{-t} \\ 2e^{-t} \end{bmatrix}}_{\mathbf{b}} \qquad \mathbf{x}_0 = \begin{bmatrix} 1 \\ 1 \end{bmatrix}.$$

The matrix $A$ is already in real canonical form, and using $\alpha = -1$ and $\beta = -1$ in Proposition 2.1.26, we find that

$$e^{tA} = e^{-t} \begin{bmatrix} \cos(-t) & -\sin(-t) \\ \sin(-t) & \cos(-t) \end{bmatrix} = e^{-t} \begin{bmatrix} \cos t & \sin t \\ -\sin t & \cos t \end{bmatrix}.$$

Here, we have used the fact that $\cos(-t) = \cos t$ and $\sin(-t) = -\sin t$. Replacing $t$ with $-s$ and using these same trigonometric facts, we have

$$e^{-sA} = e^{s} \begin{bmatrix} \cos s & -\sin s \\ \sin s & \cos s \end{bmatrix}.$$

Therefore, the integrand in formula (2.18) is

$$e^{-sA}\mathbf{b}(s) = e^{s} \begin{bmatrix} \cos s & -\sin s \\ \sin s & \cos s \end{bmatrix} \begin{bmatrix} e^{-s} \\ 2e^{-s} \end{bmatrix} = \begin{bmatrix} \cos s - 2\sin s \\ \sin s + 2\cos s \end{bmatrix}.$$

Integrating each component of this vector separately,

$$\int_0^t e^{-sA}\mathbf{b}(s)\, ds = \int_0^t \begin{bmatrix} \cos s - 2\sin s \\ \sin s + 2\cos s \end{bmatrix} ds = \begin{bmatrix} \sin s + 2\cos s \\ -\cos s + 2\sin s \end{bmatrix}\Bigg|_0^t$$

$$= \begin{bmatrix} \sin t + 2\cos t - 2 \\ -\cos t + 2\sin t + 1 \end{bmatrix}.$$

Putting all of this together, Equation (2.18) tells us that the solution of the initial value problem is

$$\begin{bmatrix} x_1 \\ x_2 \end{bmatrix} = e^{tA}\mathbf{x}(0) + e^{tA} \int_0^t e^{-sA}\mathbf{b}(s)\, ds$$

$$= e^{-t} \begin{bmatrix} \cos t & \sin t \\ -\sin t & \cos t \end{bmatrix} \begin{bmatrix} 1 \\ 1 \end{bmatrix} + e^{-t} \begin{bmatrix} \cos t & \sin t \\ -\sin t & \cos t \end{bmatrix} \begin{bmatrix} \sin t + 2\cos t - 2 \\ -\cos t + 2\sin t + 1 \end{bmatrix}.$$

Expanding these products, our overall result is

$$\begin{aligned} x_1(t) &= -e^{-t}\cos t + 2e^{-t}\sin t + 2e^{-t} \\ x_2(t) &= e^{-t}\sin t + 2e^{-t}\cos t - e^{-t}. \end{aligned}$$

Readers interested in learning more about the types of linear systems of ODEs for which exact solution is possible are encouraged to explore other resources such as the references listed in the bibliography. The special classes of linear systems we have discussed up to this point will be sufficient for our purposes in the remainder of this text.

---

## Exercises

1. If $\lambda$ is an eigenvalue of an invertible matrix $A$, show that $1/\lambda$ is an eigenvalue of $A^{-1}$. What are the eigenvectors of $A^{-1}$ associated with $1/\lambda$?
2. An $n \times n$ matrix $A$ is called *idempotent* if $A^2 = A$. Show that each eigenvalue of an idempotent matrix is either 0 or 1.
3. Show that if $N$ is a nilpotent matrix (see Definition 2.1.32), then zero is the only eigenvalue of $N$.
4. Give an example of a $3 \times 3$ matrix that is diagonalizable but not invertible. Then, give an example of a $3 \times 3$ matrix that is invertible but not diagonalizable.
5. Square matrices $A$ and $B$ are called *simultaneously diagonalizable* if they are diagonalizable by the same invertible matrix $P$. Show that if $A$ and $B$ are simultaneously diagonalizable, then $A$ and $B$ commute. That is, $AB = BA$.
6. The cosine of an $n \times n$ square matrix $A$ is defined in terms of the Maclaurin series representation of the cosine function:

$$\cos(A) = \sum_{k=0}^{\infty} (-1)^k \frac{A^{2k}}{(2k)!} = I - \frac{A^2}{2!} + \frac{A^4}{4!} + \cdots$$

   (a) Suppose that $D = \text{diag}\{\lambda_1, \lambda_2, \ldots \lambda_n\}$ is a diagonal matrix. What is $\cos(D)$?

   (b) Suppose that $A$ is a diagonalizable matrix and $A = PDP^{-1}$ where $D$ is diagonal. Show that $\cos(A) = P[\cos(D)]P^{-1}$.

(c) Find $\cos(A)$ if

$$A = \begin{bmatrix} -3 & 6 \\ 4 & -1 \end{bmatrix}.$$

7. Use the Maclaurin series representation of the sine function to define $\sin(A)$, where $A$ is an $n \times n$ matrix. Use your definition to compute (a) the sine of the $2 \times 2$ zero matrix, and (b) the sine of the diagonal matrix $D = \text{diag}\{\pi/2, \pi/2, \pi/2\}$.

8. Consider the two matrices

$$A = \begin{bmatrix} 1 & 0 \\ 0 & 0 \end{bmatrix} \qquad \text{and} \qquad B = \begin{bmatrix} 0 & 1 \\ 0 & 0 \end{bmatrix}.$$

(a) Show that A and B do *not* commute.

(b) Show that B is *not* diagonalizable.

(c) Show that $e^{A+B} \neq e^A e^B$.

9. For each of the following matrices $A$, compute $e^{tA}$:

$$\text{(a)} \quad \begin{bmatrix} 2 & 25/8 \\ 2 & 2 \end{bmatrix} \qquad \text{(b)} \quad \begin{bmatrix} 0 & -1 \\ -1 & 2 \end{bmatrix} \qquad \text{(c)} \quad \begin{bmatrix} 2 & -1 \\ 2 & 0 \end{bmatrix}.$$

10. Solve the system

$$\frac{dx_1}{dt} = 3x_1 + x_2 \qquad \frac{dx_2}{dt} = x_1 + 3x_2,$$

with $x_1(0) = -1$ and $x_2(0) = 4$.

11. Find the general solution of the system

$$\frac{dx_1}{dt} = x_1 + 2x_2 + 3x_3, \qquad \frac{dx_2}{dt} = 2x_2 + 8x_3, \qquad \frac{dx_3}{dt} = 3x_3.$$

HINT: Do you *really* need to exponentiate a matrix here?

12. Find the general solution of the system

$$\frac{dx_1}{dt} = x_1 - 5x_2 \qquad \frac{dx_2}{dt} = 5x_1 + x_2.$$

13. Solve the initial value problem

$$\begin{aligned} \frac{dx_1}{dt} &= 5x_1 + 10x_2, & x_1(0) &= 1 \\ \frac{dx_2}{dt} &= -x_1 + 3x_2, & x_2(0) &= 1. \end{aligned}$$

14. Solve the initial value problem

$$\frac{dx_1}{dt} = x_1 - 5x_2 \qquad \frac{dx_2}{dt} = 2x_1 + 3x_2,$$

with $x_1(0) = 1$ and $x_2(0) = 1$.

15. Find the general solution of the system $\mathbf{x}' = A\mathbf{x}$ where $A$ is the matrix

$$A = \begin{bmatrix} 0 & 1 \\ -1 & 2 \end{bmatrix}.$$

16. Find the general solution of the system $\mathbf{x}' = A\mathbf{x}$ where $A$ is the matrix

$$A = \begin{bmatrix} 1 & 0 & 0 \\ 2 & 1 & 0 \\ 3 & 2 & 1 \end{bmatrix}.$$

17. Find the general solution of the system $\mathbf{x}' = A\mathbf{x}$ where $A$ is the matrix

$$A = \begin{bmatrix} 1 & 0 & 0 \\ -1 & 2 & 0 \\ 1 & 0 & 2 \end{bmatrix}.$$

18. Solve the initial value problem $\mathbf{x}' = A\mathbf{x}$ where

$$A = \begin{bmatrix} 0 & 1 & 0 \\ 0 & 0 & 1 \\ -8 & -12 & -6 \end{bmatrix}$$

and $x_1(0) = 1$, $x_2(0) = 2$, and $x_3(0) = 3$.

19. For each of the following matrices $A$, sketch the phase portrait for the linear system $\mathbf{x}' = A\mathbf{x}$. In each case, indicate whether the origin is a stable node,

unstable node, stable focus, unstable focus, saddle, or center.

$$A = \begin{bmatrix} 2 & 0 \\ 0 & 2 \end{bmatrix} \qquad A = \begin{bmatrix} -1 & 0 \\ 0 & 3 \end{bmatrix} \qquad A = \begin{bmatrix} 1 & -3 \\ 3 & 1 \end{bmatrix}.$$

**20.** For the following matrices $A$, **carefully** sketch the phase portrait for the linear system $\mathbf{x}' = A\mathbf{x}$. In each case, identify the stable subspace $E^s$ and the unstable subspace $E^u$.

$$A = \begin{bmatrix} 1 & 2 \\ 4 & 4 \end{bmatrix} \qquad A = \begin{bmatrix} 0 & -1 \\ 9 & 0 \end{bmatrix}.$$

Hint: Try to mimic examples that appear in the text.

**21.** (Trace and determinant.) Without finding the eigenvalues of the coefficient matrices of the following systems, determine whether the origin is a stable node, unstable node, stable focus, unstable focus, or center.

(a) $\mathbf{x}' = \begin{bmatrix} -1 & 1 \\ 3 & 2 \end{bmatrix} \mathbf{x}$ (b) $\mathbf{x}' = \begin{bmatrix} 2 & 1 \\ 1 & 2 \end{bmatrix} \mathbf{x}$

(c) $\mathbf{x}' = \begin{bmatrix} -2 & 1 \\ 1 & -1 \end{bmatrix} \mathbf{x}$ (d) $\mathbf{x}' = \begin{bmatrix} 2 & -1 \\ 5 & -2 \end{bmatrix} \mathbf{x}.$

**22.** The trace of *any* $n \times n$ matrix $A$ is equal to the sum of its eigenvalues; the purpose of this exercise is to prove this statement for *diagonalizable* matrices $A$.

(a) Suppose $A$ and $B$ are $n \times n$ matrices. Show that $\text{tr}(AB) = \text{tr}(BA)$.

(b) Using the result from Part (a), show that if $A$ and $B$ are similar, then $\text{tr}A = \text{tr}B$.

(c) Suppose that $A$ is diagonalizable and that $A = PDP^{-1}$ where $D$ is diagonal. Use the result from Part (b) to explain why the trace of $A$ is equal to the sum of its eigenvalues.

**23.** Consider the constant-coefficient system $\mathbf{x}' = A\mathbf{x}$ where

$$A = \begin{bmatrix} 4 & 0 & -3 \\ 0 & -2 & 0 \\ 3 & 0 & 4 \end{bmatrix}.$$

Determine the stable, unstable and center subspaces $E^s$, $E^u$ and $E^c$ associated with the equilibrium at the origin.

24. Consider the system $\mathbf{x}' = A\mathbf{x}$, where

$$A = \begin{bmatrix} -3 & 0 & 0 \\ 0 & 2 & -4 \\ 0 & 4 & 2 \end{bmatrix}.$$

Identify $E^s$, $E^u$ and $E^c$ for this system, and provide a rough sketch of the phase portrait.

25. Determine $E^s$, $E^u$ and $E^c$ for the following system, and describe the flow starting from initial conditions within each of these subspaces.

$$\mathbf{x}' = \begin{bmatrix} 0 & -3 & 0 \\ 3 & 0 & 0 \\ 0 & 0 & -1 \end{bmatrix} \mathbf{x}.$$

26. Each of the following systems contains an unspecified constant $\alpha$. For each system, (i) determine the eigenvalues of the coefficient matrix in terms of $\alpha$; (ii) find the critical values of $\alpha$ at which the qualitative nature of the phase portrait experiences a dramatic change; and (iii) sketch the phase portrait for different choices of $\alpha$: one just below each critical value, and one just above each critical value.

(a) $\mathbf{x}' = \begin{bmatrix} 1 & \alpha^2 \\ 4 & 1 \end{bmatrix} \mathbf{x}$ (b) $\mathbf{x}' = \begin{bmatrix} -1 & 1 \\ 1 & \alpha \end{bmatrix} \mathbf{x}$ (c) $\mathbf{x}' = \begin{bmatrix} -1 & \alpha \\ 1 & 2 \end{bmatrix} \mathbf{x}$.

27. Solve the initial value problem

$$\frac{dy}{dx} = (-\tan x)y + \sec x, \quad y\left(\frac{\pi}{4}\right) = \sqrt{2}.$$

28. Solve the initial value problem

$$\frac{dy}{dx} = -\frac{y}{x} + e^{x^2}, \quad y(1) = 3.$$

**29.** Solve the initial value problem

$$\mathbf{x}' = \begin{bmatrix} -1 & 2 \\ 0 & 1 \end{bmatrix} \mathbf{x} + \begin{bmatrix} 1 \\ 2t \end{bmatrix}, \qquad \mathbf{x}_0 = \begin{bmatrix} 1 \\ 0 \end{bmatrix}.$$

**30.** Solve the initial value problem

$$\begin{aligned} \frac{dx}{dt} &= x + 2t & x(0) &= 2 \\ \frac{dy}{dt} &= -y + e^t & y(0) &= 1. \end{aligned}$$

**31.** Solve the initial value problem

$$\begin{aligned} \frac{dx_1}{dt} &= -x_1 - 2x_2 + \cos 2t, & x_1(0) &= 3 \\ \frac{dx_2}{dt} &= 2x_1 - x_2 + \sin 2t, & x_2(0) &= 3. \end{aligned}$$

# CHAPTER 3

## Nonlinear Systems: Local Theory

We now turn our attention to nonlinear systems of ODEs, which present a host of new challenges. Obtaining exact analytical solutions to such systems is usually impossible, so we must settle for qualitative descriptions of the dynamics. On the other hand, nonlinear systems can exhibit a wide variety of behaviors that linear systems cannot. Moreover, most dynamical phenomena in nature are inherently nonlinear.

Consider a general system of the form $\mathbf{x}' = f(\mathbf{x}, t)$, where $\mathbf{x} \in \mathbb{R}^n$ is a vector of unknowns and

$$f(\mathbf{x}, t) = \begin{bmatrix} f_1(x_1, x_2, \ldots x_n, t) \\ f_2(x_1, x_2, \ldots x_n, t) \\ \vdots \\ f_n(x_1, x_2, \ldots x_n, t) \end{bmatrix}$$

is a vector-valued function $f : \mathbb{R}^{n+1} \to \mathbb{R}^n$. In what follows, we shall work only with *autonomous* systems—those of the form $\mathbf{x}' = f(\mathbf{x})$ where $f : \mathbb{R}^n \to \mathbb{R}^n$ does not explicitly involve the independent variable $t$. This is actually not a severe restriction at all, because non-autonomous systems can be converted to autonomous ones by introducing an extra dependent variable. For example, the non-autonomous system

$$\begin{bmatrix} x_1' \\ x_2' \end{bmatrix} = \begin{bmatrix} \cos(x_1 t) + x_2 \\ t^2 + x_1^2 \end{bmatrix}$$

becomes autonomous if we introduce a third dependent variable $x_3 = t$ and corresponding ODE $x_3' = 1$. The result is the autonomous system

$$\begin{bmatrix} x_1' \\ x_2' \\ x_3' \end{bmatrix} = \begin{bmatrix} \cos(x_1 x_3) + x_2 \\ x_3^2 + x_1^2 \\ 1 \end{bmatrix},$$

whose order is one larger than the original non-autonomous system. Hence, it will be sufficient to consider autonomous systems only.

For autonomous systems $\mathbf{x}' = f(\mathbf{x})$, it is straightforward to determine whether the system is linear or nonlinear.

**Definition 3.0.5.** Consider the system $\mathbf{x}' = f(\mathbf{x})$, where $f : \mathbb{R}^n \to \mathbb{R}^n$. The system of ODEs is *linear* if the function $f$ satisfies $f(\alpha\mathbf{x} + \mathbf{y}) = \alpha f(\mathbf{x}) + f(\mathbf{y})$ for all vectors $\mathbf{x}, \mathbf{y} \in \mathbb{R}^n$ and all scalars $\alpha \in \mathbb{R}$. Otherwise, the system of ODEs is called *nonlinear*.

**Example 3.0.6.** The right-hand side of the ODE $\frac{dx}{dt} = x^2$ is $f(x) = x^2$, and

$$f(x+y) = (x+y)^2 = x^2 + 2xy + y^2 \neq x^2 + y^2 = f(x) + f(y).$$

Therefore, the ODE is nonlinear.

**Warning:** Be careful when classifying ODEs as autonomous/non-autonomous or linear/nonlinear. These concepts are independent of one another. Here are several one-dimensional examples to reinforce this point:

☞ $\frac{dx}{dt} = x$ is linear and autonomous.

☞ $\frac{dx}{dt} = x + \cos t$ is linear and non-autonomous.

☞ $\frac{dx}{dt} = \cos x$ is nonlinear and autonomous.

☞ $\frac{dx}{dt} = t + \cos x$ is nonlinear and non-autonomous.

There are very few analytical techniques for solving nonlinear ODEs. Separation of variables is one such method, but it has very limited scope.

**Example 3.0.7.** Consider the nonlinear, autonomous ODE $\frac{dx}{dt} = x - x^2$ with initial condition $x(0) = 1/2$. The fact that the ODE is autonomous makes separating the variables easy:

$$\frac{1}{x - x^2}\frac{dx}{dt} = 1.$$

Integrating both sides with respect to $t$ yields

$$\int \frac{1}{x - x^2}\,dx = \int dt,$$

and a partial fractions decomposition simplifies the left-hand side as

$$\int \frac{1}{x}\,dx + \int \frac{1}{1-x}\,dx = \int dt.$$

Hence, $\ln|x| - \ln|1-x| = t + C$ where $C$ is an integration constant. In the vicinity of the initial condition $x(0) = 1/2$, we are dealing with $x$ values for which both $x$ and $1 - x$ are positive quantities, allowing us to drop the absolute value bars. Moreover, the initial condition tells us that $C = 0$. Solving for $x$ then gives

$$x(t) = \frac{e^t}{1 + e^t}.$$

Notice that as $t \to \infty$, the solution satisfies $x \to 1$ and as $t \to -\infty$, we have $x \to 0$. The linear systems we considered in the previous chapter cannot exhibit this sort of behavior. In fact, the only systems $\mathbf{x}' = A\mathbf{x}$ whose non-constant solutions remain bounded (i.e., trapped in a finite interval) for all time $t$ are those for which all eigenvalues of $A$ have zero real part.

The remarks in the preceding example allude to the fact that nonlinear systems can exhibit very different behavior from the linear ones we have considered up to now. The next two examples help drive home this point.

**Example 3.0.8.** In the previous chapter, we learned that the linear system $\mathbf{x}' = A\mathbf{x} + \mathbf{b}(t)$ with initial condition $\mathbf{x}(0) = \mathbf{x}_0$ has a *unique* solution given by the variation of parameters formula (2.18). By contrast, the nonlinear system

$$\frac{dx}{dt} = 3x^{2/3} \qquad x(0) = 0$$

has at least *two* solutions. Clearly one solution is the constant function $x(t) = 0$. Separation of variables produces a second solution $x(t) = t^3$, and therefore this initial value problem does not have a unique solution.

**Example 3.0.9.** Consider the nonlinear system

$$\frac{dx}{dt} = x^2 \qquad x(0) = 1.$$

By separation of variables, you can show that $x(t) = (1-t)^{-1}$ is the solution. Unfortunately, the solution *blows up* as $t \to 1^-$. Although the *function* $x(t) = (1-t)^{-1}$ is defined for $t > 1$, it does not make sense to regard this as the *solution* of the initial value problem for $t \geq 1$ since something went horribly wrong with our system at $t = 1$. Note that solutions of the linear equations we studied in the previous chapter cannot blow up in finite time.

**Moral:** For practical purposes, we need a way of determining whether a nonlinear system $\mathbf{x}' = f(\mathbf{x})$ has a *unique* solution, and whether solutions are well-behaved. In practice, if we obtain multiple solutions to an initial value problem which is supposed to model some natural phenomenon, then which solution (if any) is physically relevant? To address the issue of whether a nonlinear system has a unique solution, we will typically approximate the dynamics using linear systems.

---

## 3.1. Linear Approximations of Functions of Several Variables

One approach toward tackling nonlinear systems $\mathbf{x}' = f(\mathbf{x})$ involves making successive linear approximations, a process known as Picard iteration. Before explaining that procedure, we expand upon some of the ideas covered in multi-variable calculus. The purpose of this section is to explain how to obtain a linear approximation of a function $f : \mathbb{R}^n \to \mathbb{R}^n$.

For functions $f(x)$ of a single variable, the derivative provides a means for writing down a linear approximation. Namely, the best linear approximation of $f(x)$ near a point $x = a$ is given by the tangent line approximation

$$f(x) \approx f(a) + f'(a) \cdot (x - a).$$

In multi-variable calculus, you learned that graphs of functions $f : \mathbb{R}^2 \to \mathbb{R}$ of two variables are *surfaces* in $\mathbb{R}^3$. Given a point $(x_0, y_0) \in \mathbb{R}^2$, the best linear approximation of $f$ near that point is given by the tangent plane approximation

$$f(x,y) \approx f(x_0,y_0) + \frac{\partial f}{\partial x}(x_0,y_0) \cdot (x - x_0) + \frac{\partial f}{\partial y}(x_0,y_0) \cdot (y - y_0).$$

Recall that if $f(x,y)$ is a function of two variables, its gradient $\nabla f$ is the vector

$$\nabla f(x,y) = \left(\frac{\partial f}{\partial x}, \frac{\partial f}{\partial y}\right).$$

Hence, an equivalent way of writing the equation for the tangent plane approximation is

$$f(x,y) \approx f(x_0,y_0) + \underbrace{\nabla f(x_0,y_0) \bullet (x - x_0, y - y_0)}_{\text{dot product}}.$$

For functions $f : \mathbb{R}^n \to \mathbb{R}$, linear approximation is simply an extension of the tangent plane approximation. If $(\overline{x}_1, \overline{x}_2, \ldots \overline{x}_n)$ is some point in $\mathbb{R}^n$, then the linear approximation of $f$ near that point is given by

$$\begin{aligned} f(x_1, x_2, \ldots x_n) \approx\ & f(\overline{x}_1, \overline{x}_2, \ldots \overline{x}_n) \\ & + \nabla f(\overline{x}_1, \overline{x}_2, \ldots \overline{x}_n) \bullet (x_1 - \overline{x}_1, x_2 - \overline{x}_2, \ldots x_n - \overline{x}_n), \end{aligned}$$

where

$$\nabla f(x_1, x_2, \ldots x_n) = \left(\frac{\partial f}{\partial x_1}, \frac{\partial f}{\partial x_2}, \cdots \frac{\partial f}{\partial x_n}\right).$$

We are now ready to explain how to obtain linear approximations for functions $f : \mathbb{R}^n \to \mathbb{R}^n$. If $f$ is such a function, we can write $f$ in terms of its component functions

$$f(x_1, x_2, \ldots x_n) = \begin{bmatrix} f_1(x_1, x_2, \ldots x_n) \\ f_2(x_1, x_2, \ldots x_n) \\ \vdots \\ f_n(x_1, x_2, \ldots x_n) \end{bmatrix}. \tag{3.1}$$

Since each component function $f_i$ is a function from $\mathbb{R}^n \to \mathbb{R}$, we know how to write down their linear approximations near a specific point $(\overline{x}_1, \overline{x}_2, \ldots \overline{x}_n) \in \mathbb{R}^n$. Namely, the right hand side of Equation (3.1) is approximated by

$$\begin{bmatrix} f_1(\overline{x}_1, \overline{x}_2, \ldots \overline{x}_n) + \nabla f_1(\overline{x}_1, \overline{x}_2, \ldots \overline{x}_n) \bullet (x_1 - \overline{x}_1, x_2 - \overline{x}_2, \ldots x_n - \overline{x}_n) \\ f_2(\overline{x}_1, \overline{x}_2, \ldots \overline{x}_n) + \nabla f_2(\overline{x}_1, \overline{x}_2, \ldots \overline{x}_n) \bullet (x_1 - \overline{x}_1, x_2 - \overline{x}_2, \ldots x_n - \overline{x}_n) \\ \vdots \\ f_n(\overline{x}_1, \overline{x}_2, \ldots \overline{x}_n) + \nabla f_n(\overline{x}_1, \overline{x}_2, \ldots \overline{x}_n) \bullet (x_1 - \overline{x}_1, x_2 - \overline{x}_2, \ldots x_n - \overline{x}_n) \end{bmatrix}. \tag{3.2}$$

Finally, we re-write the vector (3.2) in the equivalent form

$$\underbrace{\begin{bmatrix} f_1(\overline{x}_1, \overline{x}_2, \ldots \overline{x}_n) \\ f_2(\overline{x}_1, \overline{x}_2, \ldots \overline{x}_n) \\ \vdots \\ f_n(\overline{x}_1, \overline{x}_2, \ldots \overline{x}_n) \end{bmatrix}}_{\text{vector in } \mathbb{R}^n} + \underbrace{\begin{bmatrix} \nabla f_1(\overline{x}_1, \overline{x}_2, \ldots \overline{x}_n) \\ \nabla f_2(\overline{x}_1, \overline{x}_2, \ldots \overline{x}_n) \\ \vdots \\ \nabla f_n(\overline{x}_1, \overline{x}_2, \ldots \overline{x}_n) \end{bmatrix}}_{n \times n \text{ matrix}} \underbrace{\begin{bmatrix} x_1 - \overline{x}_1 \\ x_2 - \overline{x}_2 \\ \vdots \\ x_n - \overline{x}_n \end{bmatrix}}_{\text{vector in } \mathbb{R}^n}. \tag{3.3}$$

Here, it is understood that the gradients in (3.3) are written out as row vectors. Equation (3.3) is the best linear approximation for a function $f : \mathbb{R}^n \to \mathbb{R}^n$ near the point $(\overline{x}_1, \overline{x}_2, \ldots \overline{x}_n)$. The matrix appearing in (3.3) will be important throughout our study of nonlinear ODEs, and it has a special name.

**Definition 3.1.1.** Suppose $f : \mathbb{R}^n \to \mathbb{R}^n$ is written in terms of its components as in (3.1), and further suppose that the first partial derivatives of each component function exist. That is, $\partial f_i / \partial x_j$ exists for $1 \le i, j \le n$. Then the *Jacobian of* $f$ is the matrix

$$Jf(x_1, x_2, \ldots x_n) = \begin{bmatrix} \frac{\partial f_1}{\partial x_1} & \frac{\partial f_1}{\partial x_2} & \cdots & \frac{\partial f_1}{\partial x_n} \\ \frac{\partial f_2}{\partial x_1} & \frac{\partial f_2}{\partial x_2} & \cdots & \frac{\partial f_2}{\partial x_n} \\ \vdots & \vdots & \ddots & \vdots \\ \frac{\partial f_n}{\partial x_1} & \frac{\partial f_n}{\partial x_2} & \cdots & \frac{\partial f_n}{\partial x_n} \end{bmatrix}. \tag{3.4}$$

Evaluating a Jacobian matrix $Jf$ at a point $(\overline{x}_1, \overline{x}_2, \ldots \overline{x}_n) \in \mathbb{R}^n$ is perfectly analogous to evaluating the derivative of a single-variable function $f$ at a point $x = a$. Readers who plan to take advanced calculus/analysis courses will learn that, if all partial derivatives in a Jacobian matrix are continuous, then the Jacobian matrix represents the derivative of a function $f : \mathbb{R}^n \to \mathbb{R}^n$.

**Example 3.1.2.** Consider the function $f : \mathbb{R}^2 \to \mathbb{R}^2$ defined by

$$f(x, y) = \begin{bmatrix} x^2 + y^3 \\ x\cos(xy) \end{bmatrix}.$$

Then

$$Jf(x, y) = \begin{bmatrix} \partial f_1/\partial x & \partial f_1/\partial y \\ \partial f_2/\partial x & \partial f_2/\partial y \end{bmatrix} = \begin{bmatrix} 2x & 3y^2 \\ \cos(xy) - xy\sin(xy) & -x^2\sin(xy) \end{bmatrix}.$$

In particular,

$$Jf(0,0) = \begin{bmatrix} 0 & 0 \\ 1 & 0 \end{bmatrix}.$$

---

## 3.2. Fundamental Existence and Uniqueness Theorem

The notion of Jacobian matrices will become important when we state conditions which guarantee that a general nonlinear system $\mathbf{x}' = f(\mathbf{x})$ with $\mathbf{x}(0) = \mathbf{x}_0$ has a unique solution. By Lemma 2.5.2, this initial value problem can be written as an integral equation

$$\mathbf{x}(t) = \mathbf{x}_0 + \int_0^t f(\mathbf{x}(s))\,\mathrm{d}s.$$

Picard's method of successive approximations states that, under reasonable assumptions on $f$, the solution of this integral equation can be constructed recursively by setting $\mathbf{u}_0(t) = \mathbf{x}_0$ and defining the functions

$$\mathbf{u}_{k+1}(t) = \mathbf{x}_0 + \int_0^t f(\mathbf{u}_k(s))\,\mathrm{d}s$$

for each $k \geq 0$. Provided that $f$ is reasonably well-behaved, the sequence of functions $\{\mathbf{u}_k(t)\}$ converges to $\mathbf{x}(t)$, the solution of the initial value problem, on some time interval $t \in (-\alpha, \alpha)$. We illustrate the Picard iteration procedure via an example.

**Example 3.2.1.** Consider the one-dimensional system $x' = ax$ with initial condition $x(0) = x_0$. In this case, $f(x) = ax$ and we know in advance that the solution of this problem is simply $x = x_0 e^{at}$. To start the process of Picard iteration, we define $u_0(t) = x_0$, a constant. Then

$$u_1(t) = x_0 + \int_0^t f(u_0(s))\,\mathrm{d}s = x_0 + \int_0^t ax_0\,\mathrm{d}s = x_0 + ax_0t = x_0(1+at).$$

The next iterate is

$$u_2(t) = x_0 + \int_0^t f(u_1(s))\,\mathrm{d}s = x_0 + \int_0^t ax_0(1+as)\,\mathrm{d}s = x_0\left(1 + at + \frac{(at)^2}{2}\right).$$

After computing a few more iterates, we are led to conjecture that

$$u_k(t) = x_0 \sum_{j=0}^{k} \frac{(at)^j}{j!},$$

which can be proved by straightforward induction. Letting $k \to \infty$, we find that

$$\lim_{k\to\infty} u_k(t) = x_0 \sum_{j=0}^{\infty} \frac{(at)^j}{j!} = x_0 e^{at}.$$

Here, we have recognized the infinite series as the Maclaurin series representation of the exponential function $e^{at}$. Notice that the limit of our sequence of functions $u_k(t)$ is $x(t)$, the solution of the initial value problem. In this example, the solution is valid for all real $t$.

We now state a criterion under which the Picard iteration process will actually produce the *unique* solution of our initial value problem.

**Theorem 3.2.2. (Fundamental Existence & Uniqueness Theorem):** Consider the initial value problem $\mathbf{x}' = f(\mathbf{x})$ and $\mathbf{x}(0) = \mathbf{x}_0$, where $f : \mathbb{R}^n \to \mathbb{R}^n$. Suppose that all partial derivatives in the Jacobian matrix $Jf(\mathbf{x})$ are continuous for all $\mathbf{x}$ in the vicinity of the initial condition $\mathbf{x}_0$. Then there exists a positive number $\alpha$ such that the initial value problem has a *unique solution* $\mathbf{x}(t)$ in the interval $[-\alpha, \alpha]$.

**Remark.** (i) This Theorem is *local*. It does not guarantee that a unique solution will exist for all time $t$. Conditions for global existence and uniqueness will be provided later. (ii) It is possible to state a stronger version of this theorem which does not require that $Jf(\mathbf{x})$ consist of continuous functions. See Perko [8] for details. (iii) There are several ways to prove this theorem (see [7]). The most intuitive proof requires familiarity with the notions of uniform convergence and the contraction mapping principle.

**Example 3.2.3.** Consider the initial value problem

$$\begin{aligned} \frac{dx_1}{dt} &= 2x_1 - 3x_1x_2 \qquad & x_1(0) = 3 \\ \frac{dx_2}{dt} &= x_1x_2 - 4x_2 \qquad & x_2(0) = 8. \end{aligned}$$

In this case, $f : \mathbb{R}^2 \to \mathbb{R}^2$ is given by

$$f(x_1, x_2) = \begin{bmatrix} f_1(x_1, x_2) \\ f_2(x_1, x_2) \end{bmatrix} = \begin{bmatrix} 2x_1 - 3x_1x_2 \\ x_1x_2 - 4x_2 \end{bmatrix}.$$

The corresponding Jacobian matrix is

$$Jf(x_1, x_2) = \begin{bmatrix} 2 - 3x_2 & -3x_1 \\ x_2 & x_1 - 4 \end{bmatrix}.$$

Not only are the entries of this matrix continuous at the point

$$\mathbf{x}_0 = \begin{bmatrix} 3 \\ 8 \end{bmatrix},$$

they are continuous at all points $(x_1, x_2) \in \mathbb{R}^2$. Therefore, there exists $\alpha > 0$ such that this initial value problem has a unique solution on the interval $[-\alpha, \alpha]$.

**Example 3.2.4.** Recall the initial value problem

$$\frac{dx}{dt} = 3x^{2/3}, \qquad x(0) = 0,$$

for which we found two solutions: $x(t) = 0$ and $x(t) = t^3$. In this one-dimensional example, we have $f(x) = 3x^{2/3}$, and the "Jacobian" of $f$ is simply its usual derivative. That is, $Jf(x) = f'(x) = 2x^{-1/3}$. Notice that $f'(x)$ is *NOT* continuous at $x = 0$, which is precisely the value of $x$ assigned by the initial condition. Hence, we have not satisfied the conditions of the Fundamental Existence and Uniqueness Theorem. If the initial condition had been $x(0) = 1$, then our problem *would* have a unique solution. (Warning: Simply failing to satisfy the conditions of Theorem 3.2.2 is not enough to logically conclude that the initial value problem does *not* have a unique solution. It merely tells us that we cannot immediately conclude that our problem *does* have a unique solution.)

---

## 3.3. Global Existence, Dependence on Initial Conditions

Before jumping into a discussion of qualitative theory of nonlinear systems, we address two questions which follow upon the theory in the previous section. If we slightly change our initial conditions, can this make a profound impact on solutions of our initial value problem? Under what circumstances can we guarantee *global* existence of solutions as opposed to the local existence that Theorem 3.2.2 ensures? We will answer these questions after recalling some basic facts from single-variable calculus.

**Lemma 3.3.1. (Extreme Value Theorem).** If a function $f(x)$ is continuous on a closed interval $[a, b]$, then $f(x)$ achieves both a minimum value $m$ and a maximum value $M$ on this interval.

For example, on the closed interval $[0, 3\pi/4]$, the function $f(x) = \sin x$ achieves a maximum value of 1 at $x = \pi/2$ and a minimum value of 0 at $x = 0$. By contrast, on the open interval $(0, 1)$, the function $g(x) = x^2$ attains neither a maximum nor a minimum value.

**Lemma 3.3.2. (Mean Value Theorem).** Suppose $f(x)$ is continuous on $[a, b]$ and differentiable on $(a, b)$. Then there exists at least one point $c \in (a, b)$ such that

$$f'(c) = \frac{f(b) - f(a)}{b - a}.$$

Geometrically, the mean value theorem is easy to explain. Refer to Figure 3.1, which shows a graph of a differentiable function $f(x)$ on an interval $[a, b]$. The dashed line has slope

$$\frac{f(b) - f(a)}{b - a}.$$

In the figure, there happen to be two $x$ values in the interval $(a, b)$ at which the slope of the tangent line (solid line segments) is exactly equal to the slope of the dashed line. The mean value theorem guarantees that at least one such $x$ value will always exist.

**Lemma 3.3.3.** Suppose that $f(x)$ is continuously differentiable on a closed interval $[a, b]$. (That is, $f'(x)$ exists and is continuous on that interval.) Then there exists a positive number $K$ such that $|f(x) - f(y)| \leq K|x - y|$ for all $x, y \in [a, b]$.

*Proof.* Since $f'(x)$ is continuous on the closed interval $[a, b]$, Lemma 3.3.1 tells us that $f'(x)$ achieves both a minimum $m$ and a maximum $M$ somewhere on this interval. Let $K = \max\{|m|, |M|\}$. Suppose $x, y \in [a, b]$ and, without loss of generality, $x < y$. The the mean value theorem 3.3.2 says that there exists a

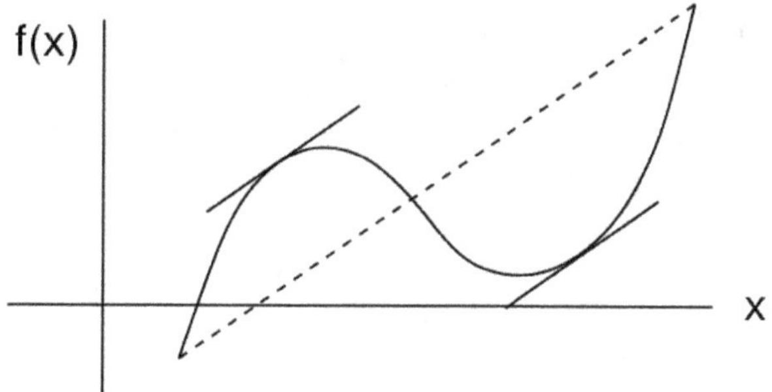


**Figure 3.1.** Illustration of the mean-value theorem. See text for details.

number $c \in (x, y)$ such that

$$f'(c) = \frac{f(y) - f(x)}{y - x}.$$

Taking absolute values, we find that

$$\left| \frac{f(y) - f(x)}{y - x} \right| = |f'(c)| \leq K.$$

Therefore, $|f(y) - f(x)| \leq K|y - x|$ regardless of $x$ and $y$. □

**Lemma 3.3.4.** Suppose $f$ is continuous on the interval $[a, b]$. Then

$$\left| \int_a^b f(x)\, dx \right| \leq \int_a^b |f(x)|\, dx.$$

*Proof.* Exercise. □

**Lemma 3.3.5. (Gronwall's Inequality).** Suppose $f$ is non-negative and continuous on $\mathbb{R}$, and suppose there exist positive constants $C$ and $K$ such that

$$f(t) \leq C + K \int_0^t f(s)\, ds.$$

for all $t \in [0, a]$. Then $f(t) \leq Ce^{Kt}$ for all $t \in [0, a]$.

*Proof.* Exercise. □

Our final Lemma will reference a definition that the reader may be familiar with.

**Definition 3.3.6.** If $\mathbf{x}$ is a vector in $\mathbb{R}^n$, then the *Euclidean norm* of $\mathbf{x}$ is defined as

$$\|\mathbf{x}\|_2 = \sqrt{x_1^2 + x_2^2 + \cdots + x_n^2}.$$

If $\mathbf{x}$ and $\mathbf{y}$ are two points in $\mathbb{R}^n$, the *Euclidean distance* between these two points is given by

$$\|\mathbf{x} - \mathbf{y}\|_2 = \sqrt{(x_1 - y_1)^2 + (x_2 - y_2)^2 + \cdots + (x_n - y_n)^2}.$$

The Euclidean norm of a vector is a generalization of our usual notion of the length of a vector, and Euclidean distance is a generalization of our usual notion of distance in two or three dimensions.

**Lemma 3.3.7. (Triangle Inequality).** If $\mathbf{x}, \mathbf{y} \in \mathbb{R}^n$, then $\|\mathbf{x} + \mathbf{y}\|_2 \le \|\mathbf{x}\|_2 + \|\mathbf{y}\|_2$. In the special case $n = 1$, this means $|x + y| \le |x| + |y|$.

We now use the above lemmas to prove an important theorem which tells us how changing the initial condition may affect the solution of an initial value problem.

**Theorem 3.3.8. (Dependence on Initial Conditions).** Suppose $f$ is continuously differentiable on $\mathbb{R}$ and consider the two initial value problems

$$\begin{aligned} x' = f(x)x(0) &= x_0 \\ y' = f(y)y(0) &= y_0 = x_0 + C. \end{aligned}$$

Then the solutions of these two problems separate at most exponentially fast as $t$ increases (at least over the time interval in which the solutions exist). More exactly, there exists a positive constant $K$ such that $|x(t) - y(t)| \le |C|e^{Kt}$.

*Proof.* Notice that $x$ and $y$ satisfy the *same* ODE, but with different initial conditions. We write both of these initial value problems as integral equations

$$\begin{aligned} x(t) &= x_0 + \int_0^t f(x(s))\, ds \\ y(t) &= y_0 + \int_0^t f(y(s))\, ds = x_0 + C + \int_0^t f(y(s))\, ds. \end{aligned}$$

Measuring the gap between the solutions,

$$\begin{aligned} |y(t) - x(t)| &= \left| x_0 + C + \int_0^t f(y(s))\, ds - \left( x_0 + \int_0^t f(x(s))\, ds \right) \right| \\ &= \left| C + \int_0^t f(y(s)) - f(x(s))\, ds \right|. \end{aligned}$$

By Lemma 3.3.7, we have

$$|y(t) - x(t)| \leq |C| + \left| \int_0^t f(y(s)) - f(x(s))\, ds \right|,$$

from which Lemma 3.3.4 tells us that

$$|y(t) - x(t)| \leq |C| + \int_0^t |f(y(s)) - f(x(s))|\, ds.$$

Since $f$ is continuously differentiable, Lemma 3.3.3 says that there exists a positive constant $K$ such that $|f(y(s)) - f(x(s))| \leq K|y(s) - x(s)|$ which provides the estimate

$$|y(t) - x(t)| \leq |C| + K \int_0^t |y(s) - x(s)|\, ds.$$

Finally, Gronwall's Inequality (Lemma 3.3.5) gives

$$|y(t) - x(t)| \leq |C| e^{Kt}.$$

This means that solutions of the two initial value problems we started with can separate at most exponentially fast. □

With Theorem 3.3.8 in mind, we give an important definition.

**Definition 3.3.9.** An initial value problem

$$x' = f(x) \qquad x(0) = x_0$$

is called *well-posed* if each of the following criteria is satisfied.

☞ *Existence:* The problem has at least one solution.

☞ *Uniqueness:* The problem has at most one solution.

☞ *Dependence on Initial Conditions:* A slight change in initial conditions does not profoundly impact the solution. In other words, solutions of "nearby" initial value problems do not separate faster than exponentially.

In practice, one always hopes to work with well-posed problems. Non-existence of a solution has obvious negative implications, and typically raises questions as to how best to create an "approximate" solution. Non-uniqueness of solutions can sometimes be remedied by imposing additional requirements that solutions must satisfy (e.g., boundary conditions or initial conditions) allowing us to single out one particular solution of interest. Sensitive dependence on initial conditions can be a severe problem. If the behavior of a system is extremely sensitive to the choice of initial conditions, then the solutions may be completely unreliable in making predictions of future behavior. Mathematical models of the weather tend to suffer from this latter drawback, making it difficult to forecast beyond a few days ahead.

We now turn our attention to the question of global existence of solutions of the initial value problem $x' = f(x)$ and $x(0) = x_0$, where $f$ is continuously differentiable. The Fundamental Existence and Uniqueness Theorem 3.2.2 and Theorem 3.3.8 guarantee that this problem is well-posed, but they only ensure that the solution will exist *locally* (i.e., in some interval $(\alpha, \beta)$ containing $t = 0$).

**Lemma 3.3.10.** Consider the initial value problem

$$\mathbf{x}' = f(\mathbf{x}) \qquad \mathbf{x}(0) = \mathbf{x}_0$$

where $f$ is continuously differentiable on $\mathbb{R}^n$. Then there is a *maximal* interval $J = (\alpha, \beta)$ over which the initial value problem has a unique solution $\mathbf{x}(t)$. This *maximal interval of existence* $J$ is an open interval containing $t = 0$.

*Proof.* See Section 2.4 of Perko [8]. □

**Remark.** (i) Let us clarify what we mean by the *maximal* interval of existence. Lemma 3.3.10 states that if the initial value problem has a solution $\mathbf{y}(t)$ on some interval $I$, then $I$ must be a subset of $J$, and $\mathbf{y}(t) = \mathbf{x}(t)$ for all $t \in I$. (ii) If the maximal interval of existence if $(-\infty, \infty)$, then we say that the solution exists globally. (iii) The statement of Lemma 3.3.10 need not require that $f$ be continuously differentiable on all of $\mathbb{R}^n$, only on some open set containing the initial condition $\mathbf{x}_0$. Our less general formulation of 3.3.10 will facilitate the statement and proof of the next lemma, as well as our main Theorem (see 3.3.13 below).

**Lemma 3.3.11.** Suppose that the initial value problem in Lemma 3.3.10 has a maximum interval of existence $(\alpha, \beta)$ where $\beta < \infty$. Then given any positive number $M$ (no matter how large), there exists some $t \in (\alpha, \beta)$ such that $\|\mathbf{x}(t)\|_2 > M$.

*Proof.* See Section 2.4 of Perko [8]. □

In words, Lemma 3.3.11 tells us that if the solution of the initial value problem fails to exist for all positive time $t$, then the solution curve cannot possibly remain confined within some fixed distance $M$ of the origin throughout the entire interval of existence.

**Lemma 3.3.12. (Strong Gronwall Inequality).** Suppose $g \geq 0$ is continuous on $\mathbb{R}$ and that there exist positive constants $C$, $B$, and $K$ such that

$$g(t) \leq C + Bt + K \int_0^t g(s)\, ds$$

for $t \geq 0$. Then

$$g(t) \leq Ce^{Kt} + \frac{B}{K}\left(e^{Kt} - 1\right).$$

*Proof.* Straightforward extension of the proof of the basic Gronwall inequality 3.3.5. □

**Theorem 3.3.13. (Global Existence).** Consider the initial value problem

$$\mathbf{x}' = f(\mathbf{x}) \qquad \mathbf{x}(0) = \mathbf{x}_0,$$

where $f : \mathbb{R}^n \to \mathbb{R}^n$ is continuously differentiable. If there exist positive constants $K$ and $B$ such that $\|f(\mathbf{x})\|_2 \leq K\|\mathbf{x}\|_2 + B$ for all $\mathbf{x} \in \mathbb{R}^n$, then the initial value problem has a unique solution valid for all real $t$.

**Remark.** The condition $\|f(\mathbf{x})\|_2 \leq K\|\mathbf{x}\|_2 + B$ essentially means that the "size" of the vector $\|f(\mathbf{x})\|_2$ grows no faster than linearly as $\mathbf{x}$ moves away from the origin.

*Proof.* We give the proof for $0 < t < \infty$, arguing by contradiction. Suppose the maximal interval of existence is $(\alpha, \beta)$ where $\beta < \infty$. Let $g(t) = \|\mathbf{x}\|_2$, where $\mathbf{x} = \mathbf{x}(t)$ denotes the solution of the initial value problem. Then $g$ is continuous, but not necessarily differentiable. Since

$$\mathbf{x}(t) = \mathbf{x}_0 + \int_0^t f(\mathbf{x}(s))\,\mathrm{d}s,$$

we make the estimate

$$\begin{aligned} g(t) = \left\| \mathbf{x}_0 + \int_0^t f(\mathbf{x}(s))\,\mathrm{d}s \right\|_2 &\le \|\mathbf{x}_0\|_2 + \left\| \int_0^t f(\mathbf{x}(s))\,\mathrm{d}s \right\|_2 \\ \le \|\mathbf{x}_0\|_2 + \int_0^t \|f(\mathbf{x}(s))\|_2\,\mathrm{d}s &\le \|\mathbf{x}_0\|_2 + \int_0^t K\,\|\mathbf{x}(s)\|_2 + B\,\mathrm{d}s \\ = \|\mathbf{x}_0\|_2 + \int_0^t B\,\mathrm{d}s + K\int_0^t \|\mathbf{x}(s)\|_2\,\mathrm{d}s &= \|\mathbf{x}_0\|_2 + Bt + K\int_0^t g(s)\,\mathrm{d}s. \end{aligned}$$

By the Strong Gronwall Inequality 3.3.12, we conclude that

$$g(t) \le \|\mathbf{x}_0\|_2 e^{Kt} + \frac{B}{K}\left(e^{Kt} - 1\right).$$

This means that $\|\mathbf{x}(t)\|_2$ can grow at most exponentially fast. Returning to our indirect assumption that $\beta < \infty$, Lemma 3.3.11 tells us that our solution $\mathbf{x}(t)$ will never remain confined to a given bounded region throughout its maximal interval of existence. In particular, suppose we choose

$$M = \|\mathbf{x}_0\|_2 e^{\beta K} + \frac{B}{K}\left(e^{\beta K} - 1\right)$$

in the statement of Lemma 3.3.11. Then the Lemma implies that

$$g(t) = \|\mathbf{x}(t)\|_2 > M$$

for some $t \in (\alpha, \beta)$. This contradicts our above inequality for $g(t)$. Therefore, $\beta = \infty$ and the solution of the initial value problem exists for all positive $t$. Extending this proof to establish existence for all negative $t$ is straightforward. □

**Example 3.3.14.** The right hand side of the planar system

$$\begin{aligned} \frac{\mathrm{d}x}{\mathrm{d}t} = f_1(x,y) &= \sin(x^2 + y^2) \\ \frac{\mathrm{d}x}{\mathrm{d}t} = f_2(x,y) &= \cos(x^2 + y^2) \end{aligned}$$

consists of functions that are continuously differentiable. Since the sine and cosine functions remain bounded between -1 and 1, we have $\|f(x,y)\|_2 \leq \sqrt{1^2+1^2} = \sqrt{2}$. Therefore, $f$ obeys an estimate of the form $\|f\|_2 \leq K\|x\|_2 + B$, where $K = 0$ and $B = \sqrt{2}$. The global existence theorem 3.3.13 guarantees that any solution of this system of ODEs will exist for all time $t$.

---

## 3.4. Equilibria and Linearization

When approximating nonlinear systems with linear ones, one typically works in the vicinity of equilibrium solutions (for reasons that we shall soon reveal). We begin this section with some definitions which will be used frequently throughout the remainder of this chapter.

**Definition 3.4.1.** An *equilibrium* solution of $\mathbf{x}' = f(\mathbf{x})$ is any constant vector $\mathbf{x}^*$ such that $f(\mathbf{x}^*) = 0$.

**Definition 3.4.2.** Let $\epsilon$ be a fixed, positive number and suppose $\mathbf{x} \in \mathbb{R}^n$. The *open ball* of radius $\epsilon$ centered at $\mathbf{x}$ is the set of all points whose distance from $\mathbf{x}$ is less than $\epsilon$. We will use the notation

$$B(\mathbf{x},\epsilon) = \{\mathbf{y} \in \mathbb{R}^n \text{ such that } \|\mathbf{x}-\mathbf{y}\|_2 < \epsilon\}.$$

**Definition 3.4.3.** An equilibrium $\mathbf{x}^*$ of $\mathbf{x}' = f(\mathbf{x})$ is called *isolated* if there exists a positive number $\epsilon$ such that the open ball $B(\mathbf{x}^*,\epsilon)$ contains no equilibria other than $\mathbf{x}^*$.

**Example 3.4.4.** Find all equilibria of the system

$$x_1' = x_1 - 2x_1x_2 \qquad x_2' = x_1x_2 - 3x_2.$$

**Solution:** Setting both $x_1' = 0$ and $x_2' = 0$, we seek points which simultaneously satisfy

$$\begin{aligned} 0 = x_1 - 2x_1x_2 &= x_1(1-2x_2) \\ 0 = x_1x_2 - 3x_2 &= x_2(x_1-3). \end{aligned}$$

If we set $x_1 = 0$ in the first equation, this would force $x_2 = 0$ in order for the second equation to hold. Likewise, if we set $x_2 = 1/2$ in the first equation, this forces $x_1 = 3$ in the second equation. It follows that there are exactly two solutions of this system:

$$\begin{bmatrix} x_1 \\ x_2 \end{bmatrix} = \begin{bmatrix} 0 \\ 0 \end{bmatrix} \quad \text{and} \quad \begin{bmatrix} x_1 \\ x_2 \end{bmatrix} = \begin{bmatrix} 3 \\ \frac{1}{2} \end{bmatrix}.$$

These are the equilibrium solutions, and both of them are isolated.

**Remark.** (i) Finding equilibria of nonlinear systems is usually impossible to do algebraically. There is no straightforward extension of the matrix algebra techniques you learned for handling linear, constant-coefficient systems. (ii) Whereas matrix systems $A\mathbf{x} = \mathbf{b}$ have either 0, 1, or infinitely many solutions, nonlinear systems can have *any* number of solutions. In the above example, there were *two* equilibria (which could never happen for the linear systems discussed in the previous chapter).

As we might expect, if we start out in equilibrium of a system $\mathbf{x}' = f(\mathbf{x})$, then we are stuck there forever. More exactly, suppose the initial condition is $\mathbf{x}(0) = \mathbf{x}^*$, where $\mathbf{x}^*$ is an equilibrium. Clearly the constant function $\mathbf{x}(t) = \mathbf{x}^*$ is a solution of the initial value problem, because $\mathbf{x}' = 0$ and $f(\mathbf{x}) = 0$. Assuming that all partial derivatives in $Jf(\mathbf{x})$ are continuous in the vicinity of $\mathbf{x}^*$, then the Fundamental Existence and Uniqueness Theorem 3.2.2 guarantees that the initial value problem has a *unique* solution in some interval containing $t = 0$. Since we have already produced the solution $\mathbf{x}(t) = \mathbf{x}^*$, this constant solution must be the only solution.

**Linearization.** Our earlier remarks about linear approximation of functions allude to how we will approximate nonlinear systems of ODEs with linear ones. For the sake of illustration, we introduce the process of linearization for two-variable systems

$$x_1' = f_1(x_1, x_2), \qquad x_2' = f_2(x_1, x_2), \tag{3.5}$$

and remark that the following is easily extended to $n$-variable systems. Suppose that $(x_1^*, x_2^*)$ is any point in $\mathbb{R}^2$ and we wish to approximate the behavior of the nonlinear system (3.5) near this point. Replacing the right-hand sides of the equations (3.5) with their tangent plane approximations at $(x_1^*, x_2^*)$, we obtain

$$\begin{bmatrix} x_1' \\ x_2' \end{bmatrix} \approx \underbrace{\begin{bmatrix} f_1(x_1^*, x_2^*) \\ f_2(x_1^*, x_2^*) \end{bmatrix}}_{\text{vector in } \mathbb{R}^2} + \underbrace{\begin{bmatrix} \nabla f_1(x_1^*, x_2^*) \\ \nabla f_2(x_1^*, x_2^*) \end{bmatrix}}_{2\times 2 \text{ Jacobian}} \underbrace{\begin{bmatrix} x_1 - x_1^* \\ x_2 - x_2^* \end{bmatrix}}_{\text{vector in } \mathbb{R}^2}.$$

This is nothing more than a restatement of Equation (3.3) in $n = 2$ dimensions. If we use vector notation

$$\mathbf{x} = \begin{bmatrix} x_1 \\ x_2 \end{bmatrix} \quad \text{and} \quad \mathbf{x}^* = \begin{bmatrix} x_1^* \\ x_2^* \end{bmatrix},$$

then the linear approximation of the system (3.5) becomes

$$\mathbf{x}' \approx f(\mathbf{x}^*) + Jf(\mathbf{x}^*)(\mathbf{x} - \mathbf{x}^*).$$

We give this approximation a name.

**Definition 3.4.5.** The system

$$\mathbf{x}' = f(\mathbf{x}^*) + Jf(\mathbf{x}^*)(\mathbf{x} - \mathbf{x}^*) \tag{3.6}$$

is called the *linearization* of the system $\mathbf{x}' = f(\mathbf{x})$ at the point $\mathbf{x} = \mathbf{x}^*$.

The linearization is a linear, constant-coefficient system which can be solved using the techniques in the previous chapter. If $\mathbf{x}^*$ happens to be an equilibrium of the system, then $f(\mathbf{x}^*) = 0$ and the linearization takes the particularly convenient form

$$\mathbf{x}' = Jf(\mathbf{x}^*)(\mathbf{x} - \mathbf{x}^*).$$

**Example 3.4.6.** Consider our earlier example

$$x_1' = x_1 - 2x_1x_2, \qquad x_2' = x_1x_2 - 3x_2,$$

for which we found two equilibria:

$$\begin{bmatrix} 0 \\ 0 \end{bmatrix} \quad \text{and} \quad \begin{bmatrix} 3 \\ \frac{1}{2} \end{bmatrix}.$$

To obtain the linearization at the origin, we begin by computing the Jacobian

$$Jf(\mathbf{x}) = \begin{bmatrix} 1 - 2x_2 & -2x_1 \\ x_2 & x_1 - 3 \end{bmatrix}.$$

Evaluating the Jacobian at the first equilibrium gives

$$Jf(0,0) = \begin{bmatrix} 1 & 0 \\ 0 & -3 \end{bmatrix},$$

and therefore the linearization or our system at $(0,0)$ is

$$\begin{bmatrix} x_1' \\ x_2' \end{bmatrix} = \begin{bmatrix} 1 & 0 \\ 0 & -3 \end{bmatrix} \begin{bmatrix} x_1 \\ x_2 \end{bmatrix}.$$

We immediately see that the origin is a *saddle* for the linearized system and the solution is $x_1(t) = c_1 e^t$ and $x_2(t) = c_2 e^{-3t}$. The linearization about the other equilibrium point is handled in a similar way. Referring to the Jacobian matrix $Jf(\mathbf{x})$ above, we calculate that

$$Jf\left(3, \frac{1}{2}\right) = \begin{bmatrix} 0 & -6 \\ \frac{1}{2} & 0 \end{bmatrix}.$$

Notice that the characteristic equation of this matrix is $\lambda^2 + 3 = 0$, which means the eigenvalues are $\lambda = \pm\sqrt{3}i$. This would suggest that the equilibrium $\mathbf{x}^* = (3, 1/2)$ is a *center*, at least for the linearized system

$$\begin{bmatrix} x_1' \\ x_2' \end{bmatrix} = \begin{bmatrix} 0 & -6 \\ \frac{1}{2} & 0 \end{bmatrix} \begin{bmatrix} x_1 - 3 \\ x_2 - \frac{1}{2} \end{bmatrix} = \begin{bmatrix} 0 & -6 \\ \frac{1}{2} & 0 \end{bmatrix} \begin{bmatrix} x_1 \\ x_2 \end{bmatrix} + \begin{bmatrix} 3 \\ -\frac{3}{2} \end{bmatrix}.$$

Solving this (barely inhomogeneous) system is straightforward.

When linearizing about an equilibrium, one would hope that the linearized system would mimic the behavior of the nonlinear system, at least in the vicinity of the equilibrium. This is usually, but not always, the case. In the previous example, one of the equilibria was hyperbolic (no eigenvalues of the Jacobian had zero real part), and the other was non-hyperbolic. In the next section, we shall formally state the Hartman-Grobman Theorem, which tells us that if we linearize about a *hyperbolic* equilibrium, then the linearization exhibits the same qualitative behavior as the original nonlinear system. This need not be true for non-hyperbolic equilibria.

---

## 3.5. The Hartman-Grobman Theorem

In the previous section, we introduced the notion of isolated equilibria for systems of ODEs.

**Example 3.5.1.** For a linear, homogeneous, constant-coefficient system $\mathbf{x}' = A\mathbf{x}$, equilibria $\mathbf{x}^*$ must satisfy $A\mathbf{x}^* = 0$. If $A$ is invertible, then the only equilibrium is the origin, $\mathbf{x}^* = 0$, which is clearly isolated. If $A$ is not invertible, then solutions of $A\mathbf{x}^* = 0$ form a subspace of $\mathbb{R}^n$ of dimension at least 1 (i.e., the nullspace of $A$ is non-trivial). This implies that there are infinitely many equilibria, none of which are isolated.

In this section, we will assume that all equilibria are isolated. We also need one more definition concerning equilibria of nonlinear systems.

**Definition 3.5.2.** An equilibrium $\mathbf{x}^*$ of the system $\mathbf{x}' = f(\mathbf{x})$ is called *hyperbolic* if all eigenvalues of the Jacobian $Jf(\mathbf{x}^*)$ have non-zero real part.

**Example 3.5.3.** By algebra, you can show that the system

$$x' = x - xy \qquad y' = -y + xy$$

has exactly two equilibria: $(x, y) = (0, 0)$ and $(x, y) = (1, 1)$. The Jacobian matrix is

$$Jf(x, y) = \begin{bmatrix} 1 - y & -x \\ y & -1 + x \end{bmatrix},$$

from which we calculate that

$$Jf(0,0) = \begin{bmatrix} 1 & 0 \\ 0 & -1 \end{bmatrix} \quad \text{and} \quad Jf(1,1) = \begin{bmatrix} 0 & -1 \\ 1 & 0 \end{bmatrix}.$$

Since the eigenvalues of $Jf(0,0)$ are 1 and $-1$, the equilibrium $(0,0)$ is hyperbolic. However, the eigenvalues of $Jf(1,1)$ are $\lambda = \pm i$, both of which have zero real part. Thus, $(1,1)$ is a non-hyperbolic equilibrium.

We now state one of our most important theorems about the qualitative behavior of solutions of nonlinear systems of ODEs.

**Theorem 3.5.4. (Hartman-Grobman).** Suppose $\mathbf{x}^*$ is an isolated equilibrium of a nonlinear system $\mathbf{x}' = f(\mathbf{x})$. Then in the vicinity of $\mathbf{x}^*$, the linearization $\mathbf{x}' = Jf(\mathbf{x}^*)(\mathbf{x} - \mathbf{x}^*)$ about that equilibrium has the same qualitative behavior as the original nonlinear system.

More precisely, there exists a positive number $\epsilon$ such that, within the open ball $B(\mathbf{x}^*, \epsilon)$, the phase portraits of the nonlinear system and its linearization are *topologically equivalent*. In other words, there is a continuous, one-to-one correspondence between phase portrait trajectories which preserves the orientation of corresponding trajectories. We illustrate this via an example.

**Example 3.5.5.** Although the system

$$x' = -x \qquad\qquad y' = x^2 + y$$

is nonlinear, it can be solved exactly, one equation at a time. Solving the first equation yields

$$x(t) = x_0 e^{-t},$$

where $x_0 = x(0)$. Substituting this into the second equation and using the variation of parameters formula, we find that

$$y(t) = y_0 e^t + \frac{1}{3}x_0^2 \left(e^t - e^{-2t}\right).$$

With the assistance of a computer, we could sketch the phase portrait for the nonlinear system by choosing various initial conditions $(x_0, y_0)$ and plotting the parametrized curve $(x(t), y(t))$ given by the above formulas. The phase portrait is sketched in the left panel of Figure 3.2. Let us compare the dynamics of the nonlinear system with that of its linearization. Clearly the only equilibrium of this system is $(x, y) = (0, 0)$. We linearize by computing the Jacobian

$$Jf(x, y) = \begin{bmatrix} -1 & 0 \\ 2x & 1 \end{bmatrix},$$

from which

$$Jf(0, 0) = \begin{bmatrix} -1 & 0 \\ 0 & 1 \end{bmatrix}.$$

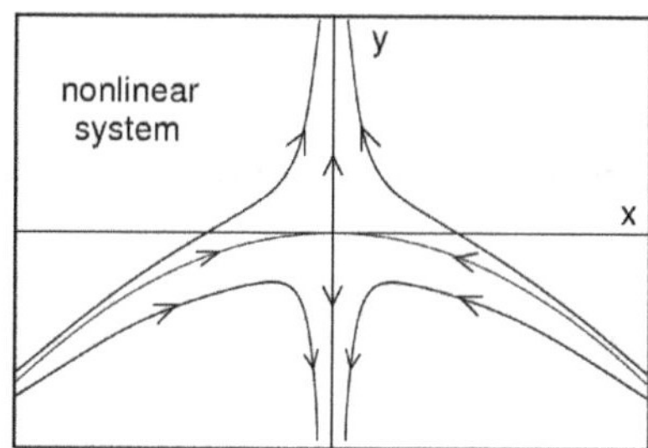


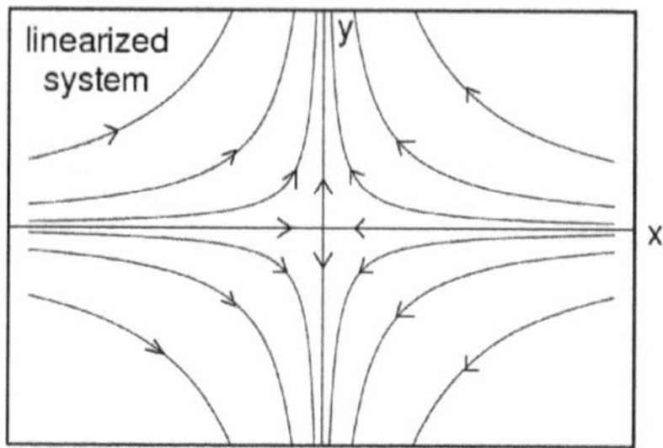


**Figure 3.2.** Comparing the phase portrait of a nonlinear system (left panel) with that of its linearization (right panel).

The eigenvalues of $Jf(0,0)$ are $-1$ and $1$, neither of which has zero real part. Hence, the origin is an isolated, hyperbolic equilibrium point. Moreover, since the eigenvalues are real and of opposite sign, the origin is a saddle. The solution of the linearized system

$$\begin{bmatrix} x' \\ y' \end{bmatrix} = \begin{bmatrix} -1 & 0 \\ 0 & 1 \end{bmatrix} \begin{bmatrix} x \\ y \end{bmatrix}$$

is $x = x_0 e^{-t}$ and $y = y_0 e^t$. You may check that the stable subspace $E^s$ is the x-axis and the unstable subspace $E^u$ is the y-axis. The phase portrait for this saddle is shown in the right panel of Figure 3.2. Comparing the two panels in Figure 3.2 serves to illustrate the Hartman-Grobman Theorem. The phase portrait for the nonlinear system is a "warped" version of the phase portrait of its linearization. However, if we "zoom in" very close the origin, the two phase portraits would be indistinguishable. Notice that the unstable subspace $E^u$ for the linearization also happens to form a separatrix for the nonlinear system. However, the nonlinear system "bends" the stable subspace $E^s$ into a parabola—this parabolic trajectory does *not* form a subspace of $\mathbb{R}^2$. In the next section, we explain that the nonlinear counterparts of stable, unstable, and center subspaces are called stable, unstable, and center *manifolds*.

---

## 3.6. The Stable Manifold Theorem

A *manifold* is a generalization of the flat objects (lines, planes, and hyperplanes) that you learned about in linear algebra. The precise definition of a manifold is somewhat technical and is not required for our purposes; instead we give some examples. A smooth curve such as a circle in the plane is an example of a one-dimensional manifold. If we "zoom in" on any point within the circle, what we see would be indistinguishable from a small interval on the (one-dimensional) real line. Likewise, the surfaces in $\mathbb{R}^3$ you learned about in multi-variable calculus would be examples of two-dimensional manifolds. If we suitably magnify any tiny region on the graph of the paraboloid $f(x, y) = x^2 + y^2$, then the result would be virtually indistinguishable from a region within the "flat" space $\mathbb{R}^2$. We shall deal only with differentiable manifolds (imagine a surface which does not have any sharp corners or ridges).

Below, we shall find that near hyperbolic equilibria, we expect nonlinear systems to have stable and unstable manifolds which have the same dimensions and invariance properties of $E^s$ and $E^u$ for the linearized systems. First, we review the concept of the *flow*.

Suppose that $\mathbf{x}' = f(\mathbf{x})$ and assume that all partial derivatives in $Jf(\mathbf{x})$ are continuous for all $\mathbf{x} \in \mathbb{R}^n$. Given an initial condition $\mathbf{x}(0) = \mathbf{x}_0$, Theorem 3.2.2 tells us that the initial value problem has a unique solution which exists in some open interval containing $t = 0$. Let $\phi_t(\mathbf{x}_0)$ denote the solution of this initial value problem, defined on the largest time interval containing $t = 0$ over which the solution exists.

**Definition 3.6.1.** The set of all such functions $\phi_t(\mathbf{x}_0)$ (i.e., for all possible choices of $\mathbf{x}_0$) is called the *flow* of the ODE $\mathbf{x}' = f(\mathbf{x})$.

Notice that for each choice of $\mathbf{x}_0$, the function $\phi_t(\mathbf{x}_0)$ defines a parametrized curve in $\mathbb{R}^n$ (parametrized by $t$), crossing the point $\mathbf{x}_0$ at time $t = 0$. For the linear systems we discussed in the previous chapter, we know that the phase portraits may contain saddles, foci, nodes, and centers. The flows for nonlinear systems may exhibit much more interesting behavior.

**Definition 3.6.2.** A subset $E \subset \mathbb{R}^n$ is called *invariant* with respect to the flow if for each $\mathbf{x}_0 \in E$, the curve $\phi_t(\mathbf{x}_0)$ remains inside $E$ for all time $t$ over which the solution actually exists.

For linear, homogeneous, constant-coefficient systems $\mathbf{x}' = A\mathbf{x}$, the subspaces $E^s$ $E^c$, and $E^u$ associated with the equilibrium at the origin are all invariant with respect to the flow. If we start from an initial condition inside any one of these subspaces, our solution trajectory remains confined to that subspace for all time $t$.

We now state another major theorem which, together with the Hartman-Grobman Theorem, provides much of our basis for understanding the qualitative behavior of solutions of nonlinear ODEs.

**Theorem 3.6.3. (Stable Manifold Theorem).** Suppose that $\mathbf{x}' = f(\mathbf{x})$ is a system of ODEs for which the Jacobian matrix $Jf(\mathbf{x})$ consists of continuous functions. Further suppose that this system has an isolated, hyperbolic equilibrium point at the origin, and that the Jacobian $Jf(0)$ has $k$ eigenvalues with negative real part and $(n-k)$ eigenvalues with positive real part. Then

☞ There is a $k$-dimensional differentiable manifold $W^s$ which is (i) tangent to the stable subspace $E^s$ of the linearized system $\mathbf{x}' = Jf(0)\mathbf{x}$ at the origin; (ii) is invariant with respect to the flow; and (iii) for all initial conditions $\mathbf{x}_0 \in W^s$, we have

$$\lim_{t\to\infty} \phi_t(\mathbf{x}_0) = 0.$$

☞ There is an $(n-k)$-dimensional differentiable manifold $W^u$ which is (i) tangent to the unstable subspace $E^u$ of the linearized system $\mathbf{x}' = Jf(0)\mathbf{x}$ at the origin; (ii) is invariant with respect to the flow; and (iii) for all initial conditions $\mathbf{x}_0 \in W^u$, we have

$$\lim_{t\to-\infty} \phi_t(\mathbf{x}_0) = 0.$$

Here, $W^s$ and $W^u$ are called the *stable and unstable manifolds*, respectively.

**Remark.** Although the statement of this Theorem is a bit wordy, it is not difficult to understand what is going on. Basically, it says that near a *hyperbolic*, isolated equilibrium point, nonlinear systems produce objects $W^s$ and $W^u$ which are "curvy versions" of the subspaces $E^s$ and $E^u$ for the linearized system. The manifold $W^s$ is tangent to the subspace $E^s$ at the equilibrium. In the previous example, $W^s$ was a parabola which was tangent to $E^s$ (the x-axis) at the origin (see Figure 3.2). If we start from an initial condition inside $W^s$, we will stay inside $W^s$ for all time $t$, and we will approach the equilibrium as $t \to \infty$. Similar statements hold for the unstable manifold.

**Example 3.6.4.** In the example we did in the previous section

$$x' = -x, \qquad y' = x^2 + y,$$

we noted that this nonlinear system has an exact solution

$$x(t) = x_0 e^{-t}, \qquad y(t) = \left(y_0 + \frac{1}{3}x_0^2\right)e^t - \frac{1}{3}x_0^2 e^{-2t}.$$

The origin is the only equilibrium solution, and we already established that it is a saddle. To determine $W^s$, we need to determine which special choices of initial conditions $(x_0, y_0)$ would cause us to approach the origin as $t \to \infty$. In the equation for $x(t)$, we find that

$$\lim_{t\to\infty} x_0 e^{-t} = 0$$

regardless of our choice for $x_0$. This imposes no restrictions on our choice of initial conditions. Taking the same limit in the equation for $y(t)$ is much more interesting: in order to guarantee that

$$\lim_{t\to\infty}\left[\left(y_0 + \frac{1}{3}x_0^2\right)e^t - \frac{1}{3}x_0^2 e^{-2t}\right] = 0$$

we must insist that

$$y_0 + \frac{1}{3}x_0^2 = 0,$$

because the exponential function $e^t$ will increase without bound as $t \to \infty$. No other restrictions are necessary, because $e^{-2t} \to 0$ as $t \to \infty$. Therefore, we conclude that the one-dimensional stable manifold is given by the parabola $y = -x^2/3$. A similar argument shows that $W^u$ is the $y$-axis. Notice that $W^s$ is tangent to the stable subspace $E^s$ (the $x$-axis) at the equilibrium point.

**Example 3.6.5.** Consider the nonlinear system

$$x' = -x + 3y^2, \qquad y' = -y, \qquad z' = 3y^2 + z.$$

Solve this system and compute the stable and unstable manifolds for the equilibrium at the origin.

**Solution:** This is very similar to the previous example. By computing the linearization, you should convince yourself that the origin really is a *hyperbolic* equilibrium. We see immediately that $y(t) = y_0 e^{-t}$. Substituting this expression into the first equation, we obtain a linear, inhomogeneous equation

$$x' + x = 3y_0^2 e^{-2t}.$$

The variation of parameters technique applies. Using

$$e^{\int 1\, dt} = e^t$$

as an integrating factor, we find that

$$e^t (x' + x) = 3y_0^2 e^{-t},$$

or equivalently,

$$\frac{d}{dt}\left(e^t x\right) = 3y_0^2 e^{-t}.$$

Integrating both sides,

$$e^t x = e^t x\big|_{t=0} + \int_0^t 3y_0^2 e^{-s}\, ds = x_0 + 3y_0^2 \left(-e^{-s}\right)\big|_0^t = x_0 + 3y_0^2(1 - e^{-t}).$$

Multiplying both sides by $e^{-t}$,

$$x(t) = x_0 e^{-t} + 3y_0^2\left(e^{-t} - e^{-2t}\right).$$

The equation for $z$ is solved in a similar fashion, and the general solution of the system is

$$\begin{aligned} x &= \left(x_0 + 3y_0^2\right) e^{-t} - 3y_0^2 e^{-2t} \\ y &= y_0 e^{-t} \\ z &= \left(z_0 + y_0^2\right) e^{t} - y_0^2 e^{-2t}. \end{aligned}$$

The stable manifold $W^s$ consists of all initial conditions $(x_0, y_0, z_0)$ such that the flow guides us to the origin as $t \to \infty$. That is,

$$\lim_{t\to\infty} \phi_t(x_0, y_0, z_0) = (0, 0, 0).$$

Notice that all of the exponential functions in the solution are decaying *except* for the $e^t$ term in the $z(t)$ equation. In order to guarantee that we approach the

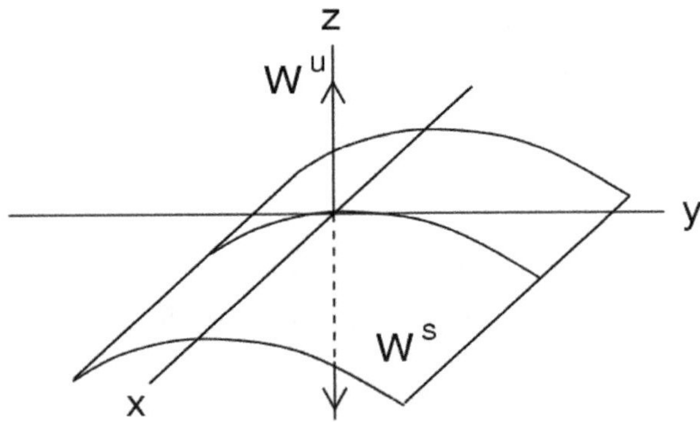


**Figure 3.3.** Sketch of the two-dimensional stable manifold $z = -y^2$ and the one-dimensional unstable manifold (z-axis) from the example in the text.

origin as $t \to \infty$, we need to coefficient of $e^t$ to be 0. This forces

$$z_0 + y_0^2 = 0.$$

If we graph $z = -y^2$ in $\mathbb{R}^3$, the result is a two-dimensional manifold: a parabolic sheet (see Figure 3.3). Similarly, solutions in the unstable manifold must approach the origin as $t \to -\infty$. In this limit, we have $e^t \to 0$ but $e^{-t} \to \infty$ and $e^{-2t} \to \infty$. Requiring the coefficients of both $e^{-t}$ and $e^{-2t}$ to be 0, it must be the case that $x_0 = 0$ and $y_0 = 0$, while $z_0$ remains free. Therefore, the unstable manifold $W^u$ consists of all points on the $z$-axis.

We remark that when calculating $W^s$ and $W^u$, it was important to express the solution $(x, y, z)$ of the ODEs in terms of the initial conditions $(x_0, y_0, z_0)$ as opposed to introducing purely arbitrary constants $C_1$, $C_2$, and $C_3$.

Calculating $W^s$ and $W^u$ by hand is usually impossible. There are methods for obtaining successive approximations of these manifolds, but such techniques can be very tedious.

---

## 3.7. Non-Hyperbolic Equilibria and Lyapunov Functions

The Hartman-Grobman and Stable Manifold Theorems tell us that near an isolated, hyperbolic equilibrium point $\mathbf{x}_0$, the behavior of the nonlinear system

$\mathbf{x}' = f(\mathbf{x})$ is qualitatively similar to that of its linearization $\mathbf{x}' = Jf(\mathbf{x}_0) \cdot (\mathbf{x} - \mathbf{x}_0)$. What about non-hyperbolic equilibria?

**Example 3.7.1.** We claim that the system

$$x' = -y - x(x^2 + y^2) \qquad y' = x - y(x^2 + y^2)$$

has exactly one equilibrium (the origin). To verify this, we set both $x' = 0$ and $y' = 0$ to obtain

$$y = -x(x^2 + y^2) \quad \text{and} \quad x = y(x^2 + y^2).$$

Combining these two equations, we have

$$x = -x(x^2 + y^2)^2.$$

Clearly $(x^2 + y^2) \geq 0$ regardless of $x$ and $y$, with equality only if $x = y = 0$. If it were the case that $x \neq 0$, we could divide both sides of the latter equation by $x$ to obtain $1 = -(x^2 + y^2)^2 < 0$, which is absurd. Therefore, it must be the case that $x = 0$, from which it follows that $y = 0$ as well.

Next, we claim that the equilibrium $(0, 0)$ is non-hyperbolic. The Jacobian matrix associated with our system is

$$Jf(x, y) = \begin{bmatrix} -3x^2 - y^2 & -1 - 2xy \\ 1 - 2xy & -x^2 - 3y^2 \end{bmatrix},$$

and evaluating this matrix at the equilibrium point yields

$$Jf(x, y) = \begin{bmatrix} 0 & -1 \\ 1 & 0 \end{bmatrix}.$$

The linearization of our original system is

$$\begin{bmatrix} x' \\ y' \end{bmatrix} = \begin{bmatrix} 0 & -1 \\ 1 & 0 \end{bmatrix} \begin{bmatrix} x \\ y \end{bmatrix},$$

and the eigenvalues of the coefficient matrix are $\lambda = \pm i$. Both eigenvalues have zero real part, implying that the equilibrium is non-hyperbolic and that the origin is a *center* for the linearized system (with circular trajectories in the phase plane). Is this an accurate representation of the behavior of the nonlinear system?

**Sneaky observation:** The original system is easy to analyze if we introduce a "radial" variable $u(t) = x(t)^2 + y(t)^2$. Notice that $u$ measures the square of the Euclidean distance from the origin for points on our solution curve $(x(t), y(t))$. Taking the derivative,

$$\begin{aligned}\frac{du}{dt} &= 2x\frac{dx}{dt} + 2y\frac{dy}{dt} = 2x[-y - x(x^2+y^2)] + 2y[x - y(x^2+y^2)] \\ &= -2x^2(x^2+y^2) - 2y^2(x^2+y^2) = -2(x^2+y^2)(x^2+y^2) = -2u^2.\end{aligned}$$

The equation $u' = -2u^2$ is easy to solve by separation of variables, and we find that

$$u(t) = \frac{1}{2t+C},$$

where $C$ is an arbitrary constant which we could solve for if given an initial condition $u_0$. Since $u$ is, by definition, a non-negative quantity, we know that $u_0 \geq 0$. In fact, we may assume that $u_0 > 0$ because otherwise we would have $x_0 = y_0 = 0$, placing us at the equilibrium for all time $t$. Setting $t = 0$ in the equation for $u(t)$, we find that $C = u_0^{-1} > 0$ because $u_0 > 0$. Since $C$ is positive, the solution $u(t)$ exists for all positive $t$ because the denominator $2t + C$ is never 0. Taking the limit $t \to \infty$, we conclude that

$$\lim_{t\to\infty} u(t) = 0$$

*regardless of our choice of initial conditions.* This means that all solutions trajectories for the nonlinear system approach the equilibrium at the origin as $t \to \infty$. Consequently, the origin is an *attractor* in spite of the fact that the linearized system has a *center* at the origin. Figure 3.4 shows a sample trajectory in the phase plane for the nonlinear system. The phase portrait for the linearized system is exactly as in Figure 2.8e.

**Definition 3.7.2.** In the above example, the origin is called a *weak sink* or *weak attractor*.

The reason for using the word "weak" in this definition is because trajectories approach the origin slower than those of "normal" attractors, which approach exponentially fast.

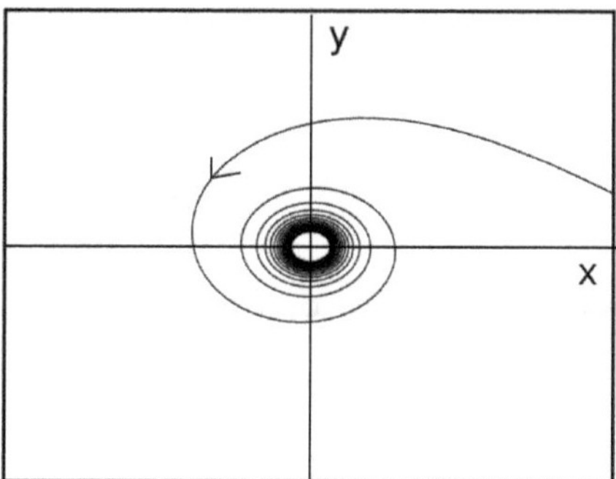


**Figure 3.4.** Phase plane of a nonlinear system for which the origin is a weak attractor (see text).

**Example 3.7.3.** Changing signs in the previous example, the system

$$x' = -y + x(x^2 + y^2) \qquad y' = x + y(x^2 + y^2)$$

still has a non-hyperbolic equilibrium at the origin. This time, the origin is unstable (a *repeller*), whereas the origin is a [stable] *center* for the linearized system. Defining $u(t)$ as before, the reader is encouraged to show that $\|u(t)\|_2$ actually blows up to $\infty$ in a *finite* amount of time for any choice of initial conditions (excluding the equilibrium itself). This is a major contrast with the behavior of solutions of the linearized system, which exist and remain bounded for all real $t$.

**Moral:** The two examples above should convince you that near non-hyperbolic equilibria, we need a completely different approach for classifying stability. Earlier, we gave an intuitive, but somewhat vague definition of what it means for an equilibrium to be stable. We now give a precise mathematical definition for stability of equilibria. To make the definition slightly less technical to state, we will assume that the flow $\phi_t$ is defined for all positive $t$. You may wish to consult Definition 3.4.2 for a reminder of how the open ball $B(\mathbf{x}^*, \epsilon)$ is defined.

**Definition 3.7.4.** Suppose $\mathbf{x}^*$ is an isolated equilibrium of the ODE $\mathbf{x}' = f(\mathbf{x})$ and let $\mathbf{x}_0$ denote an initial condition. Then $\mathbf{x}^*$ is called

☞ *locally stable* if given any number $\epsilon > 0$ there exists a $\delta > 0$ such that whenever $\mathbf{x}_0 \in B(\mathbf{x}^*, \delta)$, then the solution satisfies $\mathbf{x}(t) \in B(\mathbf{x}^*, \epsilon)$ for all $t > 0$.

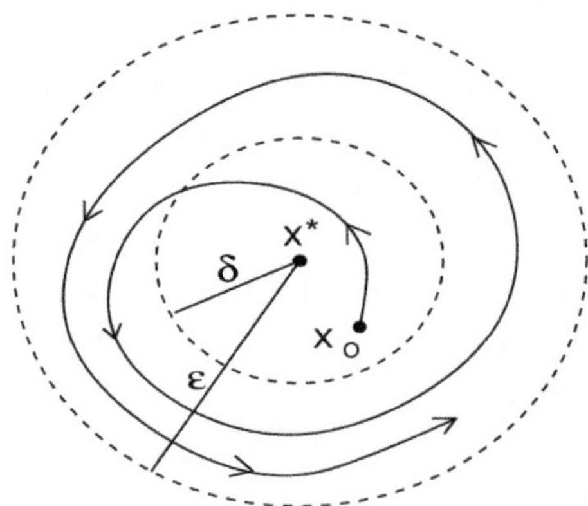


**Figure 3.5.** Illustration of local stability of an equilibrium. The inner and outer dashed circles have radii $\delta$ and $\epsilon$, respectively, and are centered at an equilibrium $\mathbf{x}^*$.

☞ *locally asymptotically stable* if $\mathbf{x}^*$ is locally stable and there exists $\eta > 0$ such that

$$\lim_{t \to \infty} \mathbf{x}(t) = \mathbf{x}^*$$

whenever $\mathbf{x}_0 \in B(\mathbf{x}^*, \eta)$.

☞ *unstable* if $\mathbf{x}^*$ is not locally stable.

The notion of local stability is illustrated in Figure 3.5, and in words, this concept is explained as follows. Suppose we are given a positive number $\epsilon$. Then no matter how small $\epsilon$ is, we can always produce another positive number $\delta$ such that whenever we start from an initial condition whose distance from $\mathbf{x}^*$ is less than $\delta$, the solution curve $\mathbf{x}(t)$ will always have distance less than $\epsilon$ from $\mathbf{x}^*$. In some sense, this basically means that trajectories that start reasonably close to $\mathbf{x}^*$ can never "escape too far" from $\mathbf{x}^*$. Local asymptotic stability means that if the initial condition $\mathbf{x}_0$ is appropriately close to $\mathbf{x}^*$, then the solution curve $\mathbf{x}(t)$ will actually converge to $\mathbf{x}^*$ as $t \to \infty$. For linear systems, centers are examples of equilibria that are stable but not asymptotically stable. The Hartman-Grobman Theorem and Stable Manifold Theorems ensure that any hyperbolic equilibrium point of $\mathbf{x}' = f(\mathbf{x})$ is either locally asymptotically stable (if *all* eigenvalues of the Jacobian matrix $Jf(\mathbf{x}^*)$ have negative real part) or unstable (if *any* eigenvalue has positive real part). An asymptotically stable equilibrium is sometimes called a *sink* or *attractor*.

**Lyapunov Functions.** Classifying non-hyperbolic equilibria $\mathbf{x}^*$ as stable, asymptotically stable, or unstable can be incredibly difficult (and often impossible). We now describe a classification technique that was originally proposed by Russian mathematician A.M. Lyapunov in his 1892 doctoral dissertation. Consider the ODE $\mathbf{x}' = f(\mathbf{x})$, where $\mathbf{f}$ is continuously differentiable. A solution $\mathbf{x}(t)$ of this equation can be written in terms of its component functions, $\mathbf{x}(t) = [x_1(t), x_2(t), \ldots x_n(t)]$. Now suppose that $V : \mathbb{R}^n \to \mathbb{R}$ is a continuously differentiable scalar-valued function. Then by the chain rule,

$$\begin{aligned}
\frac{\mathrm{d}}{\mathrm{d}t} V(\mathbf{x}(t)) &= \frac{\mathrm{d}}{\mathrm{d}t} V(x_1(t), x_2(t), \ldots x_n(t)) \\
&= \frac{\partial V}{\partial x_1}\frac{\mathrm{d}x_1}{\mathrm{d}t} + \frac{\partial V}{\partial x_2}\frac{\mathrm{d}x_2}{\mathrm{d}t} + \cdots + \frac{\partial V}{\partial x_n}\frac{\mathrm{d}x_n}{\mathrm{d}t} \\
&= \left[\frac{\partial V}{\partial x_1}, \frac{\partial V}{\partial x_2}, \cdots \frac{\partial V}{\partial x_n}\right] \bullet \left[\frac{\mathrm{d}x_1}{\mathrm{d}t}, \frac{\mathrm{d}x_2}{\mathrm{d}t}, \ldots, \frac{\mathrm{d}x_n}{\mathrm{d}t}\right] \qquad \text{(a dot product)} \\
&= \nabla V(\mathbf{x}) \bullet \mathbf{x}'(t) \;=\; \nabla V(\mathbf{x}) \bullet f(\mathbf{x}).
\end{aligned}$$

**Observation.** This calculation tells us how the function $V$ changes as we move along a solution curve $\mathbf{x}(t)$. In particular, if we find that

$$\frac{\mathrm{d}}{\mathrm{d}t} V(\mathbf{x}(t)) < 0$$

inside some set $E \subset \mathbb{R}^n$, then the function $V$ *decreases* as we move along solution curves in $E$ in the direction of increasing $t$.

Lyapunov exploited this observation to provide a creative but intuitive way for analyzing stability of equilibria $\mathbf{x}^*$. The idea is to define a function $V$ on a set $E$ containing $\mathbf{x}^*$, where $V$ is chosen in such a way that we can tell whether the flow in $E$ is towards or away from the equilibrium. In what follows, we have in mind a system $\mathbf{x}' = f(\mathbf{x})$ with an isolated equilibrium $\mathbf{x}^*$. We assume that $f$ is continuously differentiable in some open ball $E = B(\mathbf{x}^*, \epsilon)$ of radius $\epsilon > 0$ centered at the equilibrium.

**Theorem 3.7.5** (Lyapunov). Suppose there exists a function $V : \mathbb{R}^n \to \mathbb{R}$ which is (i) defined and continuously differentiable on the set $E = B(\mathbf{x}^*, \epsilon)$; (ii) $V(\mathbf{x}^*) = 0$; and (iii) $V(\mathbf{x}) > 0$ if $\mathbf{x} \neq \mathbf{x}^*$. Then the equilibrium $\mathbf{x}^*$ is

☞ *Stable* if

$$\frac{\mathrm{d}}{\mathrm{d}t} V(\mathbf{x}(t)) \;=\; \nabla V(\mathbf{x}) \bullet f(\mathbf{x}) \leq 0$$

for all $\mathbf{x} \in E$.

☞ *Asymptotically stable* if

$$\frac{\mathrm{d}}{\mathrm{d}t}V(\mathbf{x}(t)) = \nabla V(\mathbf{x}) \bullet f(\mathbf{x}) < 0$$

for all $\mathbf{x} \in E$, except possibly at $\mathbf{x}^*$ itself.

☞ *Unstable* if

$$\frac{\mathrm{d}}{\mathrm{d}t}V(\mathbf{x}(t)) = \nabla V(\mathbf{x}) \bullet f(\mathbf{x}) > 0$$

for all $\mathbf{x} \in E$, except possibly at $\mathbf{x}^*$ itself.

**Definition 3.7.6.** Any function $V : \mathbb{R}^n \to \mathbb{R}$ satisfying the conditions of Theorem 3.7.5 is called a *Lyapunov function.*

Some geometric considerations may help convince you that Lyapunov's Theorem is intuitively plausible. Figure 3.6 illustrates the shape of a typical Lyapunov function for a two-variable system of ODEs with an equilibrium at the origin. In this case, the graph of the Lyapunov function $V(x, y)$ is a surface in $\mathbb{R}^3$, and is positive everywhere except at the equilibrium (where it is 0). Any solution $(x(t), y(t))$ of the ODEs defines a parametrized curve in the $xy$-plane, and therefore $V(x(t), y(t))$ defines a curve on the surface $V$. If $V(x(t), y(t))$ decreases to $0$ as $t$ increases, then the corresponding solution trajectory in the $xy$-plane is "funneled" towards the origin, and we conclude that the equilibrium must be asymptotically stable. In general, finding a Lyapunov function is *very* difficult, and there is no general procedure for doing so.

**Example 3.7.7.** In a previous example, we showed that the planar system

$$x' = -y - x(x^2 + y^2) \qquad y' = x - y(x^2 + y^2)$$

has exactly one equilibrium (the origin), and its linearization about the origin is

$$\begin{bmatrix} x' \\ y' \end{bmatrix} = \begin{bmatrix} 0 & -1 \\ 1 & 0 \end{bmatrix} \begin{bmatrix} x \\ y \end{bmatrix}.$$

Since the eigenvalues of the coefficient matrix are $\lambda = \pm i$, both of which have zero real part, the origin is a non-hyperbolic equilibrium. We also proved that the origin is actually asymptotically stable, by introducing a new variable $u = x^2 + y^2$ and solving a differential equation for $u$. We now give an alternative proof using

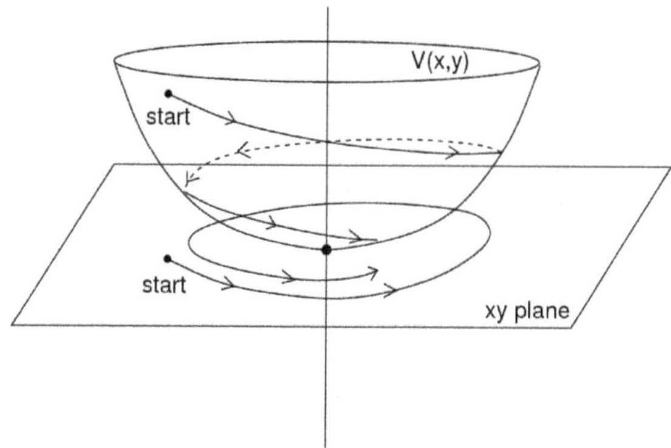


**Figure 3.6.** Illustration of a Lyapunov function for a two-variable system with an equilibrium at the origin. Corresponding to each solution of the ODE in the phase plane, there is a curve on the surface $V(x,y)$. If all such curves on the surface $V$ decrease to 0 as $t \to \infty$, then the corresponding solution trajectories in the $xy$-plane must all approach the equilibrium as $t \to \infty$, implying that the equilibrium is asymptotically stable.

Theorem 3.7.5. We claim that $V(x,y) = x^2 + y^2$ is a Lyapunov function for the equilibrium at the origin. Clearly $V(0,0) = 0$, and $V(x,y) > 0$ if $(x,y) \neq (0,0)$. Moreover, note that $\nabla V(x,y) \bullet f(x,y)$ is given by

$$\begin{aligned}(2x,2y) \bullet (-y - x(x^2+y^2), x - y(x^2+y^2)) \\ = -2xy - 2x^2(x^2+y^2) + 2xy - 2y^2(x^2+y^2) = -2(x^2+y^2)^2.\end{aligned}$$

Since $\nabla V(x,y) \bullet f(x,y) < 0$ except at $(0,0)$, we conclude from Lyapunov's Theorem that the origin is asymptotically stable.

One interesting observation regarding the previous example is that the Lyapunov function $V(x,y)$ is defined for all points $(x,y) \in \mathbb{R}^2$, and that the inequality $\nabla V(x,y) \bullet f(x,y) < 0$ is satisfied in the entire plane (except at the origin). This means that starting from *any* initial conditions, solution trajectories always approach the origin as $t \to \infty$. In other words, we may actually conclude that the origin is *globally* asymptotically stable, not just locally stable as the Theorem would guarantee. Among other things, this tells us that the above system of ODEs cannot possess any periodic solutions.

For certain types of systems of ODEs, Lyapunov functions are readily available. We now discuss one such example.

**Gradient systems.** Gradient systems are very special types of systems of ODEs which have the form

$$\frac{\mathrm{d}x_k}{\mathrm{d}t} = -\frac{\partial V}{\partial x_k} \qquad k = 1, 2, \ldots, n,$$

where $V : \mathbb{R}^n \to \mathbb{R}$ is a continuously differentiable function. In other words, the right hand sides of the ODEs in the system are obtained by taking partial derivatives of the same underlying function $V$. Now suppose that $\mathbf{x}' = f(\mathbf{x})$ is a gradient system with an isolated equilibrium $\mathbf{x}^*$, and observe that

$$\begin{aligned}
\frac{\mathrm{d}}{\mathrm{d}t} V(\mathbf{x}(t)) &= \nabla V(\mathbf{x}) \bullet f(\mathbf{x}) \\
&= \left( \frac{\partial V}{\partial x_1}, \frac{\partial V}{\partial x_2}, \ldots, \frac{\partial V}{\partial x_n} \right) \bullet \left( -\frac{\partial V}{\partial x_1}, -\frac{\partial V}{\partial x_2}, \ldots, -\frac{\partial V}{\partial x_n} \right) \\
&= -\sum_{k=1}^{n} \left( \frac{\partial V}{\partial x_k} \right)^2 = -\|\nabla V(\mathbf{x})\|_2^2,
\end{aligned}$$

which could *never* be positive. Consequently, if $V(\mathbf{x}) \geq 0$ for all $\mathbf{x} \in \mathbb{R}^n$ with equality only if $\mathbf{x} = \mathbf{x}^*$, then the function $V$ itself would be a Lyapunov function for the system. We could immediately conclude that the equilibrium $\mathbf{x}^*$ is stable.

**Example 3.7.8.** Consider the rather silly system $x' = -4x^3$ and $y' = -4y^3$, which can be easily solved by hand using separation of variables. The only equilibrium solution is at the origin $(x, y) = (0, 0)$. The Jacobian matrix associated with this system is

$$Jf(x, y) = \begin{bmatrix} -12x^2 & 0 \\ 0 & -12y^2 \end{bmatrix},$$

and evaluating this matrix at the equilibrium point yields

$$Jf(0, 0) = \begin{bmatrix} 0 & 0 \\ 0 & 0 \end{bmatrix}.$$

The linearization is completely useless, because *every* point in $\mathbb{R}^2$ is an equilibrium solution of

$$\begin{bmatrix} x' \\ y' \end{bmatrix} = \begin{bmatrix} 0 & 0 \\ 0 & 0 \end{bmatrix} \begin{bmatrix} x \\ y \end{bmatrix}.$$

Notice that $\lambda = 0$ is a double eigenvalue of the matrix $Jf(0, 0)$, and the equilibrium is non-hyperbolic. We claim that the original nonlinear system is a gradient system. Indeed, suppose that $V(x,y)$ is a function with the property that

$$-\frac{\partial V}{\partial x} = -4x^3 \quad \text{and} \quad -\frac{\partial V}{\partial y} = -4y^3.$$

Integrating the first of these equations with respect to $x$, we find that

$$V(x,y) = \int 4x^3 \, dx = x^4 + g(y),$$

where $g$ is an *arbitrary function* of $y$. Imposing the requirement that $\partial V / \partial y = 4y^3$, we calculate

$$\frac{\partial}{\partial y}\left(x^4 + g(y)\right) = 4y^3,$$

which means that $g'(y) = 4y^3$. Integrating again, we have $g(y) = y^4 + C$, where $C$ is an arbitrary constant. In order for our function $V(x,y) = x^4 + y^4 + C$ to be a Lyapunov function, we must insist that $C = 0$ so that $V(0,0) = 0$. Moreover,

$$\nabla V(x,y) \bullet f(x,y) = (4x^3, 4y^3) \bullet (-4x^3, -4y^3) = -16(x^6 + y^6) < 0$$

for all $(x,y) \neq (0,0)$. By Lyapunov's Theorem, we conclude that the origin is asymptotically stable. In fact, since the above calculations hold in the entire plane $\mathbb{R}^2$, the origin is *globally* asymptotically stable.

**Remark.** Given a function $f : \mathbb{R}^2 \to \mathbb{R}^2$, any function $V : \mathbb{R}^2 \to \mathbb{R}$ with the property that $\nabla V = f$ is called a *potential function* for $f$. It is not always possible to find a potential function for a given $f$. Most multi-variable calculus textbooks state conditions under which a potential function exists.

In your homework exercises, you will consider more "exotic" Lyapunov functions than the one used in the toy example above.

---

## Exercises

1. Consider the function $f : \mathbb{R}^2 \to \mathbb{R}^2$ defined by

$$f(x,y) = \begin{bmatrix} f_1(x,y) \\ f_2(x,y) \end{bmatrix} = \begin{bmatrix} xy\cos(\ln y) \\ \arctan\left(x^2 y\right) \end{bmatrix}.$$

Find the linear approximation for $f$ at the point $(x_0, y_0) = (1,1)$.

2. Recall from an example that the initial value problem $\frac{dy}{dx} = y^2$, $y(0) = 1$ has solution $y = (1-x)^{-1}$. Although this solution is not well-behaved, we do get one minor consolation prize: Show that the solution is unique.

3. Show that the initial value problem

$$\frac{dx}{dt} = \arctan(xy) \qquad x(0) = x_0,$$
$$\frac{dy}{dt} = xe^{-(x^2+y^2)} \qquad y(0) = y_0$$

has a unique solution regardless of the values of $x_0$ and $y_0$.

4. Suppose that $f(x)$ is continuous on some closed interval $[a, b]$. Show that

$$\left| \int_a^b f(x)\, dx \right| \leq \int_a^b |f(x)|\, dx.$$

5. This exercise will guide you through a proof of *Gronwall's Inequality*, which is stated as follows. Let $f(t)$ be a non-negative continuous function, and suppose that there exist positive constants $C$ and $K$ such that

$$f(t) \leq C + K \int_0^t f(s)\, ds$$

for all $t \in [0, a]$. Then $f(t) \leq Ce^{Kt}$ for all $t \in [0, a]$.

(a) Define the function $F(t) = C + K\int_0^t f(s)\, ds$ for $t \in [0, a]$. Explain why $F(t) \geq f(t)$ and why $F(t) > 0$ for all $t \in [0, a]$.

(b) The Fundamental Theorem of Calculus shows that $F'(t) = Kf(t)$. Combining this with part (a) above, show that, for all $t \in [0, a]$,

$$\frac{F'(t)}{F(t)} \leq K.$$

(c) Making the sneaky observation that

$$\frac{F'(t)}{F(t)} = \frac{d}{dt} \ln(F(t)),$$

use the result of part (b) above to show that

$$\ln F(t) \leq Kt + \ln F(0)$$

for all $t \in [0, a]$. Finally, exponentiate both sides of this inequality and recall from part (a) that $f(t) \leq F(t)$. Gronwall's Inequality follows.

6. Find all equilibria of the system

$$\frac{dx}{dt} = x - y^2 \qquad \frac{dy}{dt} = x^2y - 4y.$$

Linearize the system at each *hyperbolic* equilibrium, and classify these equilibria as stable/unstable nodes, saddles, or stable/unstable foci.

7. Find all equilibria of the system

$$\frac{dx}{dt} = (x^2 - 1)(y - 1) \qquad \frac{dy}{dt} = (x - 2)y.$$

Linearize the system at each *hyperbolic* equilibrium and classify these equilibria. Try to draw the phase portrait for the system. To do so, start by using the linearized systems to sketch trajectories in the vicinity of each hyperbolic equilibrium. Then, try to "interpolate", by filling in other trajectories throughout the phase plane. If you have access to computer software which can generate the true phase portrait, feel free to see how it compares to your hand sketch.

8. Consider the system

$$\frac{dx}{dt} = -y + x(\mu - x^2 - y^2) \qquad \frac{dy}{dt} = x + y(\mu - x^2 - y^2),$$

where $\mu$ is a constant.

(a) Show that the origin is the only equilibrium of this system.

(b) Find the linearization of this system at the origin.

(c) Note that the value of $\mu$ determines how the phase plane of the linearized system looks. What sort of equilibrium do we have if $\mu > 0$? What if $\mu < 0$? What if $\mu = 0$?

9. Sketch the phase planes for the following system and its linearization at the origin.

$$x' = -y + x(x^2 + y^2) \qquad y' = x + y(x^2 + y^2)$$

When dealing with the nonlinear system, you may want to define $u(t) = x(t)^2 + y(t)^2$ and proceed as in an example in the text.

**10.** Consider the system

$$x' = -3x \qquad y' = x^2 + 2y,$$

which has an isolated equilibrium at the origin. Sketch the phase portraits for this system and its linearization at the origin. Find equations for the stable and unstable manifolds $W^s$ and $W^u$.

**11.** The origin is the only equilibrium of the system

$$\frac{dx}{dt} = 2x + 5y^3, \qquad \frac{dy}{dt} = -y.$$

Find the stable and unstable manifolds $W^s(0,0)$ and $W^u(0,0)$ and sketch the phase plane.

**12.** Find all equilibria of the system

$$\frac{dx}{dt} = -x \qquad \frac{dy}{dt} = 2(x^3 + y)$$

For each equilibrium, find formulas for the stable and unstable manifolds.

**13.** This problem concerns a mass-spring system (see figure). An object of mass $m$ is attached to a spring. The object slides back and forth on a table, eventually returning to its resting position due to friction.

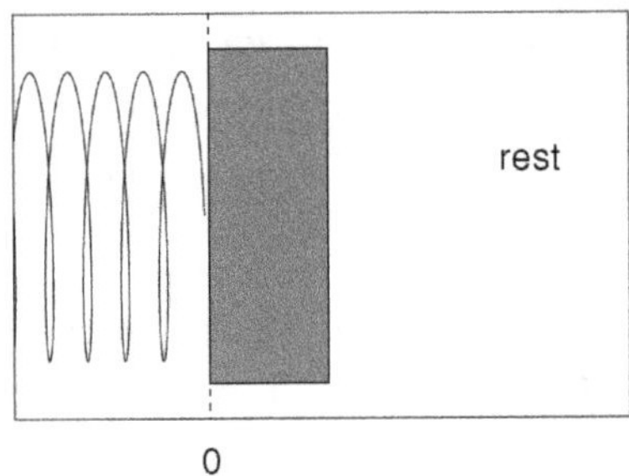


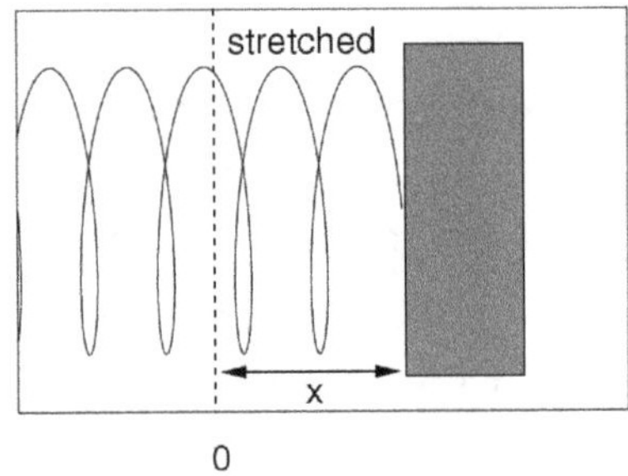


Let $x$ denote the displacement of the object from its equilibrium ($x = 0$) position, and let $v$ denote its velocity. We will use the following assumptions when modeling the motion:

☞ The force exerted on the object by the spring is proportional to the displacement $x$ of the object from equilibrium. The spring's force acts in the opposite direction of the displacement. In other words, we have assumed that

*Hooke's Law* holds: the force exerted by the spring is given by $F_{\text{spring}} = -kx$ where $k > 0$ is a positive constant.

☞ We assume that there is a frictional force exerted on the object as well, and this force acts in the direction opposite the velocity vector. For small velocity $v$, the friction is roughly proportional to the velocity, and for large $v$, the friction is proportional to the square of the velocity. That is, $F_{\text{frictional}} = -Av - Bv^2$, where $A$ and $B$ are positive constants.

☞ According to Newton's Law, $F = ma$, where $F$ is the total force acting on the object, $m$ is mass, and $a = \frac{dv}{dt}$ is acceleration. In other words,

$$ma = F_{\text{spring}} + F_{\text{frictional}} = -kx - Av - Bv^2.$$

For convenience, we will assume that mass has been scaled so that $m = 1$. Since $v = \frac{dx}{dt}$ and $a = \frac{dv}{dt}$, we get a system of two ODEs for displacement and velocity:

$$\frac{dx}{dt} = v \qquad \frac{dv}{dt} = -kx - Av - Bv^2.$$

(a) Show that this system has only one equilibrium, and find it. Give a physical interpretation of the equilibrium.

(b) Find the linearization of the system about the equilibrium.

(c) Give an intuitive physical explanation for why you would expect this equilibrium to be stable. Then, rigorously show that the equilibrium is asymptotically stable (and therefore hyperbolic), regardless of the choices of the positive constants $A$, $B$, and $k$.

(d) Solve the linearized system. If you do this correctly, you will find that the exact form of the solution depends upon the sign of $A^2 - 4k$.

(e) Give a physical interpretation of the differences between the solutions you would get if $A^2 - 4k$ is positive versus those you would get if $A^2 - 4k$ is negative. The constant $k$ is called the *spring constant* and its value determines the stiffness of the spring.

14. The Lotka-Volterra equations model the populations of two interacting species. In what follows, $x(t)$ denotes the population of a prey species, and $y(t)$ denotes the population of a predator species which depends upon the

prey for survival. The equations of this model are:

$$\frac{dx}{dt} = Ax - Bxy \qquad \frac{dy}{dt} = Cxy - Dy$$

where $A$, $B$, $C$, and $D$ are all positive constants. As usual, denote the right hand side of this system by $f(x,y)$, which maps $\mathbb{R}^2 \to \mathbb{R}^2$.

(a) Note that the origin $(x,y) = (0,0)$ is an equilibrium. Find the linearization about this equilibrium and solve the linearized system. Classify the equilibrium as stable/unstable and determine whether it is a saddle, node, focus, or center.

(b) There is one other equilibrium $(x^*, y^*)$ of this system. Find it and show that it is non-hyperbolic. Do not bother to solve the linearized system.

(c) Let $V(x,y) = Cx - D\ln x + By - A\ln y + E$, where $E$ is a constant. Show that $\nabla V \cdot f = 0$.

(d) If $(x^*, y^*)$ is the equilibrium you found in part (b) above, how should you choose the constant $E$ so that $V(x^*, y^*) = 0$?

(e) Use the second derivative test to show that $V(x,y)$ has a local minimum at the equilibrium you found in part (b). Conclude that $V(x,y)$ is a Lyapunov function for this equilibrium and, using your result from part (c), state what you have learned about the stability of this equilibrium.

15. Consider the system

$$\frac{dx}{dt} = 4x(x-1)\left(x - \frac{1}{2}\right) \qquad \frac{dy}{dt} = 2y.$$

(a) Find all equilibria of this system.

(b) Show that $V(x,y) = x^2(x+1)^2 + y^2$ is a Lyapunov function for this system at the origin.

(c) Use $V(x,y)$ to classify the stability of the origin.

16. Given that

$$\frac{dx}{dt} = x^2 y - \frac{xy^2}{2} \qquad \frac{dy}{dt} = x^2\left(\frac{x}{3} - \frac{y}{2}\right)$$

is a gradient system, find a Lyapunov function $V(x,y)$ such that $\frac{dx}{dt} = -\partial V/\partial x$ and $\frac{dy}{dt} = -\partial V/\partial y$. Then, use $V(x,y)$ to prove that the origin is asymptotically stable.

**17.** Consider the nonlinear system

$$\frac{dx}{dt} = y \qquad \frac{dy}{dt} = -a \sin x,$$

where $a$ is a positive constant.

(a) Show that the origin is a non-hyperbolic equilibrium.

(b) Define $V(x, y) = a(1 - \cos x) + \frac{1}{2}y^2$. Show that this is a Lyapunov function for the system (at least in some appropriately chosen region of the $xy$-plane which contains the origin).

(c) What can you conclude about the stability of the origin?

**18.** Consider the nonlinear system

$$\frac{dx}{dt} = -y + x(x^2 + y^2) \qquad \frac{dy}{dt} = x + y(x^2 + y^2),$$

which looks suspiciously like an example in the text (hint, hint). Notice that the system has a unique equilibrium (the origin), and this equilibrium is non-hyperbolic. Find a Lyapunov function and use it to determine whether the origin is stable or not.

**19.** The *nullclines* of a nonlinear system of ODEs provide a useful way of understanding the dynamics. For a planar system of the form

$$\frac{dx}{dt} = f_1(x, y) \qquad \frac{dy}{dt} = f_2(x, y),$$

the $x$-nullcline is the curve defined by $f_1(x, y) = 0$ and the $y$-nullcline is the curve defined by $f_2(x, y) = 0$. The intersections of these nullclines are the equilibria of the system.

Consider the system

$$\frac{dx}{dt} = -\alpha x + y \qquad \frac{dy}{dt} = \frac{5x^2}{4 + x^2} - y,$$

where $\alpha$ is a positive parameter.

(a) Find and plot the nullclines for this system with $\alpha = 2$. Since the origin is the only intersection point of the $x$- and $y$-nullclines, it is the only equilibrium. Linearize the system at origin and classify the equilibrium.

(b) Find and plot the nullclines for this system with $\alpha = 1$. How many equilibria does the system now have?

(c) Again assuming that $\alpha = 1$, find the equilibria algebraically.

(d) What types of equilibria are these?

(e) Notice that changing $\alpha$ from 1 to 2 affects the number of equilibria. In fact, there is some critical value of $\alpha$ (between 1 and 2) for which the system experiences a significant qualitative change (i.e., a change in the number of equilibria). Find this critical value of $\alpha$. This critical value is called a bifurcation point. In Chapter 5, we will explore this dependence on parameters in more detail.

# CHAPTER 4

## Periodic, Heteroclinic, and Homoclinic Orbits

In this chapter, we shift our attention away from equilibria, instead seeking more "interesting" solutions of nonlinear systems $\mathbf{x}' = f(\mathbf{x})$. Much of our discussion involves planar systems (i.e., $f : \mathbb{R}^2 \to \mathbb{R}^2$), because such systems admit particularly simple criteria for the existence of periodic solutions.

### 4.1. Periodic Orbits and the Poincaré-Bendixon Theorem

A non-equilibrium solution $\mathbf{x}$ of the system $\mathbf{x}' = f(\mathbf{x})$ is *periodic* if there exists a positive constant $p$ such that $\mathbf{x}(t+p) = \mathbf{x}(t)$ for all time $t$. The least such $p$ is called the *period* of the solution, and tells us how often the solution trajectory "repeats itself". In the phase portrait for the system of ODEs, periodic solutions (sometimes called periodic *orbits*) always appear as closed curves. On the other hand, not every closed curve corresponds to a periodic solution, as we shall see when we discuss homoclinic orbits.

For linear, constant-coefficient systems, we learned to associate pure imaginary eigenvalues with periodic solutions. Determining whether a nonlinear system has periodic solutions is less straightforward. In preparation for stating criteria for existence of periodic solutions, we review some basic notions from calculus.

Suppose that $\Gamma(t) = (\gamma_1(t), \gamma_2(t))$ is a parametrized curve in $\mathbb{R}^2$; i.e., $x = \gamma_1(t)$ and $y = \gamma_2(t)$. Assume that $\gamma_1(t)$ and $\gamma_2(t)$ are continuously differentiable. At a given time $t = t_0$, the tangent vector to the curve $\Gamma(t)$ is given by $\Gamma'(t_0) = (\gamma_1'(t_0), \gamma_2'(t_0))$.

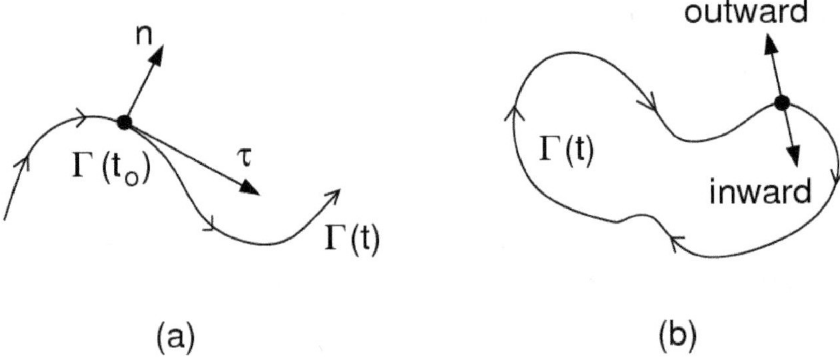


**Figure 4.1.** (a) Illustration of a tangent vector $\tau$ and normal vector $\mathbf{n}$ to a curve $\Gamma$ at a particular time $t = t_0$. (b) For the closed curves that we shall consider, each point on the curve has exactly one outward unit normal vector and one inward unit normal vector.

**Definition 4.1.1.** Any non-zero vector $\mathbf{n}(t_0)$ in $\mathbb{R}^2$ which is perpendicular to the tangent vector $\Gamma'(t_0)$ is called a *normal* vector to the curve at $t = t_0$. If $\|\mathbf{n}\|_2 = 1$, then $\mathbf{n}$ is called a *unit normal* vector.

Figure 4.1a illustrates the normal and tangent vectors to a parametrized curve $\Gamma$. When we deal with closed curves (Figure 4.1b), we will always presume that the curves are sufficiently "well-behaved" (smooth) that there is a unique inward unit normal vector and unit outward normal vectors at each point along the curve. The following Lemma formally states an observation that was made when we discussed phase portraits in Chapter 2.

**Lemma 4.1.2.** Consider a system $\mathbf{x}' = f(\mathbf{x})$ where $f$ is continuously differentiable. Then solution trajectories in the phase portrait cannot intersect each other.

*Proof.* Exercise. Convince yourself that if two trajectories did intersect, then this would violate the Fundamental Existence and Uniqueness Theorem 3.2.2. □

With the above Lemma in mind, we now give an instructive example that will motivate our main result regarding existence of periodic solutions.

**Example 4.1.3.** The system

$$\begin{aligned} x' = f_1(x,y) &= -y + x(1 - x^2 - y^2) \\ y' = f_2(x,y) &= x + y(1 - x^2 - y^2) \end{aligned}$$

has exactly one equilibrium solution: $(x, y) = (0, 0)$. The linearization about the equilibrium is given by

$$\begin{bmatrix} x' \\ y' \end{bmatrix} = \begin{bmatrix} 1 & -1 \\ 1 & 1 \end{bmatrix} \begin{bmatrix} x \\ y \end{bmatrix}.$$

The coefficient matrix is in real canonical form, and has eigenvalues $\lambda = 1 \pm i$. Since the origin is a hyperbolic equilibrium, the Hartman-Grobman Theorem tells us that our original nonlinear system has an unstable focus at the origin.

**Observation.** Even though the origin is a repeller, solution trajectories cannot escape too far from the origin as $t \to \infty$. Notice that $x^2 + y^2$ measures the square of the Euclidean distance from the origin. If $x^2 + y^2$ becomes too large, an inspection of the right hand sides of our ODEs suggests that we will be pulled back towards the origin.

To make this claim more rigorous, we will construct a curve $\Gamma$ which simultaneously encloses the origin and has the property that solution trajectories in the phase plane always cross $\Gamma$ from *outside to inside* as $t$ increases. Suppose that $\Gamma$ parameterizes a circle of radius $R$ centered at the origin. If we agree to orient the curve $\Gamma$ counterclockwise, then one natural parametrization would be

$$\Gamma(t) = (\gamma_1(t), \gamma_2(t)) = (R\cos t, R\sin t).$$

At each time $t$, the unit outward normal vector is simply $\mathbf{n}(t) = (\cos t, \sin t)$, as illustrated in Figure 4.2.

**Question:** If choose some point $(x, y) = (R\cos t, R\sin t)$ on the curve $\Gamma$, does the flow of the system of ODEs direct our motion towards the interior or exterior of the curve?

**Answer:** We claim that if $f(x, y) \bullet \mathbf{n}(t) < 0$, then the flow is directed inward, and if $f(x, y) \bullet \mathbf{n}(t) > 0$, then the flow is directed outward. To see why, we recall the geometric interpretation of the dot product of two vectors:

$$f(x, y) \bullet \mathbf{n}(t) = \|f(x, y)\|_2 \|\mathbf{n}\|_2 \cos\theta,$$

where $\theta$ is the angle between the two vectors $f(x,y)$ and $\mathbf{n}$. Clearly $\|f(x,y)\|_2$ and $\|\mathbf{n}\|_2$ are positive since they represent lengths of vectors. If $\pi/2 < \theta < 3\pi/2$, then the vector $f(x,y)$ must be oriented towards the interior of the curve $\Gamma$ (see Figure 4.2). For this range of angles $\theta$, we have $\cos\theta < 0$ and therefore $f(x,y) \bullet \mathbf{n}(t) < 0$ as well.

In the present example, our function $f(x,y)$ is given by

$$\begin{bmatrix} f_1(x,y) \\ f_2(x,y) \end{bmatrix} = \begin{bmatrix} -y + x(1-x^2-y^2) \\ x + y(1-x^2-y^2) \end{bmatrix},$$

and $\mathbf{n} = (\cos t, \sin t)$. Computing $f \bullet \mathbf{n}$ at $(x,y) = (R\cos t, R\sin t)$,

$$\begin{aligned} &f(x,y) \bullet \mathbf{n} \\ &\quad = [-R\sin t + R(\cos t)(1-R^2), R\cos t + R(\sin t)(1-R^2)] \bullet (\cos t, \sin t), \end{aligned}$$

where we have used the fact that $x^2 + y^2 = R^2$. Expanding the dot product yields

$$\begin{aligned} f(x,y) \bullet \mathbf{n} &= -R\sin t\cos t + R(\cos^2 t)(1-R^2) + R\cos t\sin t + R(\sin^2 t)(1-R^2) \\ &= R(\cos^2 t)(1-R^2) + R(\sin^2 t)(1-R^2) \;=\; R(1-R^2). \end{aligned}$$

Observe that if the radius $R$ of our circular curve $\Gamma$ is larger than 1, then $f \bullet n < 0$, implying that the flow is directed inward. Likewise, if $R < 1$, then $f \bullet \mathbf{n} > 0$, implying that the flow is directed outward. In summary, we have noted that

☞ The only equilibrium of the system is $(0,0)$, and it is a *repeller*.

☞ Excluding the equilibrium, all trajectories of this system are "funneled" towards a circle of radius 1 centered at the origin as $t$ increases.

☞ Different trajectories cannot intersect (Lemma 4.1.2).

It turns out that the only way to reconcile all three of these constraints is if the circle of radius 1 centered at the origin corresponds to a very special type of a solution: a *periodic orbit*.

In the preceding example, we constructed curves $\Gamma$ in the plane $\mathbb{R}^2$ and then determined whether the flow of the system of ODEs was directed towards the interior or exterior of $\Gamma$ as $t$ increases. The notion that trajectories may become "trapped" on the interior of some specially chosen curve $\Gamma$ motivates

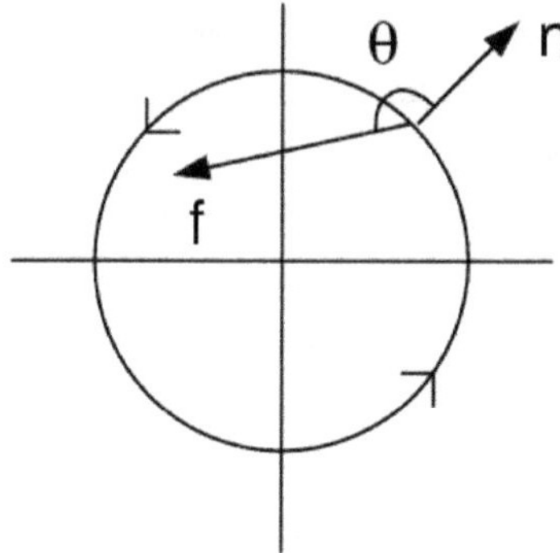


**Figure 4.2.** The outward unit normal vector for the circle $(R\cos t, R\sin t)$ is given by $\mathbf{n} = (\cos t, \sin t)$.

an important theorem. In what follows, we consider a *planar* system $\mathbf{x}' = f(\mathbf{x})$ where $\mathbf{x} = (x_1, x_2)$ and $f : \mathbb{R}^2 \to \mathbb{R}^2$ is continuously differentiable.

**Theorem 4.1.4** (Poincaré-Bendixon.). Suppose that $\Gamma$ is a continuously differentiable closed curve in $\mathbb{R}^2$ and let $\mathbf{n}$ denote an outward unit normal vector at a point $(x_1, x_2)$ on $\Gamma$. If the ODE has no equilibria on or inside $\Gamma$ and $f \bullet \mathbf{n} < 0$ at all points $(x_1, x_2)$ on $\Gamma$, then there exists at least one periodic solution of the ODE inside $\Gamma$.

**Remark.** The Poincaré-Bendixon Theorem is valid only in two dimensions. In three dimensions, the famous Lorenz ODEs have solutions which remain confined to a bounded set, but never approach an equilibrium or periodic orbit. In fact, the Lorenz equations exhibit *chaotic* behavior.

If we can construct a curve $\Gamma$ for which the conditions of the Poincaré-Bendixon Theorem are satisfied, then any solution starting from an initial condition $\mathbf{x}_0$ inside $\Gamma$ must stay trapped inside $\Gamma$ for all time $t$. As a consequence, we have

**Corollary 4.1.5.** If the conditions of the Poincaré-Bendixon Theorem are satisfied, then any solution curve starting from an initial condition $\mathbf{x}_0$ inside $\Gamma$ will exist for all time $t \geq 0$.

*Proof.* In our planar system $\mathbf{x}' = f(\mathbf{x})$, we always assume that $f : \mathbb{R}^2 \to \mathbb{R}^2$ is continuously differentiable. The function $g : \mathbb{R}^2 \to \mathbb{R}$ defined by $g(\mathbf{x}) = \|\mathbf{x}\|_2$ is also a continuous function, and therefore the composition $g(f(\mathbf{x})) = \|f(\mathbf{x})\|_2$ is continuous. If $\Omega$ denotes the region enclosed by the continuous, closed curve $\Gamma$, then $\Omega$ is a closed and bounded set. It follows that the function $\|f(\mathbf{x})\|_2$ achieves some maximum value $B$ in the region $\Omega$. Since $\|f\|_2 \leq B$ in this region, we may apply Theorem 3.3.13 (with $K = 0$) to infer global existence for all $t \geq 0$. □

When attempting to prove that a planar system $\mathbf{x}' = f(\mathbf{x})$ has a periodic solution, it can be *very* challenging to actually find a curve $\Gamma$ which satisfies the conditions of the Poincaré-Bendixon Theorem. However, there is one piece of good news—the theorem actually holds even if the curve $\Gamma$ is only *piecewise* differentiable. Consequently, instead of using smooth curves like circles, it is possible to use polygonal paths (such as squares). You may also find it helpful to use computer software to sketch the vector field defined by the function $f$ to see whether it is possible to draw a closed curve $\Gamma$ on which all arrows of the vector field are directed inward.

Like equilibria, periodic solutions of ODEs can be stable or unstable. A periodic orbit $\mathbf{p}(t)$ is called *asymptotically stable* if all nearby solution trajectories approach the periodic orbit as $t \to \infty$. This definition could be made more rigorous, as in our definition of stability of equilibria. However, we shall be content with a more intuitive understanding of stability of periodic orbits. As an example, consider an idealized frictionless pendulum which simply swings back and forth without any external forcing. The amplitude of the oscillations determines which periodic orbit the pendulum is in, and there are infinitely many possibilities. Each orbit is stable, because if we give the pendulum a tiny push, it will settle into a new periodic orbit with a slightly different amplitude. In other words, any small perturbation of our original periodic orbit will shift us to a "nearby" orbit. None of these periodic orbits is asymptotically stable, because any tiny push we apply to the pendulum will land us on a different periodic orbit than the one we started with (since the pendulum is frictionless).

**Example 4.1.6.** Consider the planar system in polar coordinates given by

$$\frac{dr}{dt} = r(1-r) \quad \text{and} \quad \frac{d\theta}{dt} = 1.$$

To envision the motion along a trajectory in the phase plane, notice that our angular variable $\theta$ continually increases with speed 1, while the radial variable obeys a logistic differential equation. By itself, the equation $\frac{dr}{dt} = r(1-r)$ has two equilibria: $r = 0$ and $r = 1$. The "Jacobian" for this 1-D equation is simply given by the derivative $Jf(r) = 1 - 2r$. For $r = 0$, we have $Jf(0) = 1 > 0$, which

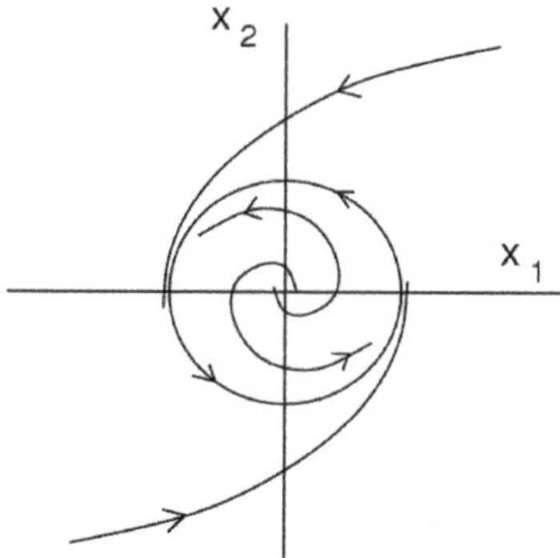


**Figure 4.3.** Phase portrait for the planar system $\frac{dr}{dt} = r(1-r)$, $\frac{d\theta}{dt} = 1$.

means that $r = 0$ is an unstable equilibrium. On the other hand, $Jf(1) = -1 < 0$ means that $r = 1$ is a stable equilibrium. The graph of $r = 1$ is a circle of radius 1, and in this case it corresponds to an asymptotically stable periodic orbit. Also, $r = 0$ corresponds to the origin, which is an unstable equilibrium. A sketch of the phase portrait for this system appears in Figure 4.3.

**Example 4.1.7.** A similar argument shows that the planar system

$$\frac{dr}{dt} = r(r-1) \quad \text{and} \quad \frac{d\theta}{dt} = 1.$$

has an *unstable* periodic orbit corresponding to the circular trajectory $r = 1$.

**Definition 4.1.8.** A periodic orbit is called a *limit cycle* if there exists a point $\mathbf{x}_0$ that does *not* lie on the periodic orbit but such that the trajectory $\phi_t(\mathbf{x}_0)$ converges to the periodic orbit either as $t \to \infty$ or as $t \to -\infty$.

In the above two examples, $r = 1$ corresponds to a limit cycle solution. In general, asymptotically stable periodic orbits and unstable periodic orbits are examples of limit cycle solutions. On the other hand, the constant-coefficient system $\mathbf{x}' = A\mathbf{x}$ where

$$A = \begin{bmatrix} 0 & -1 \\ 1 & 0 \end{bmatrix}$$

does not have any limit cycle solutions although it does have infinitely many periodic solutions. The phase portrait for that linear system consists of a family of concentric circles.

Before concluding our discussion of periodic orbits, we remark that the Poincaré-Bendixon theorem ensures that planar systems $\mathbf{x}' = f(\mathbf{x})$ (with $f$ continuously differentiable) can never exhibit behavior that is too "crazy". Specifically, such planar systems can never have *chaotic* solutions (for a proof, see Hubbard and West [5]). We now state the definition of chaos as it appears in Strogatz [11].

**Definition 4.1.9.** *Chaos* is aperiodic behavior in a deterministic system which exhibits sensitive dependence on initial conditions.

There are three important phrases in the definition of chaos. By *aperiodic* behavior, we mean that there are solutions which do not converge to a limit cycle or stable equilibrium. *Deterministic* means that the aperiodic behavior is not a result of "cheating" by artificially incorporating randomness into the system. Sensitive dependence on initial conditions means that slight changes in initial conditions can cause huge differences in how solutions behave.

**Example 4.1.10.** The Lorenz equations are given by

$$\begin{aligned} \frac{dx}{dt} &= \sigma(y - x) \\ \frac{dy}{dt} &= rx - y - xz \\ \frac{dz}{dt} &= xy - bz, \end{aligned} \tag{4.1}$$

where $\sigma$, $r$, and $b$ are constants. Clearly the origin is an equilibrium solution. It is possible to show that if $0 < r < 1$, then the origin is globally asymptotically stable. This is accomplished by using $V(x, y, z) = rx^2 + \sigma y^2 + \sigma z^2$ as a Lyapunov function. Moreover, if $r > 1$, you can show that there is a one-dimensional unstable manifold at the origin.

For the specific choices $\sigma = 10$, $r = 28$ and $b = 8/3$, this system is known to exhibit chaos. A graph of a trajectory in the phase space is shown in Figure 4.4. Although the trajectory remains confined to a bounded set, the solution is aperiodic and never converges to an equilibrium solution or limit cycle.

As we mentioned above, the Poincaré-Bendixon Theorem rules out chaos for planar systems. Apart from equilibria and limit cycle solutions, there are very few other types of interesting behavior that planar systems can exhibit. Two exceptions are discussed in the next section.

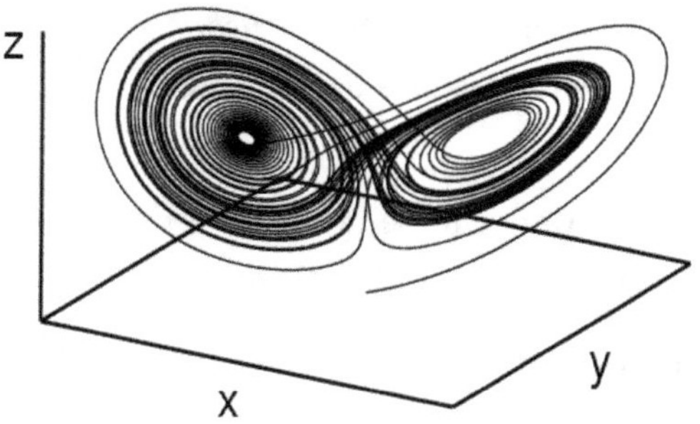


**Figure 4.4.** Chaotic solution of the Lorenz equations (4.1).

---

## 4.2. Heteroclinic and Homoclinic Orbits

Besides periodic solutions and equilibria, can there be non-constant solutions of $\mathbf{x}' = f(\mathbf{x})$ which remain bounded? The answer is yes, as we illustrate via the following examples.

**A heteroclinic orbit.** Consider the second-order ODE

$$u'' + cu' - u(u-\beta)(u-1) \ = \ 0, \tag{4.2}$$

where $c$ and $\beta$ are positive constants and $0 < \beta < \frac{1}{2}$. Introducing $w = u'$, we write the equation as a system of two first-order equations

$$\begin{aligned} u' &= w \\ w' &= u(u-\beta)(u-1) - cw \ = \ u^3 - (1+\beta)u^2 + \beta u - cw. \end{aligned} \tag{4.3}$$

Note that there are three equilibria for this system. Setting $u' = 0$ forces $w = 0$, and then setting $w' = 0$ forces $u = 0$, $u = \beta$, or $u = 1$. The Jacobian matrix for this system is

$$Jf(u,w) \ = \ \begin{bmatrix} 0 & 1 \\ 3u^2 - 2(1+\beta)u + \beta & -c \end{bmatrix}.$$

At the equilibrium $(u, w) = (0, 0)$, the Jacobian is

$$Jf(0,0) = \begin{bmatrix} 0 & 1 \\ \beta & -c \end{bmatrix},$$

which has characteristic equation $\lambda^2 + c\lambda - \beta = 0$. Since the determinant of $Jf(0,0)$ is $-\beta < 0$, we immediately conclude that $(0,0)$ is a hyperbolic equilibrium and that there is a saddle at the origin. The eigenvalues of $Jf(0,0)$ are $\lambda_{\pm} = \frac{1}{2}\left[-c \pm \sqrt{c^2 + 4\beta}\right]$, the larger of which is positive and the smaller of which is negative. To calculate an eigenvector for the positive eigenvalue $\lambda_+$, we form the matrix

$$A - \lambda_+ I = \begin{bmatrix} -\lambda_+ & 1 \\ \beta & -c - \lambda_+ \end{bmatrix}$$

which row-reduces to

$$\begin{bmatrix} -\lambda_+ & 1 \\ 0 & 0 \end{bmatrix}.$$

Eigenvectors $\mathbf{v}$ must satisfy $-\lambda_+ v_1 + v_2 = 0$, implying that $v_2 = \lambda_+ v_1$. It is convenient to treat $v_1$ as our free variable so that

$$\begin{bmatrix} v_1 \\ v_2 \end{bmatrix} = \begin{bmatrix} v_1 \\ \lambda_+ v_1 \end{bmatrix} = v_1 \begin{bmatrix} 1 \\ \lambda_+ \end{bmatrix}.$$

It follows that

$$\begin{bmatrix} 1 \\ \lambda_+ \end{bmatrix}$$

is an eigenvector for $\lambda_+$, and a similar calculation reveals that

$$\begin{bmatrix} 1 \\ \lambda_- \end{bmatrix}$$

is an eigenvector for the negative eigenvalue $\lambda_-$. The spans of these two eigenvectors form the unstable and stable subspaces, and since $\lambda_- < 0 < \lambda_+$ the orientation of these vectors is as sketched in Figure 4.5. The same sorts of calculations show that the equilibrium $(u, w) = (1, 0)$ is also a saddle and has a similar orientation as the saddle at the equilibrium (Figure 4.5).

The behavior at the equilibrium $(u, w) = (\beta, 0)$ is a bit different. The Jacobian matrix is

$$Jf(\beta, 0) = \begin{bmatrix} 0 & 1 \\ \beta^2 - \beta & -c \end{bmatrix},$$

and we note that the determinant $\beta - \beta^2$ is positive because $0 < \beta < 1$. Hence, this equilibrium is not a saddle like the other two. In fact, the eigenvalues

$$\frac{-c \pm \sqrt{c^2 - 4(\beta - \beta^2)}}{2}$$

both have negative real part. It follows that this equilibrium is either a stable node or a stable focus, depending upon the relative sizes of the constants $\beta$ and $c$.

**Interesting fact:** If the constant $c$ is chosen appropriately, we can force the stable manifold $W^s(0,0)$ at the origin to coincide with the unstable manifold $W^u(1,0)$ of the equilibrium $(1,0)$. This forms a special trajectory in the phase plane which connects the two equilibria, and there is a name for this type of solution.

**Definition 4.2.1.** Suppose that $\mathbf{x}^*$ and $\mathbf{x}^{**}$ are two distinct equilibria of the system $\mathbf{x}' = f(\mathbf{x})$. A solution $\mathbf{x}(t)$ with the property that

$$\lim_{t \to -\infty} \mathbf{x}(t) = \mathbf{x}^* \quad \text{and} \quad \lim_{t \to \infty} \mathbf{x}(t) = \mathbf{x}^{**}$$

is called a *heteroclinic orbit*.

In our example above, it is actually possible to use a special trick to find the heteroclinic orbit analytically. Indeed, consider the simpler ODE given by $u' = au(u-1)$, where $a > 0$ is a constant. Unlike our original system, this ODE has only two equilibria: $u = 0$ and $u = 1$. It is easy to check that 0 is stable and 1 is unstable. We claim that for special choices of the constants $a$ and $c$, the solutions of this simpler differential equation are also solutions of the original second-order equation (4.2). To see this, we will substitute $u' = au(u-1)$ into Equation (4.2). Since $u' = a(u^2 - u)$, we calculate that

$$u'' = a(2uu' - u') = a(2u-1)u' = a^2 u(u-1)(2u-1).$$

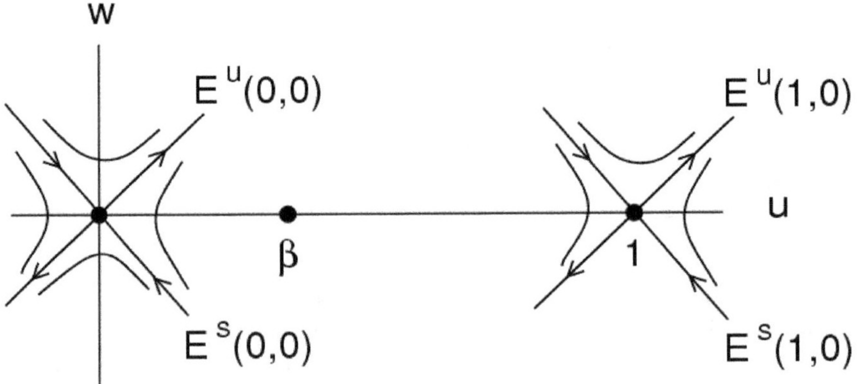


**Figure 4.5.** Two saddle equilibria for Equations (4.3).

Equation (4.2) becomes

$$a^2u(u-1)(2u-1) + cau(u-1) - u(u-\beta)(u-1) = 0$$

which, after factoring out $u(u-1)$, can be rewritten as

$$u(u-1)\left[(2a^2-1)u + (\beta + ca - a^2)\right] = 0.$$

The only way the left hand side, a function of $u$, could be identically equal to 0 is if both

$$2a^2 - 1 = 0 \quad \text{and} \quad \beta + ca - a^2 = 0.$$

The solution of this system of two equations is

$$a = \frac{1}{\sqrt{2}} \quad \text{and} \quad c = \sqrt{2}\left(\frac{1}{2} - \beta\right).$$

Note that $c > 0$ since $0 < \beta < \frac{1}{2}$. In summary, for these special choices of $c$ and $a$, the solution of the equation $u' = au(u-1)$ is also a solution of the original equation (4.2). It is straightforward to solve this equation by separation of variables. The corresponding trajectory in the $uw$-phase plane is a *heteroclinic orbit*. As $t \to -\infty$, the trajectory connects to the equilibrium $(u,w) = (1,0)$, and as $t \to \infty$, the trajectory connects to the equilibrium $(u,w) = (0,0)$. The solution exists for all time $t$, is bounded, and is neither an equilibrium nor a periodic orbit

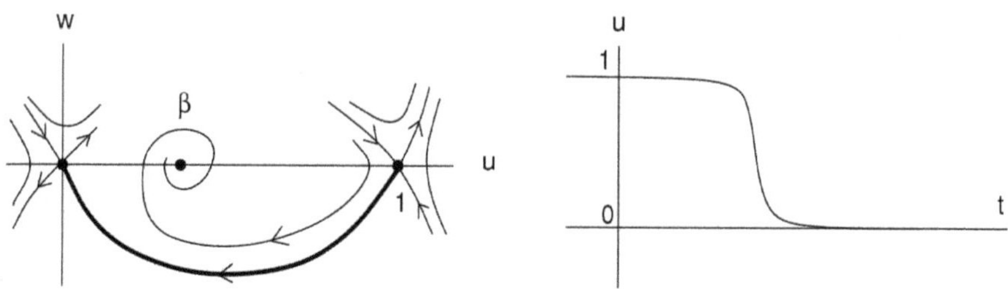


**Figure 4.6.** A heteroclinic orbit for Equation (4.2). The left panel shows the $uw$-phase plane, and the heteroclinic orbit appears in bold. The right panel shows a graph of $u$ versus $t$. Note that $u \to 0$ as $t \to \infty$ and $u \to 1$ as $t \to -\infty$.

(see Figure 4.6). The heteroclinic orbit forms a connection of the stable manifold $W^s(0,0)$ with the unstable manifold $W^u(1,0)$.

**A homoclinic orbit.** We now discuss a special type of orbit in which the stable and unstable manifolds of the *same* equilibrium are connected. The example we give is that of a *double well potential*. More exactly, consider the system

$$x' = y \quad \text{and} \quad y' = x - x^3, \tag{4.4}$$

which has three equilibria. Linearization indicates that $(x, y) = (0, 0)$ is a saddle equilibrium. The equilibria $(1, 0)$ and $(-1, 0)$ are non-hyperbolic, and are centers for the linearized systems. It is actually possible to show that they are centers for the nonlinear system as well, by defining the *energy functional* $E(x, y) = \frac{1}{2}y^2 - \frac{1}{2}x^2 + \frac{1}{4}x^4$ and arguing that $E(x, y)$ remains constant along phase plane trajectories. Figure 4.7 shows a rough sketch of the phase plane for this system. Notice that the unstable manifold $W^u(0, 0)$ at the origin happens to coincide with the stable manifold $W^s(0, 0)$. This closed loop trajectory is *not* a periodic orbit—a sketch of the corresponding solution $x(t)$ also appears in Figure 4.7. This special orbit is called a *homoclinic* orbit.

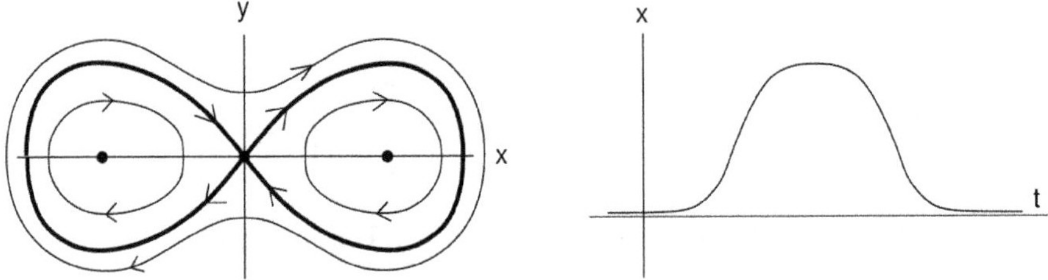


**Figure 4.7.** Two homoclinic orbits for Equation (4.4). The left panel shows the $xy$-phase plane, and the homoclinic orbits appear in bold. The right panel shows a graph of $x$ versus $t$ for one of the two homoclinic orbits. Note that $x \to 0$ as $t \to \pm\infty$.

---

## Exercises

**1.** This series of questions concerns the famous Lorenz system

$$\begin{aligned}\frac{dx}{dt} &= -\sigma x + \sigma y \\ \frac{dy}{dt} &= rx - y - xz \\ \frac{dz}{dt} &= xy - \beta z,\end{aligned}$$

where $\sigma$, $r$, and $\beta$ are *positive* parameters. The Lorenz system exhibits *chaos* if $\sigma = 10$, $r = 28$, and $\beta = 8/3$. Below, you will learn why chaos could never occur if $r < 1$.

(a) Explain why, independent of the choice of initial conditions, the initial value problem for the Lorenz system is guaranteed to have a unique solution (at least locally).

(b) Show that the origin is a *locally* asymptotically stable equilibrium if $0 < r < 1$.

(c) Show that if $r > 1$, then there will be a one-dimensional unstable manifold at the origin.

(d) The origin is actually *globally* asymptotically stable if $r < 1$, and the remaining parts of this exercise will guide you through a proof. To start, define the function $V(x,y,z) = \frac{1}{\sigma}x^2 + y^2 + z^2$. Show that

$$\nabla V \bullet f = -2\left[x - \left(\frac{r+1}{2}\right)y\right]^2 - 2\left[1 - \left(\frac{r+1}{2}\right)^2\right]y^2 - 2\beta z^2,$$

where $f$ denotes the right hand side of the Lorenz system.

(e) Show that if $r < 1$, then

$$1 - \left(\frac{r+1}{2}\right)^2 > 0.$$

Then, explain why this implies that $\nabla V \bullet f$ is strictly *less* than 0 (except at the origin) whenever $r < 1$.

(f) Finally, explain why $V$ is a Lyapunov function and why you can conclude that the origin is a *global* attractor if $r < 1$.

2. Consider the system

$$\begin{aligned} \frac{dx}{dt} &= -y - x(1 - x^2 - y^2) \\ \frac{dy}{dt} &= x - y(1 - x^2 - y^2). \end{aligned}$$

Show that this system has an unstable periodic solution and carefully sketch the phase portrait.

3. Consider the system

$$\begin{aligned} \frac{dx}{dt} &= -y + x(4 - x^2 - y^2) \\ \frac{dy}{dt} &= x + y(9 - x^2 - y^2). \end{aligned}$$

You may assume that the origin is the only equilibrium of this system. Classify the local stability of the origin. Then, show that this system has at least one stable, periodic solution.

4. Consider the system

$$\begin{aligned} \frac{dx}{dt} &= -y + x(r^2 - 6r + 8) \\ \frac{dy}{dt} &= x + y(r^2 - 6r + 8), \end{aligned}$$

where $r^2 = x^2 + y^2$. Use the Poincaré-Bendixon Theorem to prove that this system has both stable and unstable periodic orbits by following these steps:

(a) Show that the origin is the only equilibrium of this system.

(b) Using the chain rule, differentiate both sides of $r^2 = x^2 + y^2$ with respect to $t$. Then, assuming $r \neq 0$ (i.e., excluding the equilibrium solution), solve for $\frac{dr}{dt}$. You should obtain an autonomous ODE for $r$.

(c) Using the equation for $\frac{dr}{dt}$ you found in Part (b), show that $\frac{dr}{dt} > 0$ on the circle $r = 1$ and that $\frac{dr}{dt} < 0$ on the circle $r = 3$. Use the Poincaré-Bendixon Theorem to conclude that there is at least one stable periodic orbit within the annulus $1 < r < 3$.

(d) Using the equation for $\frac{dr}{dt}$ you found in Part (b), show that $\frac{dr}{dt} > 0$ on the circle $r = 5$. Combined with the fact that $\frac{dr}{dt} < 0$ on the circle $r = 3$, this seems to suggest that an *unstable* periodic orbit exists inside the annulus $3 < r < 5$. To prove this, make the substitution $t \mapsto -t$, which "reverses" the flow of time. Then, use the Poincaré-Bendixon Theorem to show that the resulting system has a *stable* periodic orbit inside the annulus $3 < r < 5$. Finally, conclude that the original system (i.e., going forward in time) has an *unstable* periodic orbit inside that annulus.

5. Here is an instance of the famous FitzHugh-Nagumo nerve membrane model:

$$\frac{dx}{dt} = -x(x-1)(x+1) - y$$
$$\frac{dy}{dt} = x - \frac{1}{2}y.$$

(a) Show that this system has exactly one equilibrium and that it is *unstable*.

(b) Consider the rectangular path $\Gamma$ with corners $(\sqrt{3}, 2\sqrt{3})$, $(-\sqrt{3}, 2\sqrt{3})$, $(-\sqrt{3}, -2\sqrt{3})$, and $(\sqrt{3}, -2\sqrt{3})$ as illustrated in Figure 4.8. By showing that the flow is directed *inward* on this path, use the Poincaré-Bendixon Theorem to conclude that these equations have at least one periodic solution. Hint: You will need to parametrize each edge of the rectangle separately, and there are many possible parameterizations.

6. Consider the second-order nonlinear ODE

$$u'' + \frac{5}{\sqrt{6}}u' + u(1-u) = 0.$$

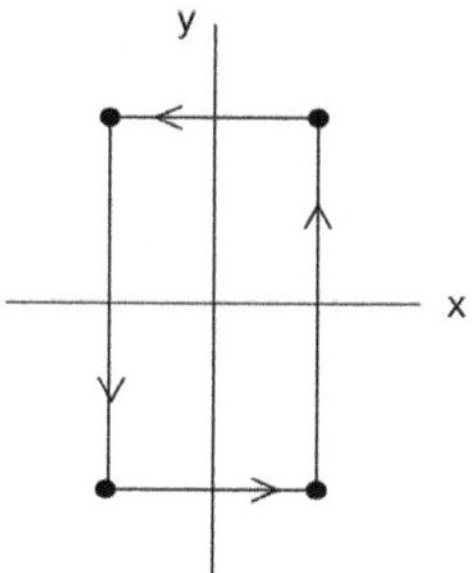


**Figure 4.8.** Rectangular path to use in exercise involving FitzHugh-Nagumo model.

(a) Write the ODE as a system of two first-order ODEs by setting $v = u'$. Show that the resulting system has two equilibria, and that one equilibrium is a stable node while the other is a saddle.

(b) Verify that

$$u(t) = \frac{1}{\left[1 + (\sqrt{2} - 1)e^{t/\sqrt{6}}\right]^2}$$

is a solution of the ODE and that it corresponds to a heteroclinic orbit. Plot $u(t)$ versus $t$ and sketch the corresponding trajectory in the $uv$-phase plane.

7. In this exercise, you will find a homoclinic orbit for a particular ODE. Consider the second-order, nonlinear equation

$$v'' + 3v^2 - \sigma v = 0$$

where $\sigma$ is a constant. This ODE could be written as a system in the usual way, by letting $w = v'$. We claim that there is a non-zero solution $v(t)$ which has the property that $v, v', v'' \to 0$ as $t \to \pm\infty$. This solution, if plotted in the $vw$-phase plane, would correspond to a homoclinic orbit.

(a) Multiply both sides of the above ODE by $v'$ and show that the resulting equation is equivalent to

$$\frac{d}{dt}\frac{(v')^2}{2} = \frac{d}{dt}\left(-v^3 + \frac{\sigma}{2}v^2\right).$$

(b) Now integrate both sides of the equation in Part (a), and let $\beta$ denote the integration constant. Explain why we may assume that $\beta = 0$.

(c) By algebraically solving for $v'$, show that

$$\frac{dv}{dt} = \pm v\sqrt{\sigma - 2v}.$$

Then, taking the minus sign for later convenience, use separation of variables to show that

$$t = -\int \frac{dv}{v\sqrt{\sigma - 2v}}.$$

(d) Recall the following definitions of hyperbolic trigonometric functions:

$$\sinh(z) = \frac{1}{2}(e^z - e^{-z}) \qquad \cosh(z) = \frac{1}{2}(e^z + e^{-z})$$

$$\tanh(z) = \frac{\sinh(z)}{\cosh(z)} \qquad \operatorname{sech}(z) = \frac{1}{\cosh(z)}$$

In the integral you wrote down in Part (c), make the (hyperbolic) trigonometric substitution

$$v = \frac{\sigma}{2}\operatorname{sech}^2(\theta) \qquad dv = -\sigma \operatorname{sech}^2(\theta)\tanh(\theta)\, d\theta$$

and note that

$$v\sqrt{\sigma - 2v} = \frac{\sigma^{3/2}}{2}\operatorname{sech}^2(\theta)\tanh(\theta).$$

Do the resulting integral to show that

$$t = \frac{2}{\sqrt{\sigma}}\theta + C$$

where $C$ is an integration constant.

(e) Use algebra to show that

$$v(t) = \frac{\sigma}{2}\operatorname{sech}^2\left[\frac{\sqrt{\sigma}}{2}(t - C)\right].$$

Then, plot the function $v(t)$ versus $t$ for the special case $\sigma = 1$ and $C = 0$ to get a sense of what the homoclinic solution would look like. Finally, set $w = v'$ and sketch the homoclinic trajectory in the $vw$-phase plane.

# CHAPTER 5

## Bifurcations

In practice, we often deal with ODEs which contain parameters (unspecified constants) whose values can profoundly influence the dynamical behavior of the system. For example, suppose we model population changes for a species. The birth and death rates of the species would be examples of parameters whose values would substantially impact that behavior of solutions of the underlying differential equation.

**Example 5.0.2.** Consider the ODE $\frac{dx}{dt} = \mu x$, where $\mu$ is a parameter. The solution of this equation is $x(t) = x_0 e^{\mu t}$, where $x_0 = x(0)$. Notice that if $\mu > 0$, the solutions exhibit exponential growth, whereas if $\mu < 0$ we observe exponential decay. If $\mu = 0$, solutions are constant. The critical value $\mu = 0$ marks the "boundary" between two very different types of dynamical behavior.

**Definition 5.0.3.** Consider a system of ODEs of the form $\mathbf{x}'(t) = f(\mathbf{x}; \mu)$, where $\mu$ is a parameter. A *bifurcation* is any major qualitative change in the dynamical behavior of the system in response to varying the parameter $\mu$.

In the previous example, we would say that a bifurcation occurs at $\mu = 0$, because the equilibrium $x = 0$ changes from stable ($\mu < 0$) to unstable ($\mu > 0$).

### 5.1. Three Basic Bifurcations

There are many ways that the qualitative behavior of a system can be drastically altered in response to changes in parameters. Equilibria and/or periodic solutions can be created or destroyed, or they can change their stability. In what follows, we will survey several common types of bifurcations. For simplicity, we will deal primarily with ODEs with only one dependent variable. An excellent reference for this material is provided by Strogatz [11].

**Saddle-node bifurcations.** First, we discuss one common mechanism for the birth or destruction of an equilibrium solution. The canonical example to keep in mind is given by the ODE

$$\frac{dx}{dt} = \mu + x^2. \tag{5.1}$$

Suppose $\mu < 0$. The equilibria satisfy $x^2 + \mu = 0$, which implies that $x = \pm\sqrt{-\mu}$ are the two equilibria of the system. To determine whether they are stable, we compute that the "Jacobian" of $f(x) = x^2 + \mu$ is simply $f'(x) = 2x$. By calculating $f'(\sqrt{-\mu}) = 2\sqrt{-\mu} > 0$, we find that $\sqrt{-\mu}$ is an unstable equilibrium. Similarly, since $f'(-\sqrt{-\mu}) = -2\sqrt{-\mu} < 0$, it follows that $-\sqrt{-\mu}$ is a stable equilibrium. Next, suppose that $\mu = 0$. There is only one equilibrium, namely $x = 0$. Although it is non-hyperbolic, it is easy to check (via separation of variables) that this equilibrium is unstable. Finally, suppose $\mu > 0$. Then $x^2 + \mu > 0$ for all $x$, implying that there are no equilibria. In summary, as $\mu$ increases from negative to positive, two equilibria (one stable and one unstable) merge and annihilate each other, leaving no equilibria if $\mu > 0$. Clearly $\mu = 0$ is a bifurcation point, and this type of bifurcation is called a *saddle-node bifurcation*. A minor adjustment of Equation (5.1) reveals that saddle-node bifurcations can also *create* new equilibria as the parameter $\mu$ increases. The equation $x' = -\mu + x^2$ experiences a saddle-node bifurcation at $\mu = 0$, creating two equilibria for $\mu > 0$.

**Bifurcation diagrams.** One very common and useful way to visualize bifurcations of a system $x' = f(x;\mu)$ is to sketch a graph of the equilibrium values of $x$ as a function of the bifurcation parameter $\mu$. As an illustration, the bifurcation diagram for Equation (5.1) appears in Figure 5.1. By convention, unstable equilibria are plotted as dashed curves and stable equilibria are plotted as solid curves. In Figure 5.1, the dashed curve corresponds to the unstable equilibrium $x = \sqrt{-\mu}$ and the solid curve corresponds to the stable equilibrium $x = -\sqrt{-\mu}$. The bifurcation diagram allows us to visualize the qualitative behavior of solutions for various choices of initial conditions and the parameter $\mu$. For example, if we start at the initial condition $x_0$ in Figure 5.1, the unstable equilibrium will repel us and $x$ will decrease towards the stable equilibrium (as indicated by an

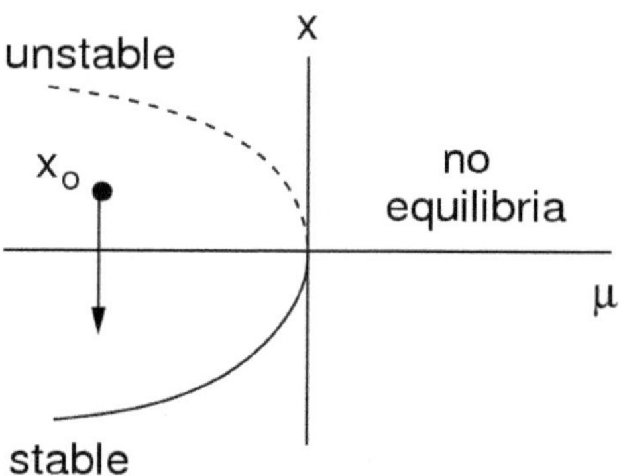


**Figure 5.1.** Saddle-node bifurcation in Equation (5.1). For $\mu < 0$, there are two equilibria: $x = \pm\sqrt{-\mu}$. The larger of these is unstable (dashed curve) and the smaller is stable (solid curve). For $\mu > 0$, there are no equilibria.

arrow in the Figure). The bifurcation diagram for $x' = -\mu + x^2$ can be obtained by reflecting the parabola in Figure 5.1 across the vertical axis.

**Transcritical bifurcations.** The saddle-node bifurcation is a very common way for equilibria to be created or destroyed as a parameter $\mu$ is varied. We now describe a mechanism by which two equilibria can exchange their stability. The canonical example of a *transcritical bifurcation* is given by the equation

$$\frac{dx}{dt} = \mu x - x^2. \tag{5.2}$$

Setting $\frac{dx}{dt} = 0$, we find that there are two equilibria: $x = 0$ and $x = \mu$. Note that the former is an equilibrium independent of the choice of $\mu$. To check the stability of these equilibria, we compute that the "Jacobian" of $f(x) = \mu x - x^2$ is simply $f'(x) = \mu - 2x$. Since $f'(0) = \mu$, we conclude that $x = 0$ is stable if $\mu < 0$ and unstable if $\mu > 0$. Similarly, since $f'(\mu) = -\mu$, we conclude that $x = \mu$ is stable if $\mu > 0$ and unstable if $\mu < 0$. In this case, $\mu = 0$ is the bifurcation point, and the two equilibria exchange their stability there. The corresponding bifurcation diagram appears as Figure 5.2, and the arrows in the figure indicate how $x$ would change starting from various choices of $x(0)$ and $\mu$. For example, if $\mu > 0$ and we choose any initial condition $x_0 \in (0, \mu)$, then $x(t) \to \mu$ as $t \to \infty$.

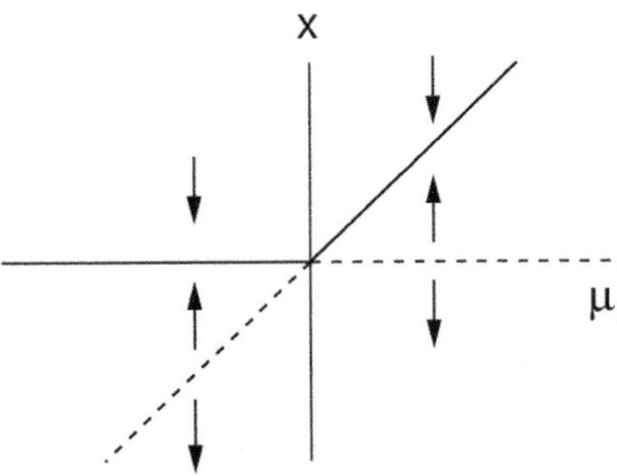


**Figure 5.2.** Transcritical bifurcation in Equation (5.2). For $\mu < 0$, there are two equilibria: $x = 0$ is stable (solid) and $x = \mu$ is unstable (dashed). A bifurcation occurs at $\mu = 0$, and the two equilibria exchange stability. Arrows indicate how $x(t)$ will change as $t$ increases, starting from various choices of $\mu$ and initial conditions $x(0)$.

**Example 5.1.1.** Equation (5.2) is similar to a well-known logistic model for population growth

$$x' = \mu x(M - x), \tag{5.3}$$

where $x$ represents population of a species, $M$ is the maximum population that the environment can sustain, and the parameter $\mu$ measures the birth rate minus the death rate. If $\mu \neq 0$, the equation has exactly two equilibria: $x = 0$ (extinction) and $x = M$ (proliferation), neither of which depend on $\mu$. If $\mu < 0$, the death rate exceeds the birth rate and $x = 0$ is a stable equilibrium while $x = M$ is unstable. If $\mu > 0$, then $x = M$ is stable and $x = 0$ is unstable, and we expect the population to approach the environment's carrying capacity as $t \to \infty$. The transcritical bifurcation at $\mu = 0$ marks the "threshold" between the species' extinction and proliferation. Figure 5.3 shows the bifurcation diagram for (5.3) (compare to Figure 5.2).

**Pitchfork bifurcations.** One type of bifurcation that is common in physical problems involving some sort of *symmetry* is the *pitchfork bifurcation*. As an example, suppose that we balance a weight on top of a vertical beam (Figure 5.4). If the weight is small, the system is in a stable equilibrium. If the weight increases beyond a certain point, the equilibrium loses its stability. Any slight deflection of the beam from a perfectly vertical position will cause the beam to buckle. If we

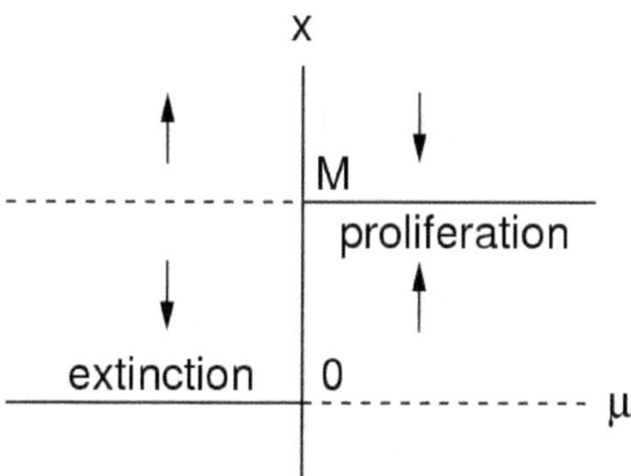


**Figure 5.3.** Transcritical bifurcation in Equation (5.3). For $\mu \neq 0$, there are two equilibria: $x = 0$ and $x = M$. The bifurcation at $\mu = 0$ causes the two equilibria to exchange stability. Arrows indicate how $x(t)$ will change as $t$ increases, starting from various choices of $\mu$ and initial conditions $x(0)$.

regard this system as "two-dimensional", then the beam will either buckle left or right. Another example is provided by a bead on a rotating hoop (Figure 5.5). For slow rotation speeds, the bead rests at the bottom of the hoop. However, this equilibrium loses its stability if the rotation speed is increased beyond a certain critical value. In these first of these examples, the bifurcation parameter $\mu$ would be the weight of the object being supported by the beam. In the latter example, the bifurcation parameter $\mu$ is the speed of rotation of the hoop.

The canonical example of a pitchfork bifurcation is provided by the equation

$$\frac{dx}{dt} = \mu x - x^3. \tag{5.4}$$

The right hand side of this equation can be factored as $f(x;\mu) = x(\mu - x^2)$. If $\mu < 0$, then the quadratic factor has no real roots and therefore $x = 0$ is the only equilibrium. On the other hand, if $\mu > 0$, then further factorization reveals that $f(x;\mu) = x(\sqrt{\mu} - x)(\sqrt{\mu} + x)$. Thus, for positive $\mu$ there are 3 equilibria: $x = 0$ and $x = \pm\sqrt{\mu}$. Using the same sort of stability analysis we used in our discussions of saddle-node and transcritical bifurcations, you should convince yourself that $x = 0$ is stable if $\mu < 0$ and unstable if $\mu > 0$. Both $x = \pm\sqrt{\mu}$ are stable equilibria for $\mu > 0$. The bifurcation diagram in Figure 5.6 should convince you that the name *pitchfork* bifurcation is indeed appropriate. As we

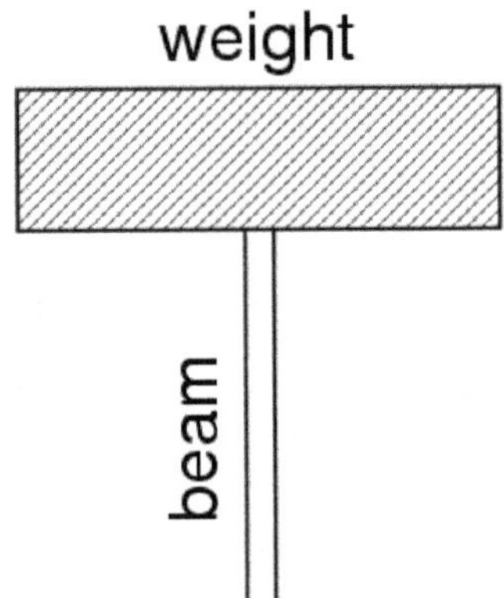


**Figure 5.4.** A weight being supported by a vertical beam.

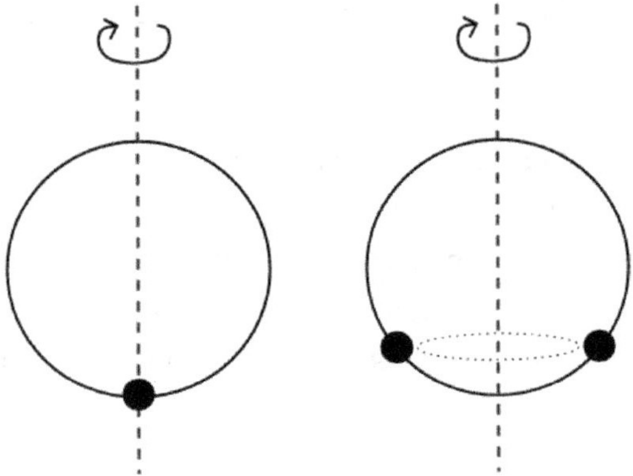

**Figure 5.5.** A bead on a rotating hoop. The vertical lines indicate the axis of rotation. For slow rotation speeds (left panel), the bead rests in a stable equilibrium at the bottom of the hoop. If the rotation speed is increased, this equilibrium loses stability and the bead settles into a new stable position (right panel).

see in the Figure, as $\mu$ increases from negative to positive, an equilibrium loses stability at the same instant that two new equilibria are born.

**Subcritical versus supercritical bifurcations.** One further way of classifying bifurcations is according to whether an equilibrium gains or loses stability as the parameter $\mu$ is increased. The pitchfork bifurcation in Equation (5.4) above is considered to be a *supercritical* bifurcation. An example of a *subcritical*

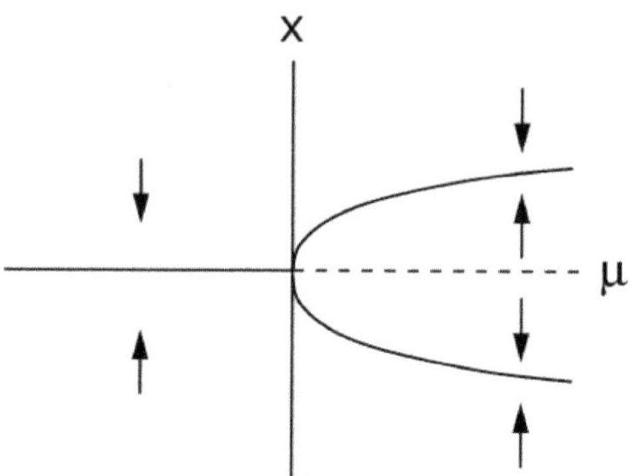


**Figure 5.6.** Pitchfork bifurcation in Equation (5.4). For $\mu < 0$, there is only one equilibrium ($x = 0$) and it is stable. For $\mu > 0$, there are three equilibria: $x = 0$ is unstable and $x = \pm\sqrt{\mu}$ are stable. Arrows indicate qualitative behavior of solutions starting from various initial conditions and choices of the parameter $\mu$.

pitchfork bifurcation would be given by the equation $x' = \mu x + x^3$. The reader is encouraged to sketch a bifurcation diagram for this equation. You should find that $x = 0$ is a stable equilibrium for $\mu < 0$ and is unstable if $\mu > 0$. For $\mu < 0$, there are two other equilibria $x = \pm\sqrt{-\mu}$, both of which are unstable. In what follows, we will rarely distinguish between subcritical and supercritical bifurcations.

**Example 5.1.2.** Let $\mu$ be a real parameter. Sketch the bifurcation diagram for the ODE

$$\frac{dx}{dt} = f_\mu(x) = 5 - \frac{\mu}{1+x^2}. \tag{5.5}$$

**Solution:** Equilibria would satisfy the equation

$$5 - \frac{\mu}{1+x^2} = 0,$$

which can be manipulated to reveal that $x^2 = \mu/5 - 1$. There are no equilibria for $\mu < 5$, but if $\mu > 5$ then there are two equilibria:

$$x = \pm\sqrt{\frac{\mu}{5} - 1}.$$

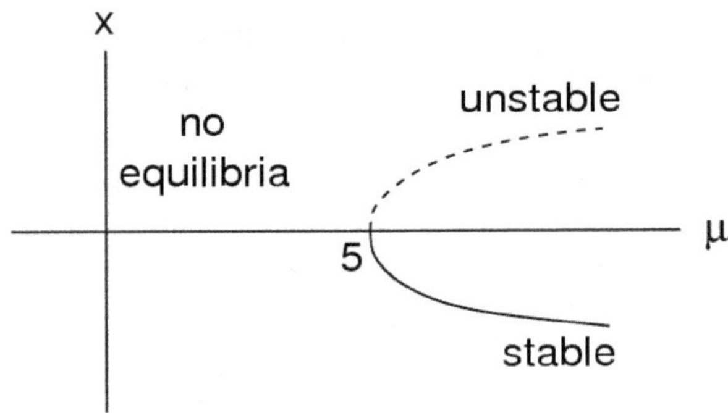


**Figure 5.7.** Bifurcation diagram for Equation (5.5). A saddle-node bifurcation occurs at $\mu = 5$.

We can already conclude that this is an example of a saddle-node bifurcation. It remains to determine which equilibria are stable. Taking the derivative of $f_\mu(x)$, we have

$$f'_\mu(x) = \frac{2\mu x}{(1+x^2)^2}.$$

It follows that

$$f'_\mu(\sqrt{\mu/5-1}) = \frac{2\mu\sqrt{\mu/5-1}}{(\mu/5)^2},$$

which is positive. It follows that for $\mu > 5$, the larger of the two equilibria is unstable. The other fixed point is stable. The bifurcation diagram appears in Figure 5.7.

**Example 5.1.3** (Courtesy of S. Lazaryan). Let $\mu$ be a real parameter. Sketch a bifurcation diagram for the ODE

$$\frac{dx}{dt} = f_\mu(x) = x(x^2-\mu)\left(e^{x^2}-\mu\right). \tag{5.6}$$

**Solution:** Equilibria must satisfy the equation

$$x(x^2-\mu)\left(e^{x^2}-\mu\right) = 0,$$

and certainly $x = 0$ is an equilibrium regardless of the value of $\mu$. If $\mu > 0$, then the factor $x^2 - \mu$ has two real roots, $x = \pm\sqrt{\mu}$. In order for $e^{x^2} - \mu = 0$ to have real solutions, we would need $\mu \geq 1$. If $\mu > 1$, algebra reveals that $x = \pm\sqrt{\ln \mu}$ are also equilibria. In total, there are five equilibria if $\mu > 1$.

To test the stability of these equilibria, we must compute the "Jacobian" of $f_\mu(x)$, taking the derivative with respect to $x$. Expanding $f_\mu(x)$ as

$$f_\mu(x) = x^3 e^{x^2} - \mu x e^{x^2} - \mu x^3 + \mu^2 x,$$

we use the chain and product rules to calculate

$$f'_\mu(x) = 3x^2 e^{x^2} + 2x^4 e^{x^2} - \mu e^{x^2} - 2\mu x^2 e^{x^2} - 3\mu x^2 + \mu^2.$$

For the equilibrium $x = 0$, we have $f'_\mu(0) = -\mu + \mu^2 = \mu(\mu - 1)$. This quantity is negative if $0 < \mu < 1$ and positive otherwise. We conclude that $x = 0$ is a stable equilibrium if $0 < \mu < 1$ and is unstable otherwise.

Next, assume $\mu > 0$ and consider the pair of equilibria $x = \pm\sqrt{\mu}$. Notice that $x^2 = \mu$ and $x^4 = \mu^2$ for both of these equilibria. Using the above formula for $f'_\mu(x)$ we have

$$\begin{aligned} f'_\mu(\pm\sqrt{\mu}) &= 3\mu e^\mu + 2\mu^2 e^\mu - \mu e^\mu - 2\mu^2 e^\mu - 3\mu^2 + \mu^2 \\ &= 2\mu e^\mu - 2\mu^2 = 2\mu\left(e^\mu - \mu\right). \end{aligned}$$

Since $e^\mu > \mu$, we conclude that $f'_\mu(\pm\sqrt{\mu}) > 0$ for all positive $\mu$. This implies that the two equilibria $x = \pm\sqrt{\mu}$ are unstable for $\mu > 0$.

Finally, assume $\mu > 1$ and consider the pair of equilibria $x = \pm\sqrt{\ln \mu}$. In this case, we have $x^2 = \ln \mu$ and $e^{x^2} = \mu$. Our formula for $f'_\mu(x)$ yields

$$\begin{aligned} f'_\mu(\pm\sqrt{\ln \mu}) &= 3\mu \ln \mu + 2\mu(\ln \mu)^2 - \mu^2 - 2\mu^2 \ln \mu - 3\mu \ln \mu + \mu^2 \\ &= 2\mu \ln \mu \left(\ln \mu - \mu\right). \end{aligned}$$

Examining the factors individually, note that $(\ln \mu - \mu) < 0$ and $\ln \mu > 0$ for $\mu > 1$. It follows that $f'_\mu(\pm\sqrt{\ln \mu}) < 0$ for all $\mu > 1$, implying that these two equilibria are stable. A bifurcation diagram summarizing the stability of the equilibria appears in Figure 5.8.

---

### 5.2. Dependence of Solutions on Parameters

In a previous chapter, we discussed how changes in initial conditions can affect how solutions behave. We even derived an estimate for how fast solution

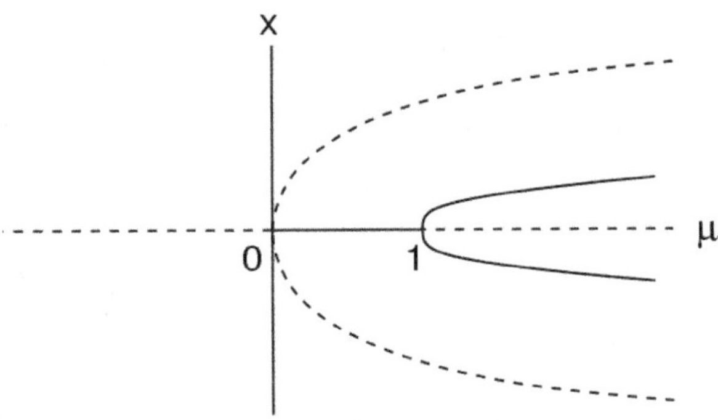


**Figure 5.8.** Bifurcation diagram for Equation (5.6). A pitchfork bifurcation occurs at $\mu = 0$, creating two unstable equilibria $x = \pm\sqrt{\mu}$ and changing $x = 0$ into a stable equilibrium. Another pitchfork bifurcation occurs at $\mu = 1$, de-stabilizing $x = 0$ and creating two new stable equilibria, $x = \pm\sqrt{\ln \mu}$.

trajectories can separate if we start from two different "nearby" initial conditions. We now perform the same sort of calculation to determine how changes in a parameter $\mu$ can affect the behavior of solutions.

Consider the two differential equations

$$x' = f(x;\mu) \quad \text{and} \quad y' = f(y;\lambda), \tag{5.7}$$

and assume that both have the same initial condition: $x(0) = a = y(0)$. In other words, we are considering two copies of the *same* initial value problem, but with two different parameter choices, $\mu$ and $\lambda$. Assume that the function $f$ is continuously differentiable with respect to *both* arguments. From a lemma that we proved when analyzing dependence of solutions on initial conditions, there exist positive constants $K_1$ and $K_2$ such that

$$|f(x;\mu) - f(x;\lambda)| \le K_1|\mu - \lambda| \quad \text{and} \quad |f(x;\lambda) - f(y;\lambda)| \le K_2|x - y|.$$

Let $K = \max\{K_1, K_2\}$. Then we can use the triangle inequality to estimate the difference between the right hand sides of the ODEs in Equation (5.7):

$$\begin{aligned} |f(x;\mu) - f(y;\lambda)| &= |f(x;\mu) - f(x;\lambda) + f(x;\lambda) - f(y;\lambda)| \\ &\le |f(x;\mu) - f(x;\lambda)| + |f(x;\lambda) - f(y;\lambda)| \end{aligned}$$

$$\leq K_1|\mu - \lambda| + K_2|x - y| \;\leq\; K\left(|\mu - \lambda| + |x - y|\right).$$

We are now in a position to measure the gap between the solutions of the two different initial value problems. Writing them as integral equations,

$$x(t) \;=\; a + \int_0^t f(x(s);\mu)\,\mathrm{d}s \quad \text{and} \quad y(t) \;=\; a + \int_0^t f(y(s);\lambda)\,\mathrm{d}s.$$

Using our above inequality, we now make the estimate

$$\begin{aligned} |y(t) - x(t)| &= \left| \int_0^t f(y(s);\lambda)\,\mathrm{d}s \;-\; \int_0^t f(x(s);\mu)\,\mathrm{d}s \right| \\ &\leq \int_0^t |f(y(s);\lambda) - f(x(s);\mu)|\,\mathrm{d}s \;\leq\; \int_0^t K\left(|\mu - \lambda| + |x - y|\right)\mathrm{d}s \\ &= K\int_0^t |\mu - \lambda|\,\mathrm{d}s \;+\; K\int_0^t |y(s) - x(s)|\,\mathrm{d}s. \end{aligned}$$

In the last line, the first integrand is simply a constant. If we introduce the abbreviation $B = K|\mu - \lambda|$ and actually evaluate the integral, our overall estimate has been written as

$$|y(t) - x(t)| \;\leq\; Bt \;+\; K\int_0^t |y(s) - x(s)|\,\mathrm{d}s.$$

This is set up perfectly for the Strong Gronwall Inequality 3.3.12, and we conclude that

$$|y(t) - x(t)| \;\leq\; \frac{B}{K}\left(\mathrm{e}^{Kt} - 1\right) \;=\; |\mu - \lambda|\left(\mathrm{e}^{Kt} - 1\right).$$

**Consequence:** Suppose the function $f$ is well-behaved in the sense that there exists a *global* constant $K$ such that

$$|f(x;\mu) - f(y;\lambda)| \;\leq\; K\left(|\mu - \lambda| + |x - y|\right) \tag{5.8}$$

for all real $x$, $y$, $\mu$, and $\lambda$. Then the solutions of the two initial value problems (corresponding to different choices of the parameter) can separate at most exponentially fast. This ensures that, at least for well-behaved problems of this sort, any bifurcations caused by changing the parameter values cannot cause a "catastrophic" change in solution behavior. For example, if solutions do not blow up to $\infty$ in finite time when we use the parameter $\mu$, then they cannot blow up in finite time if we switch our parameter value from $\mu$ to $\lambda$.

**Warning:** If we have a system $x' = f(x;\mu)$ for which $f$ does not obey inequality (5.8) *globally*, then catastrophes can happen. For example, recall our canonical example of a saddle-node bifurcation: $\frac{dx}{dt} = \mu + x^2$. Suppose $\mu < 0$ and that the initial condition $x_0$ satisfies the inequality $-\sqrt{-\mu} < x_0 < \sqrt{-\mu}$. Then the solution $x(t)$ will converge to the stable equilibrium $-\sqrt{-\mu}$ as $t \to \infty$. Now suppose we use the same initial condition, but change the parameter $\mu$ to some positive number. Then the solution of the initial value problem will blow up to $\infty$ in *finite* time! The problem is that for the function $f(x;\mu) = \mu + x^2$, there is no global constant $K$ for which inequality (5.8) will hold. More exactly, it is impossible to find one special, positive number $K$ for which the inequality

$$|x^2 - y^2| \le K|x - y|$$

for all $x$, $y$ in $\mathbb{R}$.

---

## 5.3. Andronov-Hopf Bifurcations

Up to now, we have considered bifurcations in systems with a single dependent variable. Of course, this is very restrictive because there is precisely one real "eigenvalue" associated with such systems. Higher-dimensional systems can experience other types of bifurcations that our one-dimensional systems could not. For example, what are the possible ways that a stable equilibrium of a planar system could lose its stability as we vary a parameter $\mu$? If both eigenvalues are real and one of them changes from negative to positive, the equilibrium would change from a sink (stable node) to a saddle. We could also have two real eigenvalues simultaneously change from negative to positive, converting the sink to a source (unstable node). However, there is a new possibility—we could have a pair of *complex conjugate* eigenvalues change from negative real part to positive real part, causing a transition from a stable focus to an unstable focus. We now provide an example of this sort of bifurcation.

**Example 5.3.1.** Consider the planar system

$$\frac{dx}{dt} = y \quad \text{and} \quad \frac{dy}{dt} = (\mu - x^2)y - x, \tag{5.9}$$

where $\mu$ is a parameter. It is easily checked that, independent of $\mu$, the origin $(x,y) = (0,0)$ is the only equilibrium. The Jacobian matrix associated with this

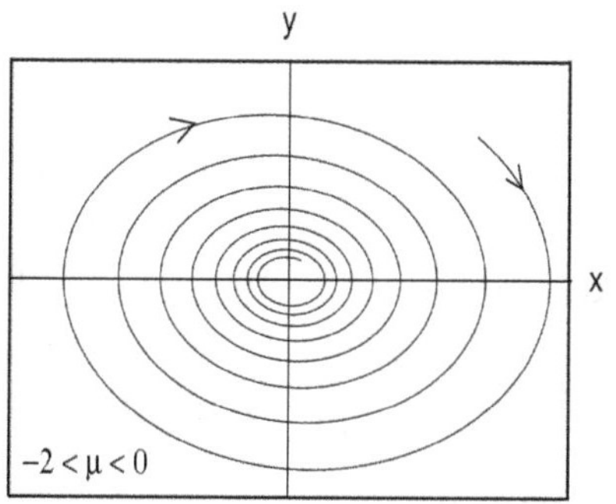


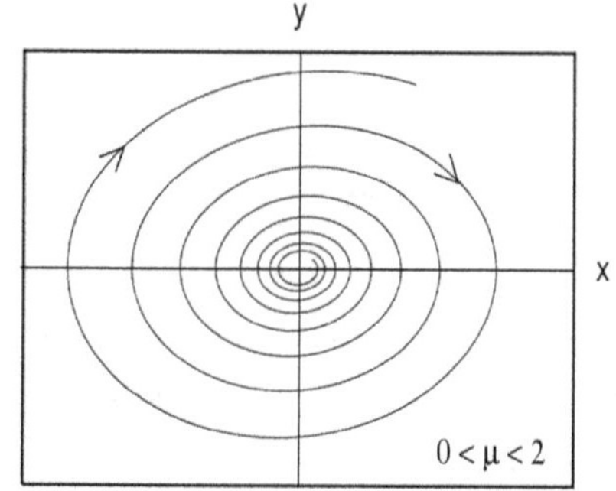


**Figure 5.9.** Example of an Andronov-Hopf bifurcation in Equation (5.9). The origin is a stable focus (left panel) if $-2 < \mu < 0$ and is an unstable focus (right panel) if $0 < \mu < 2$.

system is

$$Jf(x,y) = \begin{bmatrix} 0 & 1 \\ -1 - 2xy & \mu - x^2 \end{bmatrix},$$

from which it follows that

$$Jf(0,0) = \begin{bmatrix} 0 & 1 \\ -1 & \mu \end{bmatrix}.$$

The characteristic equation is $\lambda^2 - \mu\lambda + 1 = 0$, and the eigenvalues are

$$\lambda_{\pm} = \frac{\mu \pm \sqrt{\mu^2 - 4}}{2}.$$

Assuming that $-2 < \mu < 2$, these eigenvalues are a complex conjugate pair with real part $\mu/2$. Notice that the real part is negative if $\mu < 0$ and positive of $\mu > 0$, which means that the equilibrium $(0,0)$ changes from a stable focus to an unstable focus if $\mu$ increases from negative to positive. Figure 5.9 shows how trajectories in the phase plane reverse their orientation as we pass through the bifurcation value $\mu = 0$. This sort of bifurcation has a name.

**Definition 5.3.2.** When an equilibrium changes its stability because a pair of complex conjugate eigenvalues experiences a change in the sign of their real part, we say that an *Andronov-Hopf* bifurcation has occurred.

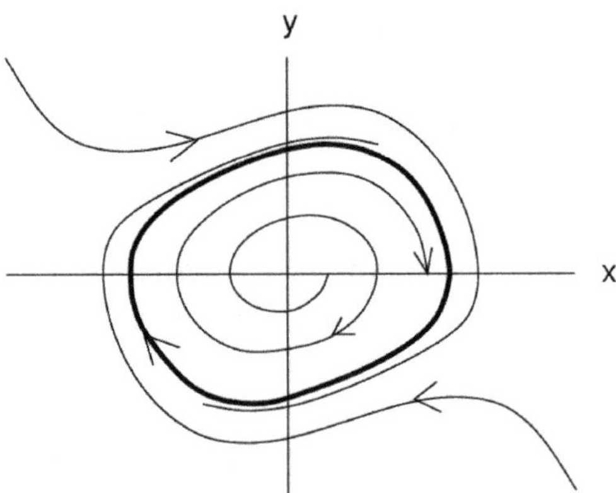


**Figure 5.10.** For $\mu = 0.3$, Equations (5.9) have a limit cycle solution (bold curve).

In the above example, the Andronov-Hopf bifurcation at $\mu = 0$ causes a dramatic change in the dynamical behavior of the system. If $-2 < \mu < 0$, the origin is a stable focus and, because the equilibrium is hyperbolic, the linearization gives a good indicator of how solutions of the nonlinear system behave. It is possible (although difficult) to prove that the origin is actually a *global* attractor by drawing appropriately chosen closed contours $\Gamma$, finding the outward normal vectors $\mathbf{n}$ to $\Gamma$, and proving the $f \bullet \mathbf{n} < 0$ everywhere on $\Gamma$. At the bifurcation point $\mu = 0$, the origin is a weakly stable equilibrium. Indeed, the distance from the origin cannot increase as we move along trajectories in the direction of increasing $t$:

$$\frac{\mathrm{d}}{\mathrm{d}t}[x^2(t) + y^2(t)] \;=\; 2xx' + 2yy' \;=\; -2x^2y^2 \leq 0.$$

Finally, we established that for $0 < \mu < 2$, the origin is an unstable equilibrium. However, more can be said. By creative use of the Poincaré-Bendixon Theorem, one may construct carefully chosen contours to find that a stable limit cycle exists for $0 < \mu < 2$. The birth of periodic solutions (limit cycles) as we pass the bifurcation value $\mu = 0$ is no accident. In fact, when Andronov-Hopf bifurcations occur, we shall find that periodic solutions are born when an equilibrium loses its stability in response to changes in $\mu$.

**Example 5.3.3.** Consider the system

$$\begin{aligned} x' &= -y + x(\mu - x^2 - y^2) \\ y' &= x + y(\mu - x^2 - y^2), \end{aligned} \tag{5.10}$$

where $\mu$ is a real parameter. We claim that, regardless of $\mu$, the origin is the only equilibrium for this system. To see why, set both $x' = 0$ and $y' = 0$ to get the two algebraic equations

$$y = x(\mu - x^2 - y^2) \quad \text{and} \quad x = -y(\mu - x^2 - y^2).$$

Substituting the second equation into the first, we find that

$$y = -y(\mu - x^2 - y^2)^2.$$

If the non-negative quantity $(\mu - x^2 - y^2)^2$ is strictly positive, then the only way the equation can be satisfied is if $y = 0$. This, in turn, implies that $x = 0$ as well. On the other hand, if $\mu - x^2 - y^2 = 0$, then the system (5.10) reduces to $x' = -y$ and $y' = x$. Again, the only possible equilibrium is $x = y = 0$.

The linearization of (5.10) about the origin is

$$\begin{bmatrix} x' \\ y' \end{bmatrix} = \begin{bmatrix} \mu & -1 \\ 1 & \mu \end{bmatrix} \begin{bmatrix} x \\ y \end{bmatrix},$$

and the eigenvalues of the coefficient matrix are $\mu \pm i$. As $\mu$ changes sign from negative to positive, the origin loses its stability via an Andronov-Hopf bifurcation.

How does this bifurcation affect the dynamical behavior of the system? To address this question, we remark that Equations (5.10) take a very convenient form if we convert to polar coordinates. Letting $R^2 = x^2 + y^2$ and $\theta = \arctan(y/x)$, we derive differential equations for $R$ and $\theta$ as follows. First,

$$\frac{\mathrm{d}}{\mathrm{d}t} R^2 = \frac{\mathrm{d}}{\mathrm{d}t}(x^2 + y^2)$$

implies that $2RR' = 2xx' + 2yy'$, where primes indicate differentiation with respect to $t$. Substituting the right hand sides of equations (5.10) for $x'$ and $y'$,

$$\begin{aligned} 2RR' &= 2x[-y + x(\mu - x^2 - y^2)] + 2y[x + y(\mu - x^2 - y^2)] \\ &= 2x^2(\mu - x^2 - y^2) + 2y^2(\mu - x^2 - y^2) \\ &= 2(x^2 + y^2)(\mu - x^2 - y^2) \ = \ 2R^2(\mu - R^2). \end{aligned}$$

Avoiding the equilibrium solution ($R = 0$), we have obtained $R' = R(\mu - R^2)$, a differential equation that describes the distance $R$ from the origin. For the angular variable $\theta$, we calculate

$$\begin{aligned} \frac{d\theta}{dt} &= \frac{d}{dt} \arctan\left(\frac{y}{x}\right) \ = \ \frac{xy' - yx'}{x^2} \cdot \frac{1}{1 + (y/x)^2} \ = \ \frac{xy' - yx'}{x^2 + y^2} \\ &= \frac{x[x + y(\mu - x^2 - y^2)] - y[-y + x(\mu - x^2 - y^2)]}{x^2 + y^2} \\ &= \frac{x^2 + y^2}{x^2 + y^2} \ = \ 1. \end{aligned}$$

In summary, converting our original system (5.10) to polar coordinates gives

$$\frac{dR}{dt} \ = \ R(\mu - R^2) \quad \text{and} \quad \frac{d\theta}{dt} \ = \ 1.$$

These two equations are un-coupled and can be handled separately. Since $\theta' = 1$, the angular variable increases with constant speed, and we move counterclockwise as we follow trajectories in the $xy$-phase plane. The radial variable $R$ measures distance from the origin. You can verify that if $\mu < 0$, then $R = 0$ is a stable equilibrium (corresponding to the origin) for the equation $R' = R(\mu - R^2)$. There are no other equilibria for the case $\mu < 0$. On the other hand, if $\mu > 0$, then there are two equilibria: $R = 0$ and $R = \sqrt{\mu}$. (Note that we exclude the equilibrium $R = -\sqrt{\mu}$ because $R$ is a non-negative quantity.) It is easily checked that $R = 0$ is unstable for $\mu > 0$, whereas $R = \sqrt{\mu}$ is stable. In the phase plane, $R = \sqrt{\mu}$ is a circle of radius $\sqrt{\mu}$, and it corresponds to a *stable limit cycle*: $x = \sqrt{\mu}\cos t$ and $y = \sqrt{\mu}\sin t$. Sketches of the phase plane for $\mu < 0$ and $\mu > 0$ appear in Figure 5.11.

The important thing to notice here is that the Andronov-Hopf bifurcation at $\mu = 0$ created a family of periodic (limit cycle) solutions. The amplitude of these limit cycles is set by $\sqrt{\mu}$. If we sketch a bifurcation diagram for the

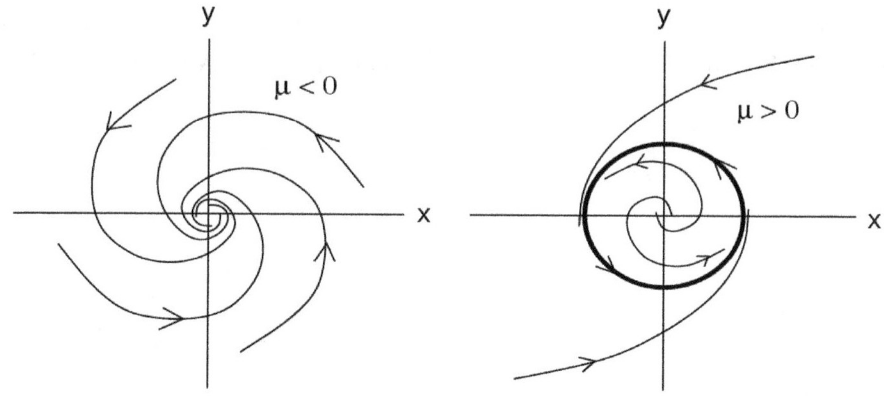


**Figure 5.11.** Phase portraits for the system (5.10) for $\mu < 0$ (left panel) and $\mu > 0$ (right panel). For $\mu < 0$, the origin is a stable focus. For $\mu > 0$, there is a limit cycle solution (indicated in bold) corresponding to a circle radius $\sqrt{\mu}$.

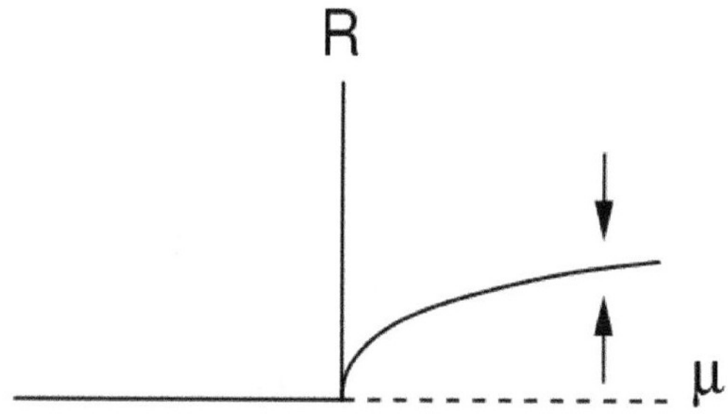


**Figure 5.12.** Bifurcation diagram for $R' = R(\mu - R^2)$. This is a "half-pitchfork" because, since the radial variable $R$ represents distance from the origin, we must exclude the possibility of negative $R$.

one-dimensional problem $R' = R(\mu - R^2)$, we obtain a "half-pitchfork"—the lower prong of the pitchfork is missing because we excluded negative values of $R$. Figure 5.12 shows this bifurcation diagram.

Plotting a bifurcation diagram for the Andronov-Hopf bifurcation in the above example is more complicated than creating a bifurcation diagram for a one-parameter system with a single dependent variable. This idea is to give a graphical rendering of the *long-term* behavior of solutions as a function of the parameter $\mu$. To do so, we must create a three-dimensional plot illustrating how the value of $\mu$ affects the behavior of $x$ and $y$ as $t \to \infty$. The bifurcation diagram

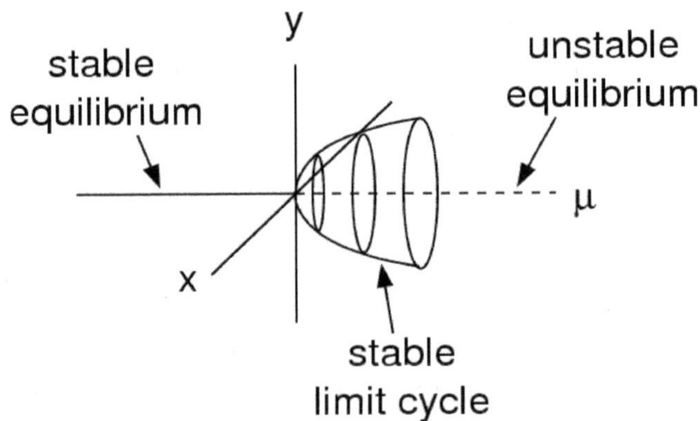


**Figure 5.13.** Bifurcation diagram for the system (5.10). For $\mu < 0$, the origin is a stable equilibrium. For each $\mu > 0$, the origin is an unstable equilibrium but there is a stable limit cycle with amplitude $\sqrt{\mu}$.

appears in Figure 5.13. Notice that the family of limit cycle solutions forms a surface (a paraboloid, to be specific).

Like other types of bifurcations, Andronov-Hopf bifurcations can be either supercritical or subcritical. The above is an example of the supercritical case, in which stable limit cycles are spawned as $\mu$ increases beyond a critical value. It is also possible to have subcritical Andronov-Hopf bifurcations in which a stable equilibrium loses stability as $\mu$ increases, and a family of unstable limit cycle solutions is destroyed in the process. Qualitatively, all Andronov-Hopf bifurcations exhibit the sort of behavior exhibited by Equations (5.10). Let $\beta \neq 0$ and consider the following linear, constant-coefficient system with a pair of complex conjugate eigenvalues $\mu \pm \beta i$:

$$\begin{bmatrix} x' \\ y' \end{bmatrix} = \begin{bmatrix} \mu & -\beta \\ \beta & \mu \end{bmatrix} \begin{bmatrix} x \\ y \end{bmatrix} = \begin{bmatrix} \mu x - \beta y \\ \beta x + \mu y \end{bmatrix}.$$

Now let us introduce nonlinearity be adding functions with "quadratic and higher-order" terms to the right hand sides of these two equations; i.e.,

$$\begin{aligned} x' &= \mu x - \beta y + p(x,y) \\ y' &= \beta x + \mu y + q(x,y). \end{aligned} \tag{5.11}$$

Here, we assume that the functions $p(x,y)$ and $q(x,y)$ are *analytic* functions of two variables. That is, $p(x,y)$ and $q(x,y)$ can be represented in terms of convergent power series expansions

$$p(x,y) = \sum_{i+j\geq 2} a_{ij}x^iy^j = (a_{20}x^2 + a_{11}xy + a_{02}y^2) \quad q(x,y) = \sum_{i+j\geq 2} b_{ij}x^iy^j = (b_{20}x^2 +$$
$$+ (a_{30}x^3 + a_{21}x^2y + a_{12}xy^2 + a_{03}y^3) + \cdots \qquad + (b_{30}x^3 + b_{21}x^2y + b_{12}x$$

Notice that we have only included quadratic and higher-order terms, and the subscripts used on the coefficients correspond to the exponents on $x$ and $y$. Roughly speaking, $p$ and $q$ are nothing more than generalizations of polynomials in two variables. By design, the system (5.11) has exactly one equilibrium (the origin), and the linearization about the origin is given by

$$\begin{bmatrix} x' \\ y' \end{bmatrix} = \begin{bmatrix} \mu & -\beta \\ \beta & \mu \end{bmatrix} \begin{bmatrix} x \\ y \end{bmatrix}.$$

If $\mu = 0$, then the origin is a center for the linearized system.

In preparation for stating a criterion for the occurrence of an Andronov-Hopf bifurcation in (5.11), we will need the following definition.

**Definition 5.3.4.** For the system (5.11), the *Lyapunov number* is defined as

$$\sigma = \frac{3\pi}{2\beta}\left[3(a_{30} + b_{03}) + (a_{12} + b_{21}) - \frac{2}{\beta}(a_{20}b_{20} - a_{02}b_{02}) \right.$$
$$\left. + \frac{a_{11}}{\beta}(a_{02} + a_{20}) - \frac{b_{11}}{\beta}(b_{02} + b_{20})\right].$$

For an explanation of where this mysterious quantity comes from, see Section 3.4 of the text of Perko [8]. Specifically, read the discussion of Poincaré maps (a test for the stability of periodic solutions).

The Lyapunov number provides a quick way of testing whether the system (5.11) actually experiences an Andronov-Hopf bifurcation at $\mu = 0$.

**Theorem 5.3.5. (Hopf Bifurcation Theorem).** If $\sigma \neq 0$, then the system (5.11) experiences an Andronov-Hopf bifurcation at $\mu = 0$. If $\sigma < 0$, a *unique, stable limit cycle* bifurcates from the origin as $\mu$ increases from negative to positive. For small, positive $\mu$, the amplitude of the limit cycle solution is approximately proportional to $\sqrt{\mu}$. Similarly, if $\sigma > 0$, then a unique unstable limit cycle bifurcates from the origin as $\mu$ decreases from positive to negative.

**Example 5.3.6.** In our previous example, we found that the system (5.10)

$$\begin{aligned} x' &= -y + x(\mu - x^2 - y^2) \\ y' &= x + y(\mu - x^2 - y^2) \end{aligned}$$

suffers an Andronov-Hopf bifurcation at $\mu = 0$. In our above notation, $\beta = 1$, $p(x,y) = -x^3 - xy^2$ and $q(x,y) = -x^2y - y^3$. Notice that there are no "quadratic" terms—i.e., terms in which the exponents of $x$ and $y$ sum to 2. This implies that $a_{20}$, $a_{11}$, $a_{02}$, $b_{20}$, $b_{11}$, and $b_{02}$ are all 0. For the "cubic" terms, we have

$$a_{30} = -1, \qquad a_{12} = -1, \qquad a_{21} = 0, \qquad a_{03} = 0,$$

and

$$b_{30} = 0, \qquad b_{12} = 0, \qquad b_{21} = -1, \qquad b_{03} = -1.$$

The Lyapunov number is

$$\sigma = \frac{3\pi}{2}[3(-1-1) + (-1-1)] = -12\pi < 0,$$

and it follows that a supercritical Andronov-Hopf bifurcation occurs at $\mu = 0$.

It is possible to state a stronger version of the Hopf Bifurcation Theorem which allows us to estimate the period and amplitude of the limit cycles that are born via an Andronov-Hopf bifurcation. Suppose that a system $\mathbf{x} = f(\mathbf{x};\mu)$ experiences an Andronov-Hopf bifurcation at some critical parameter value $\mu = \mu_c$. Let $\lambda(\mu) = \alpha(\mu) + i\omega(\mu)$ denote the eigenvalues of a complex conjugate pair of eigenvalues whose real parts change sign at $\mu = \mu_c$. At the bifurcation point, $\omega(\mu_c) \neq 0$ and $\alpha(\mu_c) = 0$. Moreover, the eigenvalues satisfy a *transversality condition* $\alpha'(\mu_c) \neq 0$, which basically just states that the eigenvalues really are *moving* from negative to positive real part (or vice-versa) as $\mu$ passes $\mu_c$. Let us consider the supercritical case in which $\alpha'(\mu_c) > 0$, so that the real parts of the pair of eigenvalues are changing from negative to positive. The Hopf Bifurcation Theorem states that for $\mu$ slightly larger than $\mu_c$, there exists a family of stable limit cycle solutions whose amplitudes are roughly proportional to $\sqrt{\mu - \mu_c}$.

Moreover, if $\mu - \mu_c$ is reasonably small, the period of the limit cycle solutions is approximately $2\pi/\omega(\mu_c)$. For details, refer to the text of Hassard et al. [4].

As an illustration, suppose that the Jacobian matrix associated with a system of ODEs has eigenvalues $(\mu - 3) \pm 2\mu i$. Then $\alpha(\mu) = \mu - 3$ and $\omega(\mu) = 2\mu$. At $\mu = \mu_c = 3$, we have $\alpha(3) = 0$ and $\omega(3) = 6 \neq 0$. The transversality condition $\alpha'(3) = 1 \neq 0$ is also satisfied, and we conclude that an Andronov-Hopf bifurcation has occurred at $\mu = 3$. The bifurcation would be supercritical because $\alpha'(3) > 0$. This means that for $\mu$ values slightly larger than 3, there exists a family of stable periodic solutions. The periods of these limit cycles are approximately $2\pi/\omega(3) = \pi/3$, and their amplitudes are roughly proportional to $\sqrt{\mu - 3}$. The bifurcation diagram for such a system would look essentially identical to the one in Figure 5.13, except that the vertex of the paraboloid-like surface would be shifted from $\mu = 0$ to $\mu = 3$.

Here, we will not attempt to provide a comprehensive list of the various types of bifurcations that can occur. Clearly, there are many ways that the qualitative behavior of a phase portrait could be suddenly and dramatically altered as a parameter $\mu$ is varied. Systems with more than one parameter offer an even greater degree of flexibility. Students interested in learning more about bifurcation theory are encouraged to take an advanced course in differential equations and dynamical systems.

---

## Exercises

1. For each of the following ODEs, $\mu$ denotes a real parameter. In each case, identify the equilibria of the ODE and determine which ones are stable and unstable. Then, produce a bifurcation diagram.

$$
\begin{aligned}
\text{(a)} \quad \frac{dx}{dt} &= (\mu - 1)x + x^2 \\
\text{(b)} \quad \frac{dx}{dt} &= \mu - 2 + 3x^2 \\
\text{(c)} \quad \frac{dx}{dt} &= \mu^2 - x^2 \\
\text{(d)} \quad \frac{dx}{dt} &= \mu x + 9x^3 \\
\text{(e)} \quad \frac{dx}{dt} &= 4x(\mu - e^x).
\end{aligned}
$$

2. For each of the following ODEs, $\mu$ denotes a real parameter. In each case, find all bifurcation points and, if possible, classify them as one of the types we discussed in this Chapter. Then, produce a bifurcation diagram.

$$\begin{aligned}
\text{(a)} \quad \frac{dx}{dt} &= (1+\mu)(x^2 - \mu x) \\
\text{(b)} \quad \frac{dx}{dt} &= \mu - x^2 + 4x^4 \\
\text{(c)} \quad \frac{dx}{dt} &= \mu^2 - x^4 \\
\text{(d)} \quad \frac{dx}{dt} &= (-x^2 + \mu^4)(x^2 - \mu^2).
\end{aligned}$$

3. Sketch the bifurcation diagram for the equation

$$\frac{dx}{dt} = (\mu - x)(\mu + x^2) = \mu^2 - \mu x + \mu x^2 - x^3,$$

where $\mu$ is a real parameter. If you do this correctly, you will discover two different bifurcation points: one transcritical, and one pitchfork.

4. Let $\mu$ be a real parameter.

   (a) Find all bifurcation points for the equation $x' = x^2 - \mu x - 2\mu^2$, and draw a bifurcation diagram.

   (b) Suppose we modify the ODE in Part (a), multiplying the right-hand side by $x$ to obtain the equation $x' = x(x^2 - \mu x - 2\mu^2)$. Sketch the new bifurcation diagram and compare it to the bifurcation diagram from Part (a).

   (c) Although the system in Part (b) has no bifurcation points, its "bifurcation diagram" still conveys useful information. In fact, the diagram completely characterizes how the system behaves depending upon the parameter $\mu$ and the initial condition $x_0$. If $\mu \neq 0$, note that there are two stable equilibria. For $\mu \neq 0$, find the *basin of attraction* of each stable equilibrium. The basin of attraction for an equilibrium $x^*$ is the set of all initial conditions $x_0$ for which $x(t) \to x^*$ as $t \to \infty$.

**Remark.** (i) Notice that the basins of attraction of the stable fixed points are "separated" by an unstable fixed point. In this respect, unstable equilibria can be *very* important, as they help dictate which stable state (if any) our system will converge to as $t \to \infty$. (ii) Suppose that $f$ is continuously differentiable and that the first-order ODE $x' = f(x;\mu)$ has multiple stable equilibria. Then there is *always* an unstable equilibrium between each pair of stable equilibria. Any solid curves in the bifurcation diagram *must* be separated by a dashed curve somewhere in between.

5. A "quad-furcation": When we discussed saddle-node bifurcations, we gave an example of an ODE that has two equilibria for $\mu < 0$ and no equilibria for $\mu > 0$. Create an example of an ODE with a single parameter $\mu$ which has no equilibria for $\mu < 0$ but *four* equilibria for $\mu > 0$. Identify which equilibria are stable/unstable.
6. A "five-pronged pitchfork": When we discussed pitchfork bifurcations, we gave an example of an ODE that has one equilibrium for $\mu < 0$ and three equilibria for $\mu > 0$. Create an example of an ODE with a single parameter $\mu$ which has one equilibrium for $\mu < 0$ but *five* equilibria for $\mu > 0$. Identify which equilibria are stable/unstable.
7. Consider the system

$$\frac{dx}{dt} = -8y + x(\mu - x^2 - y^2)$$
$$\frac{dy}{dt} = 8x + y(\mu - x^2 - y^2),$$

where $\mu$ is a parameter.

(a) Show that, regardless of $\mu$, the origin is the only equilibrium.

(b) Show that a Hopf bifurcation occurs when $\mu = 0$.

(c) Give a qualitative description of the periodic solutions which are created by this bifurcation. Estimate their period, assuming that $\mu$ is small. For small $\mu$, what can you say about the amplitude of the periodic solutions? Are these solutions stable or unstable?

8. Subcritical Hopf Bifurcations. Subcritical Hopf bifurcations can be dangerous in engineering applications because when an equilibrium loses stability as a parameter $\mu$ varies, solutions can suddenly jump to a far distant stable limit cycle. Consider, for example, the planar system in polar coordinates

$$\frac{dr}{dt} = \mu r + r^3 - r^5 \qquad \frac{d\theta}{dt} = 1.$$

where $\mu$ is a bifurcation parameter.

(a) Note that $r = 0$ is an equilibrium. Determine the range of $\mu$ values for which this equilibrium is stable/unstable.

(b) If we write the $r$ equation as $r' = r(\mu + r^2 - r^4)$, we see that other equilibria must satisfy $\mu = r^4 - r^2$. Carefully sketch this curve in the $\mu r$-plane. Hint: You may find it easier to first sketch $\mu$ versus $r$, and then flip your axes to create a plot of $r$ versus $\mu$.

(c) Show that the $r$ equation has two positive equilibria if $-\frac{1}{4} < \mu < 0$ but only one positive equilibrium if $\mu \geq 0$.

(d) Without attempting to find formulas for the positive equilibria, show that if $\mu > 0$ then the positive equilibrium is stable.

(e) For $-\frac{1}{4} < \mu < 0$, show that the smaller of the two positive equilibria is unstable.

(f) Discussion (No work required on your part!) Here is why this sort of bifurcation can be devastating. Suppose $\mu < -1/4$ and our system is in equilibrium $r = 0$. Now start gradually increasing $\mu$. When we pass $\mu = -1/4$, two limit cycles are born, corresponding to the two different positive $r$ values which satisfy the equation $\mu = r^4 - r^2$. The limit cycle with larger amplitude (larger $r$) is stable, and the other one is unstable. Meanwhile, our equilibrium $r = 0$ is still stable. However, when we reach $\mu = 0$, the unstable limit cycle is destroyed via a subcritical Hopf bifurcation, but the larger amplitude stable limit cycle still exists. As soon as the unstable limit cycle is destroyed (at $\mu = 0$), the equilibrium $r = 0$ also loses its stability. Consequently, once $\mu > 0$, any small amount of "noise" would throw us off the equilibrium and our solution would (rather dramatically) leap to a large-amplitude periodic solution. Physically, you can imagine a very well-behaved system which suddenly begins to oscillate violently as a parameter is increased. Subcritical Hopf bifurcations have been observed in mathematical models of aeroelastic flutter (vibrations in airplane wings), and a host of other physical/biological scenarios.

This system also provides a nice example of *bistability* and *hysteresis*, which can be explained as follows. Again, suppose $\mu < -1/4$ and we are at the stable equilibrium. As we increase $\mu$ past 0, our equilibrium suddenly repels nearby trajectories and suddenly forces us toward the large-amplitude stable, periodic solution. Now suppose we start to decrease $\mu$ again. This time, we have a nice, stable limit cycle until we reach $\mu = -1/4$, when we suddenly transition back to a stable equilibrium. Notice that the sudden transitions between two different stable states occurred at two different $\mu$ values. This phenomenon is known as *hysteresis*, and the fact that there are two different stable states for $-1/4 < \mu < 0$ is known as *bistability*.

9. Self-oscillations in glycolysis: For particular parameter choices, a reduced form of the famous Selkov model takes the form

$$\begin{aligned}\frac{dx}{dt} &= -x + \frac{1}{10}y + x^2y \\ \frac{dy}{dt} &= \mu - \frac{1}{10}y - x^2y.\end{aligned}$$

   (a) Show that

   $$(x, y) = \left(\mu, \frac{10\mu}{10\mu^2 + 1}\right)$$

   is an equilibrium.

   (b) Show that this equilibrium is a stable node if $\mu = 0$.

   (c) Show that this equilibrium is an unstable focus if $\mu = \sqrt{\frac{1}{2}}$.

   (d) Show that this equilibrium is a stable focus if $\mu = 1$.

   (e) Since the eigenvalues of the Jacobian matrix depend continuously on $\mu$, the above observations suggest that at least two Andronov-Hopf bifurcations have occurred: one between $\mu = 0$ and $\mu = \sqrt{\frac{1}{2}}$ and one between $\mu = \sqrt{\frac{1}{2}}$ and $\mu = 1$. Find the $\mu$ values at which the Andronov-Hopf bifurcations occur. Remark: The first bifurcation creates a family of stable, periodic solutions. The periodic solutions disappear after the second bifurcation, when the stable focus is born.

10. The purpose of this exercise is to create a bifurcation diagram for the system

$$\begin{aligned}\frac{dx}{dt} &= \mu x(1 - x) - xy \\ \frac{dy}{dt} &= x - y,\end{aligned}$$

   where $\mu > -1$ is a real parameter.

   (a) The origin is an equilibrium independent of the choice of $\mu$. Find the other equilibrium of the system.

   (b) By linearizing the system at the origin, determine the ranges of $\mu$ for which the origin is stable/unstable.

(c) Let $(x_1, y_1)$ denote the equilibrium you found in Part (a). Compute the Jacobian matrix $Jf(x_1, y_1)$ and use its determinant to find the range of $\mu$ for which the equilibrium is a saddle.

(d) When the determinant of $Jf(x_1, y_1)$ is positive, the origin is either a focus or a node. However, for the purposes of creating a bifurcation diagram, we need only determine whether the equilibrium is stable or unstable, and this is easily accomplished by inspecting the trace and determinant of $Jf(x_1, y_1)$. Show that the trace is negative for all $\mu > -1$, and conclude that if the determinant is positive, then $(x_1, y_1)$ is a stable node.

(e) Since the $x$ and $y$ coordinates of the equilibria $(0, 0)$ and $(x_1, y_1)$ happen to be equal, we need not create a three-dimensional bifurcation diagram by plotting $\mu$ versus both $x$ and $y$. Because no information is lost if we neglect the $y$ variable, sketch a bifurcation diagram of $\mu$ versus $x$. What are the bifurcation values of $\mu$? Can you classify which type(s) of bifurcations occur?

# CHAPTER 6

## Introduction to Delay Differential Equations

In this Chapter, we turn our attention to *delay differential equations* (DDEs), a major departure from the ordinary differential equations that were considered up to now. A basic reference for this material is the text of Bellman and Cooke [2]. To understand why DDEs are of mathematical interest, let us examine the simplest population growth model, which was originally proposed by Malthus. The major underlying assumption of the Malthus model is that the rate of change of population is proportional to the population itself. Mathematically, let $P(t)$ denote the population at time $t$. Then the population growth model is given by

$$\frac{\mathrm{d}P}{\mathrm{d}t} = kP,$$

where $k$ is a positive constant. The solution of this ODE is $P(t) = P(0)\mathrm{e}^{kt}$, which predicts exponential population growth as $t$ increases. However, due to the time lag between conception and birth, it may be more realistic to assume that the instantaneous rate of change of population growth is actually dependent upon the population at some fixed amount of time $\tau$ in the past. This would suggest that we adjust the above model to read

$$\frac{\mathrm{d}}{\mathrm{d}t}P(t) = kP(t-\tau). \qquad (6.1)$$

Observe that the rate of change of $P$ at time $t$ is affected by the value of $P$ at time $t-\tau$.

**Definition 6.0.7.** Suppose $x$ and $t$ are dependent and independent variables, respectively, and let $\tau$ be a positive constant. Any equation of the form

$$F(x(t), x'(t), x(t-\tau), x'(t-\tau), t) = 0$$

is called a *first-order* DDE *with a single, constant delay*. If the equation does not incorporate $x'(t-\tau)$, the DDE is called *retarded*. If the equation does incorporate $x'(t-\tau)$, the DDE is called *neutral*.

In the above definition, "first-order" refers to the fact that first derivatives are the highest-order derivatives that appear in the equation. The words "single, constant delay" refer to the fact the equation only makes reference to the present time, $t$, and one fixed time in the past, $t-\tau$.

**Example 6.0.8.** The equation

$$\frac{dx}{dt} = x^2 - (x-3) + x(t-2)$$

is a retarded first-order DDE with a single constant delay $\tau = 2$. On the right hand side, it is understood that $x^2$ means $x(t)^2$ and that $(x-3)$ means $x(t)-3$. The equation

$$x'(t) = x^2 - (x-3) + x(t-2) + x(t-4)$$

is a retarded, first-order DDE with two constant delays: $\tau_1 = 2$ and $\tau_2 = 4$. The equation $x'(t) = x(t/2)$ is a DDE with a *variable* time delay. Note that the rate of change of $x$ when $t = 1$ is influenced by the value of $x$ at time $t = 1/2$, whereas the rate of change of $x$ when $t = 6$ is influenced by the value of $x$ at time $t = 3$. Finally, the equation

$$x'(t-8) + x(t-8) + x(t) + t = 0$$

is a neutral DDE with a single constant delay $\tau = 8$.

This Chapter will focus exclusively on retarded, first-order DDEs with a single, constant time delay $\tau$.

## 6.1. Initial Value Problems

If we wish to solve a DDE such as $x'(t) = x(t-1)$, how would we specify initial data? Note that the solution at time $t-1$ influences the rate of change at time $t$. So, for example, in order to know the rate of change of $x$ for $t \in [0,1]$, we would need to know the value of $x(t)$ for $t \in [-1,0]$. That is, before we can generate the solution of this DDE on the interval $0 \leq t \leq 1$, we must require that initial data be provided as a *function* on the entire interval $-1 \leq x \leq 0$. More generally, in order to solve a retarded DDE with constant time delay $\tau$, we must specify an initial *function* $\phi(t)$ on the interval $[-\tau, 0]$.

**Method of Steps.** We illustrate one method for solving a DDE via an example. Consider the system

$$\begin{aligned} x'(t) &= x(t-1), && \text{if } t > 0 \\ x(t) &= \phi(t) \;=\; 1, && \text{if } -1 \leq t \leq 0. \end{aligned}$$

Here, we have specified the "initial function" $\phi(t)$ in a closed interval of width 1 since our time delay is $\tau = 1$. To solve the DDE, note that by the Fundamental Theorem of Calculus we have

$$x(t) \;=\; x(0) + \int_0^t x'(s)\,ds.$$

In this case, $x'(s) = x(s-1)$, and if $0 \leq s \leq 1$, then we have $-1 \leq (s-1) \leq 0$. So for $0 \leq s \leq 1$, it follows that $x(s-1) = \phi(s-1) = 1$, and the above equation reduces to

$$x(t) \;=\; x(0) + \int_0^t 1\,ds \;=\; 1 + t, \qquad (0 \leq t \leq 1).$$

Now that we have obtained the solution $x(t) = 1 + t$ on the interval $0 \leq t \leq 1$, we repeat the procedure to solve for $x(t)$ on the next interval, $1 \leq t \leq 2$. For $t \in [1,2]$, we calculate

$$x(t) = x(1) \;+\; \int_1^t x'(s)\,ds \;=\; x(1) \;+\; \int_1^t x(s-1)\,ds.$$

Making a substitution to replace $(s-1)$ with $s$, the latter integral becomes

$$x(t) = x(1) + \int_0^{t-1} 1 + s\,ds = 2 + (t-1) + \frac{(t-1)^2}{2}.$$

Therefore,

$$x(t) = \frac{t^2}{2} + \frac{3}{2} \qquad \text{if } 1 \le t \le 2.$$

Continuing in this fashion, you can calculate that for $2 \le t \le 3$,

$$x(t) = x(2) + \int_2^t x(s-1)\,ds = x(2) + \int_2^t \frac{(s-1)^2}{2} + \frac{3}{2}\,ds, \qquad \text{etc.}\ldots$$

This technique of extending the solution one interval at a time is called the *method of steps*. In this particular example, we can write the solution for $t \ge 0$ by introducing the notation $\lfloor t \rfloor$ to denote the largest integer which is smaller than or equal to $t$. By induction, one may argue that the solution of our DDE is given by

$$x(t) = \sum_{n=0}^{\lfloor t \rfloor + 1} \frac{[t-(n-1)]^n}{n!}$$

for $t \ge 0$.

---

## 6.2. Solving Constant-Coefficient Delay Differential Equations

Recall the variation of parameters formula (2.18), which says that the first-order, linear ODE $\mathbf{x}' = A\mathbf{x} + \mathbf{b}(t)$ has solution

$$\mathbf{x}(t) = e^{tA}\mathbf{x}_0 + e^{tA}\int_0^t e^{-sA}\mathbf{b}(s)\,ds.$$

This formula is easily adapted to yield the solution of the general first-order, constant-coefficient retarded DDE

$$\begin{aligned} x'(t) &= ax(t) + bx(t-\tau) + f(t) && \text{if } t > 0, \\ x(t) &= \phi(t) && \text{if } t \in [-\tau, 0]. \end{aligned} \tag{6.2}$$

Here $a$ and $b$ are constants, $\tau > 0$ is the time delay, and $f(t)$ is a continuous function. For $0 \le t \le \tau$, the solution to (6.2) is given by

$$\begin{aligned} x(t) &= e^{at}x(0) + e^{at}\int_0^t e^{-as}[bx(s-\tau) + f(s)]\,ds \\ &= e^{at}\phi(0) + e^{at}\int_0^t e^{-as}[b\phi(s-\tau) + f(s)]\,ds. \end{aligned}$$

Notice that this formula allows us to express $x(t)$ in terms of the *known* functions $\phi$ and $f$, at least on the interval $[0,\tau]$. The method of steps can then be used to extend the solution to the interval $[\tau, 2\tau]$ and so on.

We make two remarks regarding solutions of DDEs. First, notice that solutions need not be differentiable at the endpoints of consecutive intervals. In our introductory example of the method of steps, we had $x(t) = 1$ for $-1 \le t \le 0$, but $x(t) = 1 + t$ for $0 \le t \le 1$. Although the solution is continuous at $t = 0$, it is not differentiable there. Second, notice that the very construction of the above solution implies that constant-coefficient DDEs have a *unique* solution.

Provided that the functions $\phi$ and $f$ are continuous, we know that the system (6.2) has a unique solution for all $t \ge 0$, and the method of steps can be used to construct that solution. Under what conditions on $\phi$ will the solution be continuously *differentiable* for all time $t \ge 0$? First, suppose that $\phi$ is differentiable on $[-\tau, 0]$. Since $x(t) = \phi(t)$ on that interval, we also have $x'(t) = \phi'(t)$. If we want the solution to be differentiable at $t = 0$, then we must insist that left hand and right hand derivatives of $x(t)$ match at $t = 0$. Let us introduce the notation

$$x(0^-) = \lim_{t\to 0^-} x(t) \quad \text{and} \quad x(0^+) = \lim_{t\to 0^+} x(t).$$

Note that $x'(0^-) = \phi'(0^-)$ since $x'(t) = \phi'(t)$ when $t < 0$. On the interval $[0,\tau]$, the DDE tells us that

$$x'(t) = ax(t) + bx(t-\tau) + f(t) = ax(t) + b\phi(t-\tau) + f(t)$$

Taking the right hand limit, we find that

$$x'(0^+) = ax(0^+) + b\phi(-\tau^+) + f(0^+).$$

But since $f$ is continuous, it follows that $f(0^+) = f(0)$, and moreover $x(0) = \phi(0)$. Equating the left and right hand derivatives at $t = 0$, we obtain the condition

$$\phi'(0) = a\phi(0) + b\phi(-\tau) + f(0). \qquad (6.3)$$

**Theorem 6.2.1.** The solution of (6.2) is continuously differentiable for all $t \geq 0$ if and only if $\phi$ is differentiable at $t = 0$ and condition (6.3) is satisfied.

The condition (6.3) actually leads to some remarks about the "smoothness" of solutions of DDEs. For retarded DDEs with constant delays, any initial discontinuities are smoothed out as we advance forward in time. Neutral DDEs do not have this luxury—if the solution is discontinuous at $t = 0$, then the solution will also be discontinuous at the endpoint of each interval $[n\tau, (n+1)\tau]$ for each positive integer $n$. The method of steps can still be applied to neutral DDEs, but using a computer to numerically approximate solutions of such DDEs can be very challenging.

Linear, constant-coefficient DDEs can also be solved using Laplace transform methods. For details, see Bellman and Cooke [2].

---

## 6.3. Characteristic Equations

Recall from your introductory course in ODEs that for constant-coefficient problems such as $x'' + 6x' + 8x = 0$, we expect exponential solutions of the form $x = e^{\lambda t}$, where $\lambda$ is a constant. Substituting this exponential function into the ODE leads to the equation

$$\lambda^2 e^{\lambda t} + 6\lambda e^{\lambda t} + 8e^{\lambda t} = e^{\lambda t}(\lambda^2 + 6\lambda + 8) = 0.$$

Since the exponential factor could never be 0, it must be the case that $\lambda^2 + 6\lambda + 8 = 0$. This is called the *characteristic equation* for this ODE and, in this case, its roots are $\lambda = -2$ and $\lambda = -4$. Consequently, $e^{-2t}$ and $e^{-4t}$ are solutions of the ODE, as is any linear combination $C_1 e^{-2t} + C_2 e^{-4t}$.

Now consider the homogeneous, constant-coefficient DDE given by $x'(t) = ax(t) + bx(t - \tau)$, where $a$ and $b$ are constants and $\tau$ is a positive, constant time delay. As above, let us seek exponential solutions $x = e^{\lambda t}$. Then substitution yields

$$\lambda e^{\lambda t} = ae^{\lambda t} + be^{\lambda(t-\tau)} = ae^{\lambda t} + be^{\lambda t}e^{-\lambda\tau}.$$

Dividing by $e^{\lambda t}$, we obtain the characteristic equation

$$\lambda - a - be^{-\lambda\tau} = 0.$$

**Bad news:** This is a *transcendental* equation for $\lambda$, and it is impossible to algebraically solve for $\lambda$ in terms of the constants $a$, $b$, and $\tau$.

**More bad news:** Excluding the silly case $b = 0$ (in which the DDE would actually have been an ODE), the characteristic equation has *infinitely many complex-valued solutions.*

**Slightly better news:** If we could find all of the roots of the characteristic equation, then we could write the general solution of the DDE as an infinite sum of various exponential functions, provided that we can show that such series converge.

Relative to constant-coefficient ODEs, working with characteristic equations for DDEs is much more challenging. However, doing so can still be quite illuminating when we try to get a feel for the qualitative behavior of solutions. The remainder of this Chapter is dedicated to illustrating this via an example.

---

## 6.4. The Hutchinson-Wright Equation

In this section, we perform a qualitative analysis of the dynamical behavior of a nonlinear DDE. The following calculations appear in the text of Hassard et al. [4]

One of the most common models for population growth of a single species is given by the *logistic equation*

$$\frac{dP}{dt} = rP\left(1 - \frac{P}{K}\right),$$

where $r$ and $K$ are positive constants representing the population growth rate and maximum sustainable population, respectively. This ODE can be solved analytically via separation of variables. Qualitatively, it is easy to see that the ODE has two equilibria: $P = 0$ (corresponding to extinction) and $P = K$ (corresponding to the maximum population that the environment can sustain). The first of these equilibria is unstable, and the second is stable.

Now suppose that we incorporate a time delay into this equation, accounting for the lag between conception and birth. Hutchinson modified the logistic equation as

$$\frac{dP}{dt} = rP(t)\left(1 - \frac{P(t-\tau)}{K}\right), \tag{6.4}$$

where $\tau > 0$ is a positive, constant time delay. This equation is now known as the *Hutchinson-Wright equation*. It is a nonlinear DDE and cannot be solved analytically.

**Equilibria:** We begin our qualitative analysis of (6.4) by noting that there are still only two equilibria, $P = 0$ and $P = K$.

**Stability:** It is unclear how (or whether) we could extend our usual "eigenvalues of the Jacobian" test for stability from ODEs to DDEs like (6.4). The whole idea of stability analysis is to analyze how a system in equilibrium would respond to a small perturbation. First consider the equilibrium $P = 0$, and suppose that we seek a solution that is "close" to equilibrium: $P(t) = 0 + \epsilon y(t)$ where $\epsilon$ is a small, positive number. We linearize the Hutchinson-Wright equation by substituting that expression into the DDE:

$$\epsilon \frac{dy}{dt} = r\epsilon y(t)\left[1 - \frac{\epsilon y(t-\tau)}{K}\right].$$

Dividing by $\epsilon$ and expanding the right-hand side, we obtain the DDE

$$\frac{dy}{dt} = ry(t) - \frac{\epsilon r}{K} y(t) y(t-\tau).$$

Since $\epsilon$ is small by assumption, the linearization of the DDE is obtained by neglecting the latter term. The result is an ODE, namely $\frac{dy}{dt} = ry$, which is the linearization of the Hutchinson-Wright equation at the equilibrium $P = 0$. This ODE is easily solved, and we immediately see that the origin is an unstable equilibrium since $r$ is a positive constant.

The linearization at the other equilibrium $P = K$ is a bit more difficult. As before, we wish to examine a small perturbation from the equilibrium. Let $P(t) = K + \epsilon y(t)$, where $\epsilon$ is small and positive. Substituting that expression into equation (6.4),

$$\frac{d}{dt}(K+\epsilon y(t)) = r[K+\epsilon y(t)]\left[1-\frac{K+\epsilon y(t-\tau)}{K}\right].$$

By algebra,

$$\epsilon\frac{dy}{dt} = -r[K+\epsilon y(t)]\left[\frac{\epsilon y(t-\tau)}{K}\right].$$

Expanding the right-hand side and dividing by $\epsilon$ reveals that

$$\frac{dy}{dt} = -ry(t-\tau)-\epsilon\frac{r}{K}y(t-\tau)y(t).$$

Since $\epsilon$ is assumed to be small, we may obtain the linearized system by neglecting the rightmost term:

$$y'(t) = -ry(t-\tau).$$

This first-order constant-coefficient DDE is the linearization of (6.4) at the equilibrium $P = K$, and can actually be solved via the method of steps. As we shall see, it turns out that $P = K$ is a *stable* equilibrium if $r\tau < \pi/2$ and is *unstable* if $r\tau > \pi/2$. This raises the following

**Observation.** Recall that $P = K$ was always a *stable* equilibrium for the logistic ODE that we discussed at the beginning of this section. The inclusion of a time delay $\tau$ in the model has a *de-stabilizing* effect on the system, because $P = K$ becomes an unstable equilibrium if $\tau$ is appropriately large.

**Question:** What exactly happens when $r\tau = \pi/2$ that causes the equilibrium $P = K$ to suddenly lose stability? It must be some sort of bifurcation, but what type?

To answer this question and further analyze the stability of $P = K$, it will be convenient to re-scale the variables in the Hutchinson-Wright equation. First, we will re-scale the time variable in such a way that the time delay is equal to 1: let $s = t/\tau$ be our new time variable. Next, we will introduce another re-scaling which both (i) eliminates the need for the parameter $K$ and (ii) moves the equilibrium from $P = K$ to $P = 0$; namely,

$$x(s) = \frac{P(\tau s)-K}{K}.$$

With our new independent variable $s$ and dependent variable $x$, the DDE (6.4) becomes

$$\frac{d}{ds}x(s) = -(r\tau)x(s-1)[1+x(s)].$$

Finally, based upon the earlier claim that a bifurcation occurs when $r\tau = \pi/2$, it is convenient to "re-center" the bifurcation so that it will occur at 0. To accomplish this, introduce a new parameter $\mu = r\tau - \pi/2$ so that the DDE takes the form

$$x'(s) = -\left(\mu + \frac{\pi}{2}\right)x(s-1)[1+x(s)]. \tag{6.5}$$

Note that $x = 0$ is an equilibrium of this system, and it corresponds to the equilibrium $P = K$ of the original DDE (prior to re-scaling).

To see that a bifurcation really does occur when $\mu = 0$, causing the equilibrium $x = 0$ to lose stability, examine the linearization of (6.5) at that equilibrium:

$$x'(s) = -\left(\mu + \frac{\pi}{2}\right)x(s-1).$$

The characteristic equation is obtained by substituting $x = e^{\lambda s}$:

$$\lambda e^{\lambda s} = -\left(\mu + \frac{\pi}{2}\right)e^{\lambda(s-1)} = -\left(\mu + \frac{\pi}{2}\right)e^{\lambda s}e^{-\lambda}.$$

Dividing by $e^{\lambda s}$, the characteristic equation is

$$\lambda + \left(\mu + \frac{\pi}{2}\right)e^{-\lambda} = 0. \tag{6.6}$$

This is a transcendental equation, and has infinitely many complex-valued roots. To test whether any Andronov-Hopf bifurcations occur as the parameter $\mu$ is varied, we wish to determine conditions under which the characteristic equation has pure imaginary roots $\lambda = \pm i\omega$, where $\omega > 0$ is real. After all, we know that Andronov-Hopf bifurcations occur when a complex conjugate pair of eigenvalues switches from negative to positive real part, or vice-versa. Substituting $\lambda = i\omega$ into the characteristic equation reveals that

$$i\omega + \left(\mu + \frac{\pi}{2}\right)(\cos\omega - i\sin\omega) = 0, \tag{6.7}$$

where we have used Euler's identity to write $e^{-i\omega} = \cos\omega - i\sin\omega$. Equation (6.7) can only be satisfied if both the real and imaginary parts of the expressions on the left-hand side are both equal to 0. Therefore,

$$\left(\mu + \frac{\pi}{2}\right)\cos\omega \;=\; 0 \quad \text{and} \quad \omega - \left(\mu + \frac{\pi}{2}\right)\sin\omega \;=\; 0.$$

Since we originally claimed that a bifurcation occurs when $\mu = 0$, focus attention on that particular parameter choice. The two equations become

$$\frac{\pi}{2}\cos\omega \;=\; 0 \quad \text{and} \quad \omega = \frac{\pi}{2}\sin\omega.$$

Substituting the second of these into the first,

$$\frac{\pi}{2}\cos\left[\frac{\pi}{2}\sin\omega\right] \;=\; 0.$$

This requires $\sin\omega = \pm 1$, and we already assumed that $\omega > 0$. The only possible choices for $\omega$ are odd positive integer multiples of $\pi/2$. That is

$$\omega = (2n+1)\frac{\pi}{2} \qquad n = 0, 1, 2, \ldots.$$

In particular, we have found that for $\mu = 0$, the characteristic equation has pure imaginary roots $\pm\frac{\pi}{2}\mathrm{i}$. Note that due to the periodicity of the sine and cosine functions, we have

$$e^{(\pi/2)\mathrm{i}} \;=\; e^{(5\pi/2)\mathrm{i}} \;=\; e^{(9\pi/2)\mathrm{i}} \;=\; \cdots$$

and

$$e^{(-\pi/2)\mathrm{i}} \;=\; e^{(3\pi/2)\mathrm{i}} \;=\; e^{(7\pi/2)\mathrm{i}} \;=\; \cdots$$

Hence, we need only direct our attention towards the specific roots $\pm\frac{\pi}{2}$.

The fact that the characteristic equation has pure imaginary roots when $\mu = 0$ suggests that an Andronov-Hopf bifurcation may occur at that critical parameter value. According to the remarks about the Hopf Bifurcation Theorem at the end of the preceding Chapter, it remains only to verify that the transversality condition holds at $\mu = 0$; i.e., that the signs of the real parts of a pair of complex conjugate eigenvalue really does *change* from negative to positive (or vice-versa) when $\mu$ passes through 0. Indeed, we will show that the real part of $\frac{d\lambda}{d\mu}$ is non-zero when $\mu = 0$. To compute $\frac{d\lambda}{d\mu}$, we use implicit differentiation of the characteristic equation (6.6)

$$\lambda + \frac{\pi}{2}e^{-\lambda} + \mu e^{-\lambda} = 0.$$

Differentiating both sides with respect to $\mu$,

$$\frac{d\lambda}{d\mu} + \left(\frac{\pi}{2}\right)\left(-\frac{d\lambda}{d\mu}\right)e^{-\lambda} + e^{-\lambda} + \mu\left(-\frac{d\lambda}{d\mu}\right)e^{-\lambda} = 0.$$

Algebraically solving for $\frac{d\lambda}{d\mu}$ reveals that

$$\frac{d\lambda}{d\mu} = \frac{-e^{-\lambda}}{1-\left(\frac{\pi}{2}+\mu\right)e^{-\lambda}}.$$

Recall that when $\mu = 0$, we know that there is a pair of pure imaginary eigenvalue $\lambda = \pm\frac{\pi}{2}i$. Substituting these values into the last equation,

$$\left.\frac{d\lambda}{d\mu}\right|_{\mu=0} = -\frac{e^{-(\pi/2)i}}{1-\left(\frac{\pi}{2}\right)e^{-(\pi/2)i}}.$$

By Euler's identity, $e^{-(\pi/2)i} = \cos(-\pi/2) + i\sin(-\pi/2) = -i$. Thus,

$$\left.\frac{d\lambda}{d\mu}\right|_{\mu=0} = \frac{i}{1+\frac{\pi}{2}i}.$$

Multiplying and dividing the denominator of this last expression by its conjugate yields

$$\left.\frac{d\lambda}{d\mu}\right|_{\mu=0} = \frac{\frac{\pi}{2}+i}{1+\frac{1}{4}\pi^2}.$$

The real part of $\frac{d\lambda}{d\mu}$ is

$$\frac{\frac{\pi}{2}}{1+\frac{1}{4}\pi^2} > 0.$$

Consequently, as $\mu$ increases from negative to positive, the real parts of our pair of eigenvalues $\lambda$ is also changing from negative to positive. A supercritical Andronov-Hopf bifurcation occurs, and creates *stable, periodic solutions for* $\mu > 0$. According to our remarks at the end of the preceding chapter, for $\mu$ slightly larger than 0, the amplitude of these periodic solutions should be (roughly) proportional to $\sqrt{\mu}$. The period of the oscillations is estimated by dividing $2\pi$

by the imaginary part of the eigenvalues at the bifurcation point $\mu = 0$. In this case, the period would be approximately $2\pi/(\pi/2) = 4$. For our re-scaled Hutchinson-Wright equation, there are known approximations of the periodic solutions that are created when $\mu > 0$. Specifically, if $\mu$ is small and positive, then

$$x(s) = \sqrt{\frac{40}{3\pi - 2}}\sqrt{\mu}\cos\left(\frac{\pi}{2}s\right)$$

is an approximate solution, and the error in this approximation is roughly proportional to $\mu$.

**Observation.** The Hutchinson-Wright equation predicts that the equilibrium population $P = K$ will lose stability if the growth rate $r$ and/or the time delay $\tau$ are too large, resulting in oscillations in population. The fact that a *first-order* DDE can produce oscillations is noteworthy, because first-order autonomous ODEs cannot have [non-constant] periodic solutions.

To see why, consider the autonomous, first-order equation $\frac{dy}{dt} = f(y)$ and suppose that there is a periodic solution with period $\tau$. Then $y(t+\tau) = y(t)$ for all time $t$. Multiply both sides of the ODE by $\frac{dy}{dt}$ to get

$$\left(\frac{dy}{dt}\right)^2 = f(y)\frac{dy}{dt}.$$

Now integrate over one period:

$$\int_t^{t+\tau}\left(\frac{dy}{dt}\right)^2 dt = \int_t^{t+\tau} f(y)\frac{dy}{dt}\,dt = \int_{y(t)}^{y(t+\tau)} f(y)\,dy.$$

The rightmost integral is zero because $y(t) = y(t+\tau)$. On the other hand, the leftmost integral would have to be positive unless $y$ were a constant function. It follows that first-order autonomous ODEs cannot have periodic solutions. In this respect (and many others), DDEs can exhibit far richer dynamical behavior than ODEs.

---

## Exercises

1. Check that $y(t) = \sin t \ \ (t \geq 0)$ is a solution of the initial value problem

$$y'(t) = -y\left(t - \frac{\pi}{2}\right) \qquad (t \geq 0),$$

$$y(t) = \phi(t) = \sin t \qquad (t \le 0).$$

2. Let $\lfloor t \rfloor$ denote the largest integer which is less than or equal to $t$. Use induction to show that the solution of the initial value problem

$$\begin{aligned} y'(t) &= Ay(t-B) \qquad (t \ge 0), \\ y(t) &= \phi(t) = C \qquad (t \le 0) \end{aligned}$$

is given by

$$y(t) = C \sum_{n=0}^{\lfloor \frac{t}{B} \rfloor + 1} A^n \frac{[t-(n-1)B]^n}{n!}$$

for $t \ge 0$. Here, $A$, $B$, and $C$ are constants.

3. Show that the characteristic equation for the DDE $x'(t) = x(t-1)$ has exactly one real root.

# CHAPTER 7

## Introduction to Difference Equations

This Chapter concerns the dynamical behavior of systems in which time can be treated as a *discrete* quantity as opposed to a continuous one. For example, some mathematical models of the onset of cardiac arrhythmias are discrete, due to the discrete nature of the heartbeat. A more standard example involves population models for species without overlap between successive generations. If $P_n$ denotes the population of the $n$th generation, is there a functional relationship $P_{n+1} = f(P_n)$ which would allow us to predict the population of the next generation? Below, we will learn techniques for analytical and qualitative analysis of such systems. Good references for this material include the texts of Elaydi [3] and Strogatz [11].

### 7.1. Basic Notions

For the discrete systems that we shall consider, time $t$ is no longer a continuous variable as in the case of ODEs. Instead, we will typically use a non-negative integer $n$ to index our discrete time variable. If $x$ is a dependent variable, we will use subscripts $x_n$ instead of writing $x(n)$ to represent the value of $x$ at time $n$.

**Example 7.1.1.** An example of a discrete system is given by $x_{n+1} = x_n^2$. If we start with an initial condition $x_0 \in \mathbb{R}$, then we may recursively determine the values of all values in the sequence $\{x_n\}_{n=0}^{\infty}$. If $x_0 = 1/2$, then $x_1 = 1/4$, $x_2 = 1/16$, and so on.

**Definition 7.1.2.** A system of the form $x_n = f(x_{n-1}, x_{n-2}, \ldots x_{n-k})$ is an example of a *kth-order difference equation*. Such systems are sometimes called *k-dimensional mappings*.

The *solution* of a $k$th order difference equation is simply the sequence $\{x_n\}$. Notice that a $k$th-order difference equation recursively generates its iterates, and $x_n$ is affected by $x_{n-1}, x_{n-2}, \ldots x_{n-k}$. In particular, $k$ initial conditions would be required in order to start the iterative process of solving the equation.

**Example 7.1.3.** The famous *Fibonacci sequence* is generated recursively by the second-order difference equation $x_{n+1} = x_n + x_{n-1}$, with initial conditions $x_0 = 1$ and $x_1 = 1$. The next iterate is generated by summing the previous two iterates. Thus $x_2$ through $x_7$ are given by 2, 3, 5, 8, 13, and 21.

**Example 7.1.4.** Well-posedness is generally not a major issue for difference equations, because a $k$th-order difference equation with $k$ initial conditions will always generate a unique sequence of iterates, provided that $f$ is well-behaved. However, if there are restrictions on the domain of $f$, some difficulties can arise. Consider the first-order equation $x_{n+1} = \ln(x_n)$ with initial condition $x_0 = \mathrm{e}$. Then $x_1 = 1$, $x_2 = 0$, and $x_n$ is *undefined* for $n \geq 3$.

**Closed formulas.** Above we listed the first few iterates in the solution of $x_{n+1} = x_n^2$ with the initial condition $x_0 = 1/2$. Based upon the pattern exhibited by these iterates, we are led to conjecture that

$$x_n = \frac{1}{2^{2^n}},$$

which can, indeed, be proved by straightforward induction on $n$. This formula for $x_n$ is ideal in that it provides an exact formula for *all* of the iterates in the solution of the initial value problem. Such formulas are called *closed formulas* for the solution of the difference equation. Producing a closed formula for the solution of a difference equation is usually too much to hope for, but for constant-coefficient systems, closed formulas are readily available.

---

## 7.2. Linear, Constant-Coefficient Difference Equations

Recall from your course in basic differential equations that, for constant-coefficient ODEs, we typically seek exponential solutions of the form $\mathrm{e}^{\lambda t}$, where $\lambda$ is a constant that must be solved for. Substituting exponential functions into a linear, homogeneous, constant-coefficient ODE yields a polynomial equation involving $\lambda$: the so-called *characteristic equation*. By finding the roots of the characteristic polynomial, one may then build the general solution of the ODE by taking linear combinations of exponential functions. The same idea can be used to solve constant-coefficient difference equations such as the Fibonacci equation above. The only difference is that instead of seeking exponential solutions, one seeks power function solutions of the form $x_n = \lambda^n$. The "characteristic equation approach" to solving difference equations is developed in the exercises at the end of this chapter. Now, we shall introduce a more elegant method which recycles all of the techniques we learned when solving constant-coefficient systems of ODEs. First, we remark that $k$th-order difference equations, like $k$th order ODEs, can always be written as systems of $k$ *first-order* equations. The idea is to introduce new variables to represent any iterate other than the immediately preceding iterate. Let us clarify this vague remark via an example: consider the second-order difference equation $x_{n+1} = f(x_n, x_{n-1})$. If we replace $x_{n-1}$ with a new variable $y_n$, then note that $y_{n+1} = x_n$. Thus, the second-order equation we started with may be written as a system

$$x_{n+1} = f(x_n, y_n) \quad \text{and} \quad y_{n+1} = x_n.$$

The vector $(x_{n+1}, y_{n+1})$ of iterates at the $(n+1)$st time step is expressed in terms of the vector $(x_n, y_n)$ of iterates at the $n$th time step—a system of two first-order equations. Similarly, the third-order equation

$$x_{n+1} = f(x_n, x_{n-1}, x_{n-2})$$

can be written as a system by introducing $y_n = x_{n-1}$ and $z_n = x_{n-2}$. It follows that $y_{n+1} = x_n$ and $z_{n+1} = x_{n-1} = y_n$. Thus, our third-order equation can be written as a system of three first-order difference equations:

$$\begin{aligned} x_{n+1} &= f(x_n, y_n, z_n) \\ y_{n+1} &= x_n \\ z_{n+1} &= y_n. \end{aligned}$$

In this Section, we will learn to solve $k$th-order, homogeneous, constant-coefficient difference equations; i.e., equations of the form

$$x_{n+1} = a_0 x_n + a_1 x_{n-1} + \cdots + a_{k-1} x_{n-k+1}$$

where $a_0, a_1, \ldots a_{k-1}$ are constants and are independent of $n$. Some brief remarks about the special cases $k = 1$ and $k = 2$ will illuminate the solution process. If $k = 1$, then our constant-coefficient difference equation can be written as $x_{n+1} = ax_n$, where $a$ is a constant. Obtaining a closed formula for the solution is easy, because each iterate is a constant multiple of the previous iterate. The solution is $x_n = a^n x_0$. Now suppose that $k = 2$. Based upon our above remarks, any second-order constant-coefficient difference equation can be written as a system

$$x_{n+1} = a_{11}x_n + a_{12}y_n \qquad y_{n+1} = a_{21}x_n + a_{22}y_n.$$

As with systems of ODEs, it is convenient to introduce matrix/vector notation. Letting

$$A = \begin{bmatrix} a_{11} & a_{12} \\ a_{21} & a_{22} \end{bmatrix},$$

our system becomes

$$\begin{bmatrix} x_{n+1} \\ y_{n+1} \end{bmatrix} = A \begin{bmatrix} x_n \\ y_n \end{bmatrix}.$$

Notice that each vector of iterates is updated by multiplying the previous vector of iterates by the coefficient matrix $A$. Thus,

$$\begin{bmatrix} x_n \\ y_n \end{bmatrix} = A \begin{bmatrix} x_{n-1} \\ y_{n-1} \end{bmatrix} = A^2 \begin{bmatrix} x_{n-2} \\ y_{n-2} \end{bmatrix} = \cdots = A^n \begin{bmatrix} x_0 \\ y_0 \end{bmatrix},$$

which means that the closed form solution of such a system is obtained by first computing powers $A^n$ of the coefficient matrix, and then multiplying by the vector of initial conditions.

**Observation.** If we can find the appropriate canonical form for $A$, it is easy to compute powers of $A$. For example,

☞ If $A$ is diagonalizable, then we may write $A = PDP^{-1}$ where $P$ is an invertible matrix and $D$ is diagonal. In this case, $A^n = PD^nP^{-1}$.

☞ If $A$ is not diagonalizable but has real eigenvalues, then we may write $A = S + N$ where $S$ is diagonalizable, $N$ is nilpotent, and $SN = NS$. If we then write $S = PDP^{-1}$ where $D$ is diagonal, then powers of $S$ can be computed from $S^k = PD^kP^{-1}$. Since $S$ and $N$ commute, then we may use the binomial theorem to calculate

$$A^n = (S+N)^n = \sum_{k=0}^{n} \binom{n}{k} S^k N^{n-k} = \sum_{k=0}^{n} \binom{n}{k} PD^k P^{-1} N^{n-k}.$$

The fact that $N$ is nilpotent implies that many of the terms in this sum will likely vanish, because large powers of $N$ will always be 0. In principle, this sum could be computed to obtain a closed formula for the solution, although doing so would be tedious.

☞ If $A$ has complex conjugate eigenvalues, then finding the real canonical form for $A$ would facilitate computing large powers of $A$.

We now illustrate these ideas via examples.

**Example 7.2.1.** Solve the system

$$\begin{bmatrix} x_{n+1} \\ y_{n+1} \end{bmatrix} = \begin{bmatrix} 1 & -1 \\ 2 & 4 \end{bmatrix} \begin{bmatrix} x_n \\ y_n \end{bmatrix} \quad \text{where} \quad \begin{bmatrix} x_0 \\ y_0 \end{bmatrix} = \begin{bmatrix} 3 \\ -1 \end{bmatrix}.$$

**Solution:** Let $A$ denote the coefficient matrix. The characteristic equation is $\lambda^2 - 5\lambda + 6 = 0$ and, by factoring, we find that the eigenvalues are $\lambda = 2$ and $\lambda = 3$. You should check that

$$\begin{bmatrix} -1 \\ 1 \end{bmatrix} \quad \text{and} \quad \begin{bmatrix} 1 \\ -2 \end{bmatrix}$$

are eigenvectors for $\lambda = 2$ and $\lambda = 3$, respectively. Hence, we may write $A = PDP^{-1}$ where

$$P = \begin{bmatrix} -1 & 1 \\ 1 & -2 \end{bmatrix}, \quad D = \begin{bmatrix} 2 & 0 \\ 0 & 3 \end{bmatrix}, \quad P^{-1} = \begin{bmatrix} -2 & -1 \\ -1 & -1 \end{bmatrix}.$$

The powers of $A$ are given by

$$A^n = PD^nP^{-1} = P \begin{bmatrix} 2^n & 0 \\ 0 & 3^n \end{bmatrix} P^{-1} = \begin{bmatrix} 2^{n+1} - 3^n & 2^n - 3^n \\ -2^{n+1} + 2(3^n) & -2^n + 2(3^n) \end{bmatrix}.$$

Multiplying by the vector of initial conditions, the solution of the initial value problem is given by

$$\begin{bmatrix} x_n \\ y_n \end{bmatrix} = A^n \begin{bmatrix} 3 \\ -1 \end{bmatrix} = \begin{bmatrix} 3(2^{n+1}) - 3^{n+1} - 2^n + 3^n \\ -3(2^{n+1}) + 2(3^{n+1}) + 2^n - 2(3^n) \end{bmatrix}$$
$$= \begin{bmatrix} 5(2^n) - 2(3^n) \\ (-5)(2^n) + 4(3^n) \end{bmatrix}.$$

Notice that the solution involves *powers* of the eigenvalues. By contrast, we know that solutions of systems of ODEs involve exponential functions with eigenvalues appearing in the exponents.

**Example 7.2.2.** Solve the Fibonacci equation $x_{n+1} = x_n + x_{n-1}$ with $x_0 = x_1 = 1$.

*Solution:* First, we write the equation as a system by introducing $y_n = x_{n-1}$. The initial condition for $y$ would be $y_0 = x_{-1}$, which requires us to generate the iterate that precedes $x_0$, taking one step "backwards in time". From Fibonacci's equation, it must be the case that $x_{-1} + x_0 = x_1$, from which we conclude that $x_{-1} = 0$. The system now reads

$$\begin{bmatrix} x_{n+1} \\ y_{n+1} \end{bmatrix} = \begin{bmatrix} 1 & 1 \\ 1 & 0 \end{bmatrix} \begin{bmatrix} x_n \\ y_n \end{bmatrix} \quad \text{and} \quad \begin{bmatrix} x_0 \\ y_0 \end{bmatrix} = \begin{bmatrix} 1 \\ 0 \end{bmatrix}.$$

As before, we let $A$ denote the coefficient matrix. The characteristic equation is given by $\lambda^2 - \lambda - 1 = 0$, which has roots

$$\lambda_\pm = \frac{1 \pm \sqrt{5}}{2}.$$

To find an eigenvector for the positive eigenvalue $\lambda_+$, note that

$$A - \lambda_+ I = \begin{bmatrix} 1 - \lambda_+ & 1 \\ 1 & -\lambda_+ \end{bmatrix}$$

has reduced row-echelon form

$$\begin{bmatrix} 1 & -\lambda_+ \\ 0 & 0 \end{bmatrix}.$$

Eigenvectors $\mathbf{v}$ must satisfy $v_1 = \lambda_+ v_2$, so if we treat $v_2$ as a free variable we see that

$$\begin{bmatrix} \lambda_+ \\ 1 \end{bmatrix}$$

is an eigenvector for $\lambda_+$. Recycling the same computations, it follows that

$$\begin{bmatrix} \lambda_- \\ 1 \end{bmatrix}$$

is an eigenvector for $\lambda_-$. As usual, we write $A = PDP^{-1}$ where

$$P = \begin{bmatrix} \lambda_+ & \lambda_- \\ 1 & 1 \end{bmatrix}, \qquad D = \begin{bmatrix} \lambda_+ & 0 \\ 0 & \lambda_- \end{bmatrix},$$

and

$$P^{-1} = \frac{1}{\lambda_+ - \lambda_-} \begin{bmatrix} 1 & -\lambda_- \\ -1 & \lambda_+ \end{bmatrix}.$$

Multiplying $A^n$ by the vector of initial conditions,

$$\begin{aligned} \begin{bmatrix} x_n \\ y_n \end{bmatrix} = A^n \begin{bmatrix} x_0 \\ y_0 \end{bmatrix} &= PD^nP^{-1} \begin{bmatrix} 1 \\ 0 \end{bmatrix} \\ &= \frac{1}{\sqrt{5}} \begin{bmatrix} \lambda_+ & \lambda_- \\ 1 & 1 \end{bmatrix} \begin{bmatrix} \lambda_+^n & 0 \\ 0 & \lambda_-^n \end{bmatrix} \begin{bmatrix} 1 & -\lambda_- \\ -1 & \lambda_+ \end{bmatrix} \begin{bmatrix} 1 \\ 0 \end{bmatrix} \\ &= \frac{1}{\sqrt{5}} \begin{bmatrix} \lambda_+^{n+1} - \lambda_-^{n+1} \\ \lambda_+^n - \lambda_-^n \end{bmatrix}. \end{aligned}$$

Notice that the second row of the solution vector is identical to the first row, with $(n+1)$ replaced by $n$. This is not at all surprising if we recall that the variable $y_n$ was introduced to substitute for $x_{n-1}$. Only the first component of the solution vector is important for our purposes, as it provides a closed formula for $x_n$, the solution of the Fibonacci equation:

$$x_n = \frac{1}{\sqrt{5}} \left[ \left( \frac{1+\sqrt{5}}{2} \right)^{n+1} - \left( \frac{1-\sqrt{5}}{2} \right)^{n+1} \right].$$

You may find it mildly surprising that the closed formula involves powers of the *irrational* numbers $\lambda_+$ and $\lambda_-$ even though the Fibonacci sequence consists only of positive integer values.

**Example 7.2.3.** Solve the initial value problem

$$\begin{bmatrix} x_{n+1} \\ y_{n+1} \end{bmatrix} = \begin{bmatrix} 2 & 1 \\ 0 & 2 \end{bmatrix} \begin{bmatrix} x_n \\ y_n \end{bmatrix} \quad \text{and} \quad \begin{bmatrix} x_0 \\ y_0 \end{bmatrix} = \begin{bmatrix} 2 \\ 1 \end{bmatrix}.$$

**Solution:** The coefficient matrix $A$ has a repeated real eigenvalue $\lambda = 2$. By itself, this is not enough to conclude that $A$ is non-diagonalizable. However, if you try to compute eigenvectors for $A$, you will find that the eigenvalue only has geometric multiplicity 1. Hence, $A$ is not diagonalizable, and we must write $A = S + N$ where $S$ is diagonalizable and $N$ is nilpotent. This could be accomplished by finding a generalized eigenvector for $\lambda = 2$; however, in this case we may exploit the relatively simple form of $A$ in order to write $A = 2I + N$ where $I$ is the identity matrix and

$$N = \begin{bmatrix} 0 & 1 \\ 0 & 0 \end{bmatrix}$$

is nilpotent of order 2. Clearly $2I$ and $N$ commute since $I$ is the identity matrix. Now, using the binomial theorem, we may compute powers of $A$:

$$A^n = (2I + N)^n = \sum_{k=0}^{n} \binom{n}{k} (2I)^{n-k} N^k = \sum_{k=0}^{n} \binom{n}{k} 2^{n-k} N^k.$$

Since $N$ is nilpotent of order 2, we know that $N^2$ (and all higher powers of $N$) will be the zero matrix. Thus, only the first two terms of this summation survive. The relevant binomial coefficients are $\binom{n}{0} = 1$ and $\binom{n}{1} = n$, and the summation simplifies to $A^n = 2^n N^0 + n2^{n-1} N^1$. In other words,

$$A^n = \begin{bmatrix} 2^n & n2^{n-1} \\ 0 & 2^n \end{bmatrix}.$$

Finally, multiplying by the vector of initial conditions yields the closed formula

$$\begin{bmatrix} x_n \\ y_n \end{bmatrix} = \begin{bmatrix} 2^{n+1} + n2^{n-1} \\ 2^n \end{bmatrix}.$$

**Example 7.2.4.** Solve the initial value problem

$$x_{n+1} = 3x_n - 18y_n \qquad x_0 = 1$$
$$y_{n+1} = 2x_n - 9y_n \qquad y_0 = 1.$$

**Solution:** The characteristic equation associated with the coefficient matrix

$$A = \begin{bmatrix} 3 & -18 \\ 2 & -9 \end{bmatrix}$$

is $\lambda^2 + 6\lambda + 9 = 0$, and we see that $\lambda = -3$ is a repeated, real eigenvalue. Eigenvectors $\mathbf{v}$ satisfy $(A - \lambda I)\mathbf{v} = 0$. The reduced row-echelon form of

$$A - \lambda I = A + 3I = \begin{bmatrix} 6 & -18 \\ 2 & -6 \end{bmatrix}$$

is

$$\begin{bmatrix} 1 & -3 \\ 0 & 0 \end{bmatrix}.$$

It follows that

$$\mathbf{v} = \begin{bmatrix} 3 \\ 1 \end{bmatrix}$$

is an eigenvector, and the eigenspace for $\lambda = -3$ is only one-dimensional. We conclude that $A$ is non-diagonalizable, which suggests that we seek generalized eigenvectors by solving $(A - \lambda I)^2\mathbf{w} = 0$. However, because $(A + 3I)^2 = 0$, the matrix

$$N = A + 3I = \begin{bmatrix} 6 & -18 \\ 2 & -6 \end{bmatrix}$$

is nilpotent of order 2. This observation provides a useful decomposition for the matrix $A$, namely $A = -3I + N$. Since the matrix $-3I$ clearly commutes with $N$, we are allowed to use the binomial theorem to calculate powers of $A$:

$$A^n = (-3I + N)^n = \sum_{k=0}^{n} \binom{n}{k} (-3I)^{n-k} N^k.$$

Since $N$ is nilpotent of order 2, only the first two terms in this sum survive:

$$A^n = \binom{n}{0}(-3I)^n N^0 + \binom{n}{1}(-3I)^{n-1}N.$$

Here, the binomial coefficients are $\binom{n}{0} = 1$ and $\binom{n}{1} = n$, and the matrix $N^0$ is simply the identity matrix. Therefore,

$$A^n = \begin{bmatrix} (-3)^n + 6n(-3)^{n-1} & -18n(-3)^{n-1} \\ 2n(-3)^{n-1} & (-3)^n - 6n(-3)^{n-1} \end{bmatrix},$$

and the solution to the initial value problem is

$$\begin{bmatrix} x_n \\ y_n \end{bmatrix} = A^n \begin{bmatrix} x_0 \\ y_0 \end{bmatrix} = \begin{bmatrix} (-3)^n + 6n(-3)^{n-1} & -18n(-3)^{n-1} \\ 2n(-3)^{n-1} & (-3)^n - 6n(-3)^{n-1} \end{bmatrix} \begin{bmatrix} 1 \\ 1 \end{bmatrix}$$
$$= \begin{bmatrix} (-3)^n - 12n(-3)^{n-1} \\ (-3)^n - 4n(-3)^{n-1} \end{bmatrix}.$$

If the coefficient matrix associated with a constant-coefficient system of difference equations has complex conjugate eigenvalues, writing the closed-form solution can be messy. For example, consider the initial value problem

$$\begin{aligned} x_{n+1} &= -y_n & x_0 &= 1 \\ y_{n+1} &= x_n & y_0 &= 1. \end{aligned}$$

The coefficient matrix

$$A = \begin{bmatrix} 0 & -1 \\ 1 & 0 \end{bmatrix}$$

is in real canonical form. If this had been a system of ODEs, we would expect periodic solutions involving $\sin t$ and $\cos t$. In some sense, the behavior of this discrete system is similar, as we can see by computing the first few iterates:

$$\begin{bmatrix} x_1 \\ y_1 \end{bmatrix} = \begin{bmatrix} -1 \\ 1 \end{bmatrix} \qquad \begin{bmatrix} x_2 \\ y_2 \end{bmatrix} = \begin{bmatrix} -1 \\ -1 \end{bmatrix}$$
$$\begin{bmatrix} x_3 \\ y_3 \end{bmatrix} = \begin{bmatrix} 1 \\ -1 \end{bmatrix} \qquad \begin{bmatrix} x_4 \\ y_4 \end{bmatrix} = \begin{bmatrix} 1 \\ 1 \end{bmatrix}.$$

Evidently, this pattern will repeat, and the iterates will cycle through the four different vectors shown here. Writing a closed formula for the solution is straightforward, but doing so is a bit awkward because there are four cases to consider.

In general, if the coefficient matrix $A$ has a pair of complex conjugate eigenvalues $\alpha \pm \beta i$, it is useful to transform $A$ into real canonical form:

$$A = P \begin{bmatrix} \alpha & -\beta \\ \beta & \alpha \end{bmatrix} P^{-1},$$

where $P$ is a suitably-chosen invertible matrix. Letting $M$ denote the real canonical form for $A$, powers of $A$ can be computed using $A^n = PM^nP^{-1}$. To calculate powers of $M$, it is helpful to split $M$ as

$$M = \begin{bmatrix} \alpha & -\beta \\ \beta & \alpha \end{bmatrix} = \alpha I + B,$$

where

$$B = \begin{bmatrix} 0 & -\beta \\ \beta & 0 \end{bmatrix}.$$

Certainly $\alpha I$ commutes with $B$, which allows us to use the binomial theorem when computing $M^n = (\alpha I + B)^n$. Moreover, since $B^2 = -\beta^2 I$, we may calculate that $B^3 = -\beta^2 B$ and $B^4 = \beta^4 I$. Since $B^4$ is a constant multiple of the identity matrix, we might expect the same sorts of cyclic oscillations that we saw in the example above.

Finally, we remark that solving inhomogeneous constant-coefficient difference equations is straightforward, but we shall not discuss the techniques here. Indeed, for the inhomogeneous equation

$$x_{n+1} = a_1 x_n + a_2 x_{n-1} + \cdots + a_k x_{n-k+1} + g(n),$$

it is possible to state an analogue of the variation of parameters formula (2.18) for ODEs. Due to the discrete nature of difference equations, the solution contains a *summation* involving $g(n)$, as opposed to the integral in (2.18).

## 7.3. First-Order Nonlinear Equations and Stability

It is almost never possible to present a closed formula for the solution of a nonlinear difference equation. As with nonlinear ODEs, we typically settle for a qualitative understanding of how solutions behave. Our development of the qualitative analysis of nonlinear difference equations perfectly parallels the methodology we introduced for nonlinear ODEs. We will begin by analyzing constant solutions, which are analogous to equilibria for ODEs. Afterwards, we will study more exotic dynamical behavior, including periodic solutions, bifurcations, and chaos. We restrict our initial discussion to *first-order* nonlinear difference equations, later generalizing our results to higher-order systems. The material in this section is based heavily on Chapter 10 of Strogatz [11].

**Example 7.3.1.** The behavior of the iterates of the nonlinear equation $x_{n+1} = x_n^2$ depends greatly upon our choice of initial condition $x_0$. For example, if $x_0 = 2$, then $x_n \to \infty$ as $n \to \infty$. On the other hand, if $x_0 = 1/2$, then the sequence of iterates $x_n$ converges rapidly to 0. Notice also that if $x_0 = 1$, then $x_n = 1$ for all $n \geq 0$. This constant solution of the difference equation is analogous to an equilibrium for an ODE, and such solutions have a special name.

**Definition 7.3.2.** A *fixed point* of the first-order difference equation $x_{n+1} = f(x_n)$ is any number $x^*$ such that $x^* = f(x^*)$.

Notice that, by definition, if we start out by using a fixed point as our initial condition, then we will remain stuck at that fixed point for all future iterates.

**Example 7.3.3.** To find all fixed points $x$ of the difference equation $x_{n+1} = 2x_n - 2x_n^2$, we should solve the equation $x = 2x - 2x^2$. By algebra, we have $2x(x - 1/2) = 0$, a quadratic equation with two roots: $x = 0$ and $x = 1/2$. These are the two fixed points of this nonlinear difference equation.

As with equilibria of ODEs, fixed points of difference equations can be stable or unstable. Roughly speaking, a fixed point $x^*$ is locally asymptotically stable if whenever we start from an initial condition $x_0$ that is appropriately "close" to $x^*$, the sequence $\{x_n\}$ of iterates converges to $x^*$ as $n \to \infty$. Fixed points can also be repellers—i.e., the gap between $x^*$ and $x_n$ may grow as $n$ increases, no matter how close the initial condition $x_0$ is to $x^*$.

**Example 7.3.4.** Fixed points of the mapping $x_{n+1} = x_n^2$ satisfy the equation $x = x^2$. This quadratic equation has two solutions, $x = 0$ and $x = 1$. The fixed point $x = 0$ is locally asymptotically stable, because if we start from any initial condition $x_0$ that is "close" to 0, then the sequence of iterates will converge to 0. Specifically, if $x_0 \in (-1, 1)$, then $x_n \to 0$ as $n \to \infty$. In contrast, the fixed point $x = 1$ is unstable, because if we start from any initial condition other than $x_0 = 1$, the iterates will be repelled from 1.

We now devise a test to determine whether a fixed point of a difference equation is locally stable or not. As with differential equations, the stability test involves the use of Jacobian matrices. However, the conditions that eigenvalues must satisfy will be different—stability will depend upon more than just the real part of the eigenvalues.

First, consider the first-order difference equation $x_{n+1} = f(x_n)$, and assume that the function $f : \mathbb{R} \to \mathbb{R}$ is continuously differentiable. Suppose that $x^*$ is an isolated fixed point of our equation. To determine whether $x^*$ is an attractor or repeller, we need to investigate how the iterates of the mapping would behave if we start from an initial condition that is "near" $x^*$. Suppose that our initial condition is $x_0 = x^* + \epsilon_0$, where $|\epsilon_0|$ is a very small number. We will estimate the gap $\epsilon_1$ between the value of $x_1$ (the first iterate) and the fixed point $x^*$ in order to see whether $x_1$ is closer to the fixed point than $x_0$ was. More exactly, suppose $x_1 = x^* + \epsilon_1$. We also know that $x_1 = f(x_0)$ and, since $x_0 = x^* + \epsilon_0$, we may use the tangent line approximation at $x^*$ to estimate

$$x_1 = f(x^* + \epsilon_0) \approx f(x^*) + \epsilon_0 f'(x^*).$$

Recalling that $x_1 = x^* + \epsilon_1$, we equate our two expressions for $x_1$ to obtain

$$x^* + \epsilon_1 \approx f(x^*) + \epsilon_0 f'(x^*).$$

The fact that $x^*$ is a fixed point implies that $f(x^*) = x^*$, and therefore

$$x^* + \epsilon_1 \approx x^* + \epsilon_0 f'(x^*).$$

Subtracting $x^*$ from both sides and taking absolute values, we find that

$$\left| \frac{\epsilon_1}{\epsilon_0} \right| \approx |f'(x^*)|.$$

Interpreting this approximation in words will give rise to our first stability criterion. Recall that $\epsilon_0$ and $\epsilon_1$ measure the gaps $x_0 - x^*$ and $x_1 - x^*$, respectively. Thus, the left hand side of the above approximation measures the ratio of these gaps. If the fixed point is an attractor, we would need this ratio to be smaller than 1 in magnitude, implying that the gaps between iterates and the fixed point $x^*$ will shrink as we generate more iterates. Conversely, if the ratio exceeds 1 in magnitude, then the gaps between the iterates $x_n$ and the fixed point $x^*$ will grow as $n$ increases. Thus, we have provided a heuristic proof of

**Theorem 7.3.5.** Suppose $x^*$ is an isolated fixed point of the first-order difference equation $x_{n+1} = f(x_n)$, where $f$ is continuously differentiable. Then $x^*$ is locally asymptotically stable (attracting) if $|f'(x^*)| < 1$ and is unstable (repelling) if $|f'(x^*)| > 1$. If $|f'(x^*)| = 1$, this test is inconclusive.

**Warning:** Although unstable fixed points are *locally* repelling, we must exercise caution when drawing conclusions about long-term behavior of iterates (particularly if $f$ is not as smooth as required by the conditions of Theorem 7.3.5). If $f$ is merely piecewise continuous, it is possible for $x_{n+1} = f(x_n)$ to have an unstable fixed point which is *globally attracting* (see exercises).

**Example 7.3.6.** Consider the difference equation $x_{n+1} = \cos(x_n)$. We claim that this equation has exactly one fixed point. Fixed points satisfy the transcendental equation $x = \cos x$, which is impossible to solve algebraically. Equivalently, fixed points are roots of the function $g(x) = x - \cos x$. Notice that $g(x)$ is a continuous function and that $g(0) = -1$ whereas $g(\pi/2) = \pi/2$. Since $g$ is continuous and its values change from negative to positive between $x = 0$ and $x = \pi/2$, the intermediate value theorem from calculus guarantees that $g$ has *at least* one root in the interval $(0, \pi/2)$. Next, we must show that $g$ has exactly one real root. To see why, observe that $g'(x) = 1 + \sin x$ is non-negative because $\sin x \geq -1$. Thus, the function $g(x)$ is non-decreasing (it is actually a one-to-one function). It follows that the equation $g(x) = 0$ can have *at most* one root. Letting $x^*$ denote this root, we conclude that $x^*$ is the only fixed point of this difference equation.

Again, it is impossible to find the value of $x^*$ algebraically. However, the above remarks indicate that $0 < x^* < \pi/2$, and this is actually enough information for us to use Theorem 7.3.5 to test the local stability of $x^*$. Our difference equation has the form $x_{n+1} = f(x_n)$ where $f(x) = \cos x$. According to the Theorem, we should check the magnitude of $f'(x^*)$. In this case, $f'(x) = -\sin x$, from which we calculate

$$|f'(x^*)| = |-\sin(x^*)|.$$

Since $0 < x^* < \pi/2$, we have $|-\sin x^*| < 1$. Therefore, Theorem 7.3.5 guarantees that our fixed point $x^*$ is locally asymptotically *stable*. In this example, we never needed to know the exact value of $x^*$ in order to test its stability.

Remarkably, the fixed point $x^*$ of this difference equation is actually *globally* asymptotically stable. That is, for *any* choice of initial condition $x_0$, the sequence of iterates of this mapping will converge to the fixed point! You should test this out by picking any number you like and then using a calculator or computer to repeatedly take the cosine of the number you chose. Make sure your calculator is measuring angles in radians, not degrees. You will find that the value of the fixed point is $x^* = 0.739085...$, which is the only solution of the equation $x = \cos x$.

**Example 7.3.7.** Recall that the difference equation $x_{n+1} = x_n^2$ has two fixed points, 0 and 1. In this case, $f(x) = x^2$, so $f'(x) = 2x$. Since $f'(0) = 0 < 1$, we see that 0 is a locally asymptotically stable fixed point, and since $f'(1) = 2 > 1$, we conclude that 1 is an unstable fixed point.

**Example 7.3.8.** By algebra, you can check that the only fixed points of $x_{n+1} = 3x_n(1 - x_n)$ are 0 and 2/3. Here, $f(x) = 3x - 3x^2$, so $f'(x) = 3 - 6x$. Since $|f'(0)| = 3 > 1$, we see that 0 is an unstable fixed point. On the other hand, since $|f'(2/3)| = 1$, we cannot use Theorem 7.3.5 to draw any conclusions regarding the stability of that fixed point.

**Definition 7.3.9.** A fixed point $x^*$ of the equation $x_{n+1} = f(x_n)$ is called *hyperbolic* if $|f'(x^*)| \neq 1$. Otherwise, the fixed point is called *non-hyperbolic*.

Our local stability Theorem 7.3.5 can only be used to classify stability of hyperbolic fixed points. In order to determine whether a non-hyperbolic fixed point is stable, we need a finer approach. After all, the derivation of Theorem 7.3.5 was based upon linear approximation of the function $f$ in the vicinity of a fixed point $x^*$. If $f$ has a continuous third derivative, then we can obtain the following theorems regarding stability of non-hyperbolic equilibria:

**Theorem 7.3.10.** Suppose that $x^*$ is an isolated non-hyperbolic equilibrium point of $x_{n+1} = f(x_n)$ and, more specifically, that $f'(x^*) = 1$. Then $x^*$ is *unstable* if $f''(x^*) \neq 0$. If $f''(x^*) = 0$ and $f'''(x^*) > 0$ then, again, $x^*$ is *unstable*. Finally, if $f''(x^*) = 0$ and $f'''(x^*) < 0$, then $x^*$ is *locally asymptotically stable*.

In order to state the corresponding theorem for the case $f'(x^*) = -1$, it is helpful to introduce the notion of the *Schwarzian derivative.*

**Definition 7.3.11.** The *Schwarzian derivative* of a function $f$ is defined as

$$Sf(x) = \frac{f'''(x)}{f'(x)} - \frac{3}{2}\left[\frac{f''(x)}{f'(x)}\right]^2.$$

**Theorem 7.3.12.** Suppose that $x^*$ is an isolated non-hyperbolic equilibrium point of $x_{n+1} = f(x_n)$ and that $f'(x^*) = -1$. Then $x^*$ is *unstable* if $Sf(x^*) > 0$ and is *locally asymptotically stable* if $Sf(x^*) < 0$.

**Example 7.3.13.** In our previous example, we found that $x^* = 2/3$ is a non-hyperbolic fixed point of the difference equation $x_{n+1} = 3x_n(1 - x_n)$. Since $f(x) = 3x - 3x^2$, we compute that the first three derivatives of $f$ are

$$f'(x) = 3 - 6x, \qquad f''(x) = -6 \quad \text{and} \quad f'''(x) = 0.$$

Since $f'(x^*) = -1$, we may use Theorem 7.3.12. The expression for the Schwarzian derivative reduces to $Sf(x^*) = -f'''(x^*) - \frac{3}{2}f''(x^*)^2$, and we find that $Sf(x^*) = -54 < 0$. Theorem 7.3.12 tells us that the non-hyperbolic fixed point $x^* = 2/3$ is *locally asymptotically stable.*

In the preceding example, we were still able to classify the stability of the fixed point even though Theorem 7.3.5 was inconclusive. Usually, Theorems 7.3.5, 7.3.10 and 7.3.12 are enough to classify stability of fixed points, although there are cases in which all three theorems are inconclusive.

---

## 7.4. Systems of Nonlinear Equations and Stability

Fixed points for higher-order difference equations can be analyzed via techniques that are very similar to the ones we developed when considering equilibria of nonlinear systems of ODEs. To motivate our definition for fixed points of higher-order equations, let us study a specific second-order equation

$$x_{n+1} = [3 + x_n - x_{n-1}]x_n(1 - x_n).$$

As usual, we write this as a system of two first-order equations by introducing a new variable $y_n = x_{n-1}$. The resulting system is

$$x_{n+1} = [3 + x_n - y_n]x_n(1 - x_n) \quad \text{and} \quad y_{n+1} = x_n.$$

A fixed point of such a system should correspond to a *constant* solution of the difference equation. In this example, a fixed point would be any constant vector $(x^*, y^*)$ such that whenever $(x_n, y_n) = (x^*, y^*)$, we have $(x_{n+1}, y_{n+1}) = (x^*, y^*)$ as well. More generally,

**Definition 7.4.1.** Suppose $f : \mathbb{R}^m \to \mathbb{R}^m$ and that $\mathbf{x}_{n+1} = f(\mathbf{x}_n)$ is a system of $m$ first-order difference equations. A *fixed point* of the system is any vector $\mathbf{x}^* \in \mathbb{R}^m$ such that $\mathbf{x}^* = f(\mathbf{x}^*)$.

In the above example, we may solve for the fixed points by setting $x_{n+1} = x_n = x^*$ and $y_{n+1} = y_n = y^*$. The equation $y_{n+1} = x_n$ tells us that $x^* = y^*$, and substituting this into the equation for $x_{n+1}$ yields $x^* = 3x^*(1 - x^*)$. Solving this quadratic equation yields $x^* = 0$ and $x^* = 2/3$ as the two solutions. Consequently, the system has two fixed points, $(x^*, y^*) = (0, 0)$ and $(x^*, y^*) = (2/3, 2/3)$.

**Example 7.4.2.** Consider the nonlinear system

$$x_{n+1} = 2x_n - x_n y_n \quad \text{and} \quad y_{n+1} = x_n y_n.$$

Fixed points $(x, y)$ must simultaneously satisfy

$$x = 2x - xy \quad \text{and} \quad y = xy.$$

By algebra, these two equations are $x(1 - y) = 0$ and $y(x - 1) = 0$. If $x = 0$ in the first of these, then the second would force $y = 0$ as well. Similarly, if $y = 1$ in the first equation, then this would force $x = 1$ in the second equation. We have obtained precisely two equilibria, $(x, y) = (0, 0)$ and $(x, y) = (1, 1)$.

**Stability.** If $\mathbf{x}^*$ is a fixed point of a system $\mathbf{x}_{n+1} = f(\mathbf{x}_n)$ of $m$ first-order difference equations, we need a way of testing the stability of $\mathbf{x}^*$. We mimic exactly what we did when linearizing systems of ODEs at an equilibrium point. Recall that the linear approximation of a function $f : \mathbb{R}^m \to \mathbb{R}^m$ at a point $\mathbf{x}^* \in \mathbb{R}^n$ is given by

$$f(\mathbf{x}) \approx f(\mathbf{x}^*) + Jf(\mathbf{x}^*)(\mathbf{x} - \mathbf{x}^*).$$

Using this approximation in the above difference equation,

$$\mathbf{x}_{n+1} \approx f(\mathbf{x}^*) + Jf(\mathbf{x}^*)(\mathbf{x}_n - \mathbf{x}^*) \;=\; \mathbf{x}^* + Jf(\mathbf{x}^*)(\mathbf{x}_n - \mathbf{x}^*),$$

where we have used the fact that $f(\mathbf{x}^*) = \mathbf{x}^*$ since $\mathbf{x}^*$ is a fixed point. To measure the gap between iterates of the difference equation and the fixed point, define $\mathbf{y}_n = \mathbf{x}_n - \mathbf{x}^*$. Then in the vicinity of $\mathbf{x}^*$, the vectors $\mathbf{y}_n$ approximately satisfy

$$\mathbf{y}_{n+1} \;=\; Jf(\mathbf{x}^*)\mathbf{y}_n,$$

a linear, constant-coefficient system. The exact solution of this linearized system is given by

$$\mathbf{y}_n \;=\; [Jf(\mathbf{x}^*)]^n \mathbf{y}_0,$$

where the vector $\mathbf{y}_0$ is a measure of our initial gap $\mathbf{x}_0 - \mathbf{x}^*$. In order to ensure that $\mathbf{x}^*$ is locally asymptotically *stable*, we need a criterion which guarantees that the gap $\mathbf{y}_n = \mathbf{x}_n - \mathbf{x}^*$ will approach 0 as we let $n \to \infty$. By inspecting the solution of the linearized system, we must insist that the entries of the powers of the Jacobian matrix $Jf(\mathbf{x}^*)$ converge to 0 as $n \to \infty$. If this Jacobian matrix is diagonalizable, we could write $Jf(\mathbf{x}^*) = PDP^{-1}$ where $D$ is a diagonal matrix containing the eigenvalues, and $P$ is a *constant*, invertible matrix. The fact that powers of $D$ contain powers of the eigenvalues suggests a stability criterion: each eigenvalue should have "size" smaller than 1 so that their powers will converge to 0 as $n \to \infty$. For real numbers, the absolute value function gives us a notion of "size". We now generalize the concept of absolute value to include complex numbers.

**Definition 7.4.3.** If $z = x + iy$ is a complex number, then the *modulus* of $z$ is defined as $|z| = \sqrt{x^2 + y^2}$.

Notice that if $z$ is real (i.e., $y = 0$), then $|z| = \sqrt{x^2} = |x|$ and the modulus of $z$ is given by the usual absolute value function.

Now that we have a notion of "size" for complex numbers, all of our above remarks constitute a heuristic proof of the following stability theorem.

**Theorem 7.4.4.** Let $f : \mathbb{R}^m \to \mathbb{R}^m$ and suppose that all entries in the Jacobian matrix $Jf(\mathbf{x})$ are continuous. An isolated fixed point $\mathbf{x}^*$ of a system $\mathbf{x}_{n+1} = f(\mathbf{x}_n)$ of $m$ first-order difference equations is

☞ *locally asymptotically stable* if ALL eigenvalues of $Jf(\mathbf{x}^*)$ have modulus less than 1;

☞ *unstable* if ANY eigenvalue of $Jf(\mathbf{x}^*)$ has modulus greater than 1.

**Remark.** This stability criterion is different (although similar in spirit) from the one we developed when analyzing equilibria of ODEs. An equilibrium $\mathbf{x}^*$ of a system of ODEs is stable if all eigenvalues of $Jf(\mathbf{x}^*)$ have negative real part. By contrast, a fixed point $\mathbf{x}^*$ for a system of difference equations is stable if all eigenvalues of $Jf(\mathbf{x}^*)$ have modulus less than 1.

**Example 7.4.5.** Find all fixed points of the system

$$\begin{aligned} x_{n+1} &= 3x_n - x_n y_n \\ y_{n+1} &= -2y_n + x_n y_n \end{aligned}$$

and determine whether they are locally stable or unstable.

*Solution:* Fixed points $(x, y)$ of this system satisfy

$$x = 3x - xy \quad \text{and} \quad y = -2y + xy.$$

Equivalently, $x(2 - y) = 0$ and $y(x - 3) = 0$. If $x = 0$ in the first of these equations, then we are forced to set $y = 0$ in order to satisfy the second equation. Likewise, if $y = 2$ in the first equation, then we must have $x = 3$ in the second equation. Thus, there are two fixed points, $(0, 0)$ and $(3, 2)$. For this system, the function $f : \mathbb{R}^2 \to \mathbb{R}^2$ is defined by

$$f(x, y) = \begin{bmatrix} 3x - xy \\ -2y + xy \end{bmatrix},$$

and its Jacobian matrix is

$$Jf(x, y) = \begin{bmatrix} 3 - y & -x \\ y & x - 2 \end{bmatrix}.$$

At the fixed point $(0,0)$, the Jacobian matrix $Jf(0,0)$ is a diagonal matrix, namely $\text{diag}\{3,-2\}$, and the eigenvalues are the diagonal entries. The eigenvalues are real, so their moduli are simply their absolute values. Since there is an eigenvalue with modulus larger than 1 (in fact both have modulus larger than 1), Theorem 7.4.4 tells us that $(0,0)$ is an unstable equilibrium.

At the other fixed point $(3,2)$, the Jacobian matrix is

$$Jf(x,y) = \begin{bmatrix} 1 & -3 \\ 2 & 1 \end{bmatrix},$$

and the associated characteristic equation is $\lambda^2 - 2\lambda + 7 = 0$. From the quadratic formula, the eigenvalues are $\lambda_\pm = 1 \pm \sqrt{6}i$. These two eigenvalues have modulus

$$|\lambda_\pm| = \sqrt{1^2 + (\sqrt{6})^2} = \sqrt{7} > 1.$$

Again, we conclude that the fixed point is unstable because there are eigenvalues with modulus exceeding 1.

For $2 \times 2$ matrices, there is actually a quick way to determine whether all eigenvalues have modulus less than 1 without actually computing the eigenvalues. The following Lemma can be used to test the stability of fixed points for systems of two first-order difference equations.

**Lemma 7.4.6.** Suppose that $A$ is a $2 \times 2$ matrix. Then the eigenvalues of $A$ have modulus less than 1 if and only if

$$\text{tr}A + \det A > -1, \qquad \text{tr}A - \det A < 1, \quad \text{and} \quad \det A < 1.$$

The proof of Lemma 7.4.6 is relatively straightforward after you recall how to express the eigenvalues of $A$ in terms of the trace and determinant. This Lemma is a special case of the *Jury Stability Test*, which provides an iterative procedure by which one may determine whether all eigenvalues of an $n \times n$ matrix have modulus less than 1.

**Example 7.4.7.** The matrix

$$A = \begin{bmatrix} 1 & -1 \\ \frac{1}{4} & -1 \end{bmatrix}$$

has trace 0 and determinant $-3/4$. All three criteria of Lemma 7.4.6 are satisfied, which means that the eigenvalues of $A$ have modulus less than 1.

Lemma 7.4.6 tends to be especially useful for systems of difference equations which contain unspecified parameters, because it can provide inequalities that the parameters must satisfy so that equilibria will be stable. Such systems are the subject of the following section.

---

## 7.5. Period-Doubling Bifurcations

In a previous chapter, we discussed how the dynamical behavior of systems of ODEs can exhibit sudden, dramatic changes as a parameter $\mu$ is varied. The same can be said for systems of difference equations involving a parameter. Undoubtedly the most famous example of this sort is the *discrete logistic equation*, a first-order difference equation involving a single parameter $\mu$. This difference equation is given by

$$x_{n+1} = \mu x_n(1-x_n), \tag{7.1}$$

where $\mu$ is a real number. From now on, we will insist that $0 \le \mu \le 4$, because this restriction on $\mu$ endows Equation (7.1) with a convenient property. Namely, if the initial condition $x_0$ is in the interval $[0,1]$, then all subsequent iterates will remain in the interval $[0,1]$. To see why, simply inspect the graph of the function $f(x) = \mu x(1-x)$. If $\mu > 0$, then the graph of $f(x)$ is a parabola with roots at $x = 0$ and $x = 1$. The maximum value of $f$ occurs at $x = 1/2$, and the value of $f(x)$ at its maximum is $\mu/4$. Thus, if $0 \le \mu \le 4$ and $0 \le x \le 1$, then $0 \le f(x) \le 1$.

**Fixed points.** The discrete logistic equation has two fixed points. To find them, we set $x_{n+1} = x_n = x$ in Equation (7.1), obtaining $x = \mu x(1-x)$. One solution is $x = 0$, which is a fixed point independent of the value of $\mu$. The other fixed point is $x = 1 - 1/\mu$ which does, of course, change as the parameter $\mu$ changes. Theorem 7.3.5 can be used to determine whether these two fixed points are locally stable. The right-hand side of the logistic equation is defined by the function $f(x) = \mu x(1-x) = \mu x - \mu x^2$. Taking the derivative $f'(x) = \mu - 2\mu x$, we compute that $|f'(0)| = |\mu|$. According to the theorem, we need $|\mu| < 1$ in order to ensure the stability of the fixed point $x = 0$. Since we have already assumed that $0 \le \mu \le 4$, we conclude that $x = 0$ is *locally asymptotically stable* if

$0 \le \mu < 1$ and is *unstable* if $1 < \mu \le 4$. For the other fixed point $x = 1 - 1/\mu$, we calculate that $f'(1 - 1/\mu) = 2 - \mu$. Again, Theorem 7.3.5 will guarantee that the fixed point is stable provided that $|2 - \mu| < 1$. Equivalently, we have shown that $x = 1 - 1/\mu$ is *locally asymptotically stable* if $1 < \mu < 3$, and is *unstable* if either $\mu < 1$ or $\mu > 3$.

For $\mu > 3$, something very interesting happens. Both $x = 0$ and $x = 1 - 1/\mu$ are unstable fixed points, and they act as repellers. However, we have already shown that whenever we start from an initial condition $x_0 \in [0, 1]$, then all subsequent iterates $x_n$ are trapped inside the interval $[0, 1]$. How is the logistic equation (7.1) able to resolve the fact that the iterates $x_n$ are trapped inside an interval that contains two repelling fixed points? Apparently, if we start from any initial condition other than one of the two fixed points, then the system can never[1] "settle down" to equilibrium.

Suppose we choose $\mu = 3.1$ and pick any initial condition (other than one of the two fixed points) and use a computer to generate a list of the first few iterates. It turns out that the iterates converge to an *alternating* pattern—i.e., the sequence of iterates with even index $x_{2n}$ converges to one number, whereas the sequence of odd-indexed iterates $x_{2n+1}$ converges to a *different* number. The bifurcation that occurs when $\mu = 3$ gives rise to alternation, and the solutions that we obtain are analogous to periodic solutions for ODEs. For one-dimensional difference equations, when a fixed point $x^*$ loses stability because $f'(x^*)$ decreases past $-1$, we say that a *period-doubling bifurcation* has occurred. The discrete logistic equation (7.1) experiences a period-doubling bifurcation when $\mu = 3$. For the specific choice $\mu = 3.1$, the iterates eventually alternate between approximately 0.558014 and 0.764567, assuming that our initial condition $x_0$ does not coincide with one of the two unstable fixed points. This raises a natural question: "What are these numbers?"

The answer to this question is actually very sensible after we interpret what it would mean for every second iterate of the difference equation to be the same. Mathematically, it means that $x_{n+2} = x_n$ for each $n$. Moreover, since $x_{n+2} = f(x_{n+1}) = f(f(x_n))$, we are led to explore the *second iterate mapping*

$$x_{n+2} = f(f(x_n)) = \mu[\mu x_n(1 - x_n)][1 - \mu x_n(1 - x_n)]. \tag{7.2}$$

[1] Read the warning that appears after Theorem 7.3.5.

Apparently, the two mystery numbers listed above are fixed points of the second iterate mapping, but not of the original logistic equation (7.1). From Equation (7.2), fixed points $x$ of the second iterate mapping must satisfy the algebraic equation

$$x = \mu[\mu x(1-x)][1-\mu x(1-x)],$$

a fourth-degree equation for $x$. Solving this equation appears daunting until we observe that any fixed point of the original equation (7.1) is definitely a fixed point of the second iterate mapping (7.2) as well. Therefore, we already know that $x = 0$ and $x = 1 - 1/\mu$ are solutions of the above fourth-degree equation. After tedious factorization, we find that the other two roots of this equation are

$$x_{\pm} = \frac{(\mu+1) \pm \sqrt{(\mu-3)(\mu+1)}}{2\mu}, \tag{7.3}$$

which are real numbers provided that $\mu > 3$. Both of these are fixed points of the second iterate mapping but not of the logistic equation itself. We have noted that for $\mu$ slightly larger than 3, the iterates of the logistic equation settle into an alternating pattern, sometimes known as a *period-2 solution* or a *2-cycle*. Moreover, formula (7.3) tells us the two numbers that the iterates will alternate between. Interestingly, the 2-cycle is *locally asymptotically stable*: if we start from any initial condition that is "near" either $x_+$ or $x_-$, the iterates of the logistic equation will always converge to the 2-cycle solution. Testing the stability of a periodic solutions (cycles) of a difference equation is actually much more straightforward than testing stability of periodic solutions for ODEs. We know that if $x^*$ is a fixed point of a first-order difference equation $x_{n+1} = f(x_n)$, then $x^*$ is locally asymptotically stable if $|f'(x^*)| < 1$. We can apply the same criterion to the second iterate mapping (7.2): Consider the fixed point $x_-$ of the equation $x_{n+2} = f(f(x_n))$. If we define $g(x) = f(f(x))$, then the stability criterion would require that $|g'(x_-)| < 1$. By the chain rule, $g'(x_-) = f'(f(x_-))f'(x_-)$. Finally, since the iterates in the 2-cycle alternate between $x_-$ and $x_+$, we know that $f(x_-) = x_+$. It follows that the 2-cycle of the discrete logistic equation is *locally asymptotically stable* if

$$|f'(x_+)f'(x_-)| < 1.$$

After tedious algebra, this stability condition reveals that the 2-cycle solution is locally asymptotically stable if $3 < \mu < 1 + \sqrt{6}$. As $\mu$ increases beyond $1 + \sqrt{6}$, the two fixed points $x_{\pm}$ of the second iterate mapping (see Equations (7.2) and (7.3) above) lose their stability via another period-doubling bifurcation. The 2-cycle defined by $x_\pm$ still exists for $\mu > 1 + \sqrt{6}$, but it is unstable. The period-doubling bifurcation at $\mu = 1 + \sqrt{6}$ creates a stable 4-cycle. More exactly, as soon as the fixed points of the second iterate mapping lose their stability, the *fourth iterate mapping* $x_{n+4} = f(f(f(f(x_n))))$ gains *four* stable fixed points. For $\mu$ values slightly larger than $1 + \sqrt{6}$, iterates of the discrete logistic mapping will repeatedly cycle through these four values. Notice that since the function $f(x) = \mu x(1 - x)$ is quadratic, then composing it with itself four times results in an 8th-degree polynomial equation. Solving for the fixed points of the fourth-iterate mapping is possible, but is certainly not easy.

Further increasing $\mu$, the cascade of period-doubling bifurcations occurs with increasing frequency. In fact, M.J. Feigenbaum was able to prove that the lengths of intervals between successive period-doubling bifurcations decrease to 0 geometrically in the following sense: If $\mu_n$ is the value of $\mu$ at which a period-$2^n$ cycle first occurs, then

$$\lim_{n\to\infty} \frac{\mu_n - \mu_{n-1}}{\mu_{n+1} - \mu_n} = 4.669201609102990\ldots$$

Since the intervals between successive bifurcations decrease approximately geometrically by this factor, it follows that there should be some critical value of $\mu$ at which the period of the cycle becomes infinite. Indeed this is the case, and it occurs when $\mu$ is approximately 3.569946. At that critical $\mu$, the behavior of the iterates of the discrete logistic mapping becomes *chaotic* (see next section).

**Bifurcation diagram.** The cascade of period-doubling bifurcations in the logistic mapping is easiest to visualize via a bifurcation diagram. The idea is to plot the *long-term stable response* of equation (7.1) versus the parameter $\mu$. The bifurcation diagram appears in Figure 7.1, and we now summarize how to interpret the diagram. For $0 \le \mu < 1$, we know that 0 is a stable fixed point of the logistic equation, and there are no other stable fixed points. For $1 < \mu < 3$, the points in the bifurcation diagram follow the curve $x = 1 - 1/\mu$, since $1 - 1/\mu$ is a stable fixed point for that range of $\mu$ values. At $\mu = 3$, the curve branches into two curves which persist until $\mu = 1 + \sqrt{6}$. Given any $\mu \in (3, 1 + \sqrt{6})$, the long-term behavior of the iterates is alternation between the two branches, which correspond to the values of $x_\pm$ defined in (7.3). For $\mu$ slightly larger than $1 + \sqrt{6}$, there are four "branches" in the bifurcation diagram. If we select $\mu$ from this

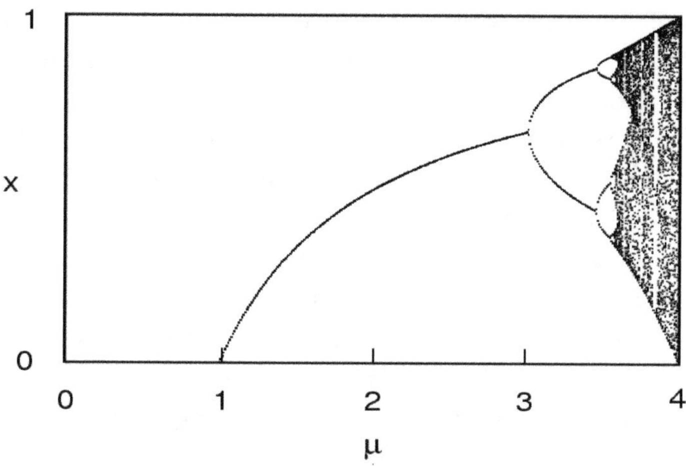


**Figure 7.1.** Bifurcation diagram for the discrete logistic equation (7.1).

"period-4 window", the iterates of the discrete logistic equation will (eventually) cycle repeatedly through the four different values indicated by the branches. For most choices of $\mu > 3.569946$, the branches in the bifurcation diagram become a blur. Apparently, the discrete logistic equation exhibits erratic behavior in that parameter regime (see next section). It is interesting to note that there is actually a period-3 cycle in the vicinity of $\mu = 3.83$, which can be seen if you look closely at the figure.

---

## 7.6. Chaos

For most values of $\mu > 3.57$, the iterates of the discrete logistic equation (7.1) form an aperiodic sequence. The fact that this simple equation can produce such erratic behavior without relying upon artificial "randomness" is rather remarkable. In fact, the discrete logistic equation is, in some sense, the simplest equation which exhibits *chaotic* dynamics. Let us recall the definition of chaos, which was stated in a previous chapter. A well-written, concise definition of *chaos* is provided in the text of Strogatz [11]; we repeat the definition exactly as it appears in his text:

**Definition 7.6.1.** "*Chaos* is aperiodic long-term behavior in a deterministic system that exhibits sensitive dependence on initial conditions."

There are three key phrases that appear in this definition. By *aperiodic long-term behavior*, we typically mean that the system has solutions that remain bounded but never converge to a fixed point or periodic orbit. By *deterministic*, we mean that the system is not allowed to incorporate any randomness in its parameters and/or inputs. The system must be able to produce erratic, aperiodic behavior on its own—artificially including randomness would be cheating. Finally, *sensitive dependence on initial conditions* means that if we start from two different initial conditions that are "nearby", then the corresponding solutions trajectories will separate exponentially fast (at least in the short term).

Testing whether a sequence of numbers is periodic can be quite challenging, because recognizing repeating patterns is difficult when the period of the oscillations is large. Fortunately, there are standard techniques for testing for periodicity within a sequence $\{x_n\}$ of real numbers. The discrete Fourier transform and, more specifically, the fast Fourier transform (FFT) are examples of such techniques. The FFT is an extremely useful way of determining the dominant frequencies of oscillations that may be hidden within sequences of numbers. We shall not cover the FFT in this text; the interested reader is encouraged to refer to any text on Fourier analysis, such as [1]. Instead, we focus on the problem of quantifying sensitive dependence on initial conditions.

**Lyapunov exponents.** We now devise a test for sensitive dependence on initial conditions for first-order difference equations. Consider the equation $x_{n+1} = f(x_n)$, where $f$ is continuously differentiable, and let $x_0$ denote an initial condition. In order to measure the discrepancy between two nearby solution trajectories, suppose we use the same difference equation $y_{n+1} = f(y_n)$ with a different initial condition $y_0 = x_0 + \epsilon_0$, where $|\epsilon_0|$ is small and positive. If $\epsilon_0 = y_0 - x_0$ represents the initial gap between our two solutions, we need to estimate the gap $\epsilon_n = y_n - x_n$ after $n$ iterations. Note that $x_n = f^{(n)}(x_0)$, where $f^{(n)}$ denotes the $n$th iterate mapping—i.e., $f^{(n)}$ represents the function $f$ composed with itself $n$ times, not the $n$th derivative of $f$. Similarly, $y_n = f^{(n)}(y_0) = f^{(n)}(x_0 + \epsilon_0)$, which implies that

$$\epsilon_n = f^{(n)}(x_0 + \epsilon_0) - f^{(n)}(x_0).$$

Dividing by the initial gap $\epsilon_0$, we obtain the approximation

$$\frac{\epsilon_n}{\epsilon_0} = \frac{f^{(n)}(x_0 + \epsilon_0) - f^{(n)}(x_0)}{\epsilon_0} \approx [f^{(n)}]'(x_0). \tag{7.4}$$

(The last approximation would be an equality if we let $\epsilon_0 \to 0$.) The derivative of the function $f^{(n)}$ at $x_0$ is actually relatively straightforward to compute:

**Lemma 7.6.2.** Assume that $x_{n+1} = f(x_n)$, where $f$ is continuously differentiable and let $f^{(n)}$ denote the composition of $f$ with itself $n$ times. Then

$$[f^{(n)}]'(x_0) = \prod_{k=0}^{n-1} f'(x_i).$$

The product notation $\Pi$ is similar to the summation notation $\Sigma$, except that terms are multiplied rather than summed.

*Proof.* It is instructive to start with the case $n = 2$, in which case $f^{(2)}(x) = f(f(x))$. Taking the derivative via the chain rule, we have

$$\frac{\mathrm{d}}{\mathrm{d}x} f(f(x)) = f'(f(x))f'(x).$$

If we set $x = x_0$, note that $f(x_0) = x_1$. Therefore, our expression becomes

$$[f^{(2)}]'(x_0) = f'(x_1)f'(x_0).$$

The remainder of the proof is a straightforward induction on $n$. □

With this Lemma in mind, Equation (7.4) can be rewritten as

$$\frac{\epsilon_n}{\epsilon_0} \approx \prod_{k=0}^{n-1} f'(x_i). \tag{7.5}$$

Since we wish to determine whether solutions of our nearby initial value problems separate exponentially fast, suppose that $|\epsilon_n| \approx |\epsilon_0| \mathrm{e}^{\lambda n}$ for some constant $\lambda$. If the gap $\epsilon_n = y_n - x_n$ obeys such a relationship, then taking absolute values in Equation (7.5) yields

$$\mathrm{e}^{\lambda n} \approx \left| \prod_{k=0}^{n-1} f'(x_i) \right|.$$

Taking logarithms, we exploit the fact that the logarithm of a product is the sum of the individual logarithms:

$$\lambda \approx \frac{1}{n} \ln \left| \prod_{k=0}^{n-1} f'(x_i) \right| = \frac{1}{n} \sum_{k=0}^{n-1} \ln |f'(x_i)|.$$

This calculation motivates the following definition:

**Definition 7.6.3.** If the limit

$$\lambda = \lim_{n\to\infty} \frac{1}{n} \sum_{k=0}^{n-1} \ln |f'(x_i)|$$

exists, it is called the *Lyapunov exponent* for the solution with initial condition $x_0$.

There are two key things to notice about the definition of the Lyapunov exponent $\lambda$. First, our definition of $\lambda$ was based upon an assumed relationship between the separation $\epsilon_n = y_n - x_n$ and the initial separation $\epsilon_0$—namely, $|\epsilon_n| = |\epsilon_0|e^{\lambda n}$. If we calculate that the Lyapunov exponent $\lambda$ is positive for some initial condition $x_0$, this suggests that if we switch to a different initial condition near $x_0$, the new solution trajectory will separate exponentially fast from the original one. In other words,

*A positive Lyapunov exponent $\lambda$ is an indicator of chaos.*

Second, notice that $\lambda$ depends upon the choice of the initial condition $x_0$.

Estimating a Lyapunov exponent by hand is typically not possible, so one typically uses a computer to estimate the value of the infinite sum. However, there are some special cases worth mentioning. Suppose that $x^*$ is a stable, hyperbolic fixed point of the difference equation $x_{n+1} = f(x_n)$, in which case we know that $|f'(x^*)| < 1$. If we start with the initial condition $x_0 = x^*$, then $x_n = x^*$ for all $n \geq 0$ as well. Finally, since $|f'(x^*)| < 1$, then $\ln |f'(x^*)| < 0$ and therefore

$$\lambda = \lim_{n\to\infty} \frac{1}{n} \sum_{k=0}^{n-1} \ln |f'(x^*)| \;=\; \lim_{n\to\infty} \frac{1}{n} \left[ n \ln |f'(x^*)| \right] \;=\; \ln |f'(x^*)| \;<\; 0.$$

It is not surprising that we get a negative Lyapunov exponent if we start at a stable fixed point as our initial condition—a constant solution that attracts nearby solution trajectories is certainly *not* chaotic.

For the discrete logistic equation (7.1), it is interesting to see how the Lyapunov exponent changes as we increase the parameter $\mu$, starting from a random initial condition $x_0 \in (0, 1)$. We would expect the Lyapunov exponent to satisfy $\lambda \leq 0$ for $\mu < 3.5699$, prior to the onset of chaos. Indeed, this is the case as illustrated in Figure 7.2. Notice that the value of $\lambda$ is actually equal to 0 at each $\mu$ value at which a period-doubling bifurcation occurs, such as $\mu = 3$ and $\mu = 1 + \sqrt{6}$.

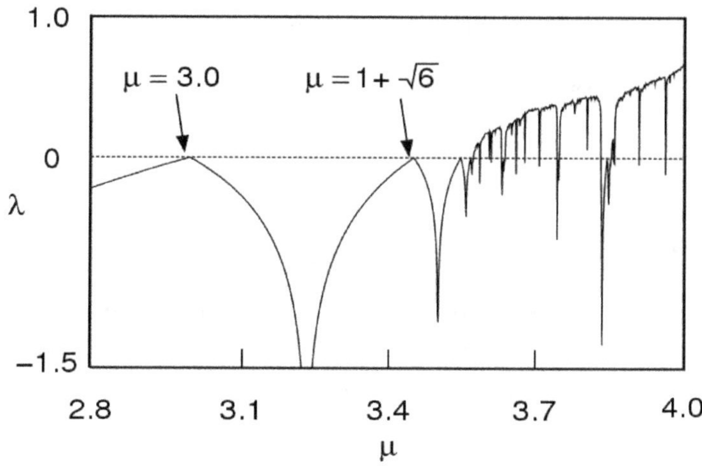


**Figure 7.2.** Lyapunov exponents for the discrete logistic equation (7.1).

The first instance of a positive Lyapunov exponent occurs at $\mu = 3.5699...$, an indication of chaos. In the vicinity of $\mu = 3.83$, the value of $\lambda$ actually becomes negative again. For $\mu$ values in that region, the logistic equation exhibits stable, period-3 behavior.

Deterministic chaos is a truly fascinating phenomenon, and readers interested in a continued development of chaos theory are encouraged to read Chapter 10 of Strogatz [11]. Despite the allure of chaos, it can be quite devastating when a system begins to behave chaotically. For example, the spatial and temporal patterns of the electrical waves in fibrillating heart ventricles (an often-fatal arrhythmia) are chaotic. We conclude this Chapter with a discussion of a feedback control algorithm for preventing the occurrence of chaos.

---

## 7.7. How to Control Chaos

The subject of chaos control was popularized in the 1990s by the work of Ott, Grebogi and Yorke. In this section, we introduce a rather intuitive chaos control algorithm known as *time-delay autosynchronization* (TDAS), which was introduced via the theory and experiments of Socolar et al. [9].

The assumption underlying TDAS control is that by making small adjustments to a parameter during each iteration, we can convert an unstable fixed point into an attractor. Now suppose that we have a system that behaves chaotically, such as the discrete logistic equation for $\mu$ appropriately chosen. In the midst of the chaotic iterates, there may be unstable fixed points. In a physical system, we would never actually see these fixed points—for example, suppose that you try to balance a broomstick on the palm of your hand, keeping your hand perfectly still. If you could position the broomstick perfectly, it would stay in the vertical position (analogous to an unstable equilibrium). Applying TDAS control would be analogous to making tiny movements of your hand in such a way that the broom remains in the vertical position.

For a more specific description of TDAS, consider the discrete logistic equation $x_{n+1} = \mu x_n(1 - x_n)$ where $\mu$ is chosen from the interval $3 < \mu < 1 + \sqrt{6}$. For that range of $\mu$ values, we know that the long-term behavior of the system will be a period-2 cycle, with the iterates alternating between two numbers $x_{\pm}$. Somewhere between these two iterates lies the unstable fixed point $x^* = 1 - 1/\mu$. The TDAS algorithm makes tiny adjustments to the system parameter $\mu$ during each iteration, in such a way that the period-2 cycle is terminated and $x^*$ becomes an attractor. Specifically, suppose that we modify $\mu$, replacing it with $\mu + \gamma(x_n - x_{n-1})$, where $\gamma$ is a positive number. The modified logistic equation is

$$x_{n+1} = [\mu + \gamma(x_n - x_{n-1})]x_n(1 - x_n). \tag{7.6}$$

Notice that we have adjusted the value of $\mu$ by an amount proportional to the gap between the previous two iterates. If $x_n > x^* > x_{n-1}$, we know that for the original logistic equation, $x_{n+1}$ would be *smaller* than $x^*$ because the iterates will alternate large-small-large-small. The modified logistic equation (7.6) effectively adjusts $\mu$ by $\gamma(x_n - x_{n-1})$, which would be *positive* if $x_n > x_{n-1}$. Hence, the modified logistic equation would compute a *larger* value for $x_{n+1}$ than the original logistic equation would have. Similarly, if $x_n < x^* < x_{n-1}$, then the original logistic equation (7.1) would generate a value of $x_{n+1}$ which is *larger* than $x^*$. The modified logistic equation (7.6) would adjust $\mu$ by $\gamma(x_n - x_{n-1})$, which is *negative* in this case. Thus, the value of $x_{n+1}$ generated by (7.6) should be smaller (and presumably closer to the fixed point $x^*$) than the value of $x_{n+1}$ generated by (7.1).

This strategy is analogous to "robbing from the rich and giving to the poor". If the iterates are alternating large-small and $x_{n+1}$ is expected to be *large*, then we will rob from it by effectively reducing $\mu$. Likewise, if $x_{n+1}$ is expected to be *small*, then we will give to it by effectively raising $\mu$.

**Definition 7.7.1.** The above technique of modifying a system parameter $\mu$ by an amount proportional to the previous two iterates is called *time-delay autosynchronization (TDAS)*.

**Stability analysis.** Above we suggested that the TDAS algorithm can sometimes successfully convert an unstable fixed point into a stable one. Let us investigate how this takes place for the discrete logistic equation. With TDAS control, we found that the logistic equation becomes

$$x_{n+1} = [\mu + \gamma(x_n - x_{n-1})]x_n(1 - x_n).$$

Observe that the introduction of the term $\gamma(x_n - x_{n-1})$ has absolutely no effect on the *values* of the fixed points. Indeed, if $x^*$ is one of the two fixed points of the original logistic equation, then $x^*$ will still be a fixed point of the modified equation because $\gamma(x^* - x^*) = 0$. Equation (7.6) is a second-order difference equation, so we write it as a system in the usual way by letting $y_{n+1} = x_n$ and $y_n = x_{n-1}$:

$$\begin{aligned} x_{n+1} = f_1(x,y) &= [\mu + \gamma(x_n - y_n)]x_n(1 - x_n) \\ y_{n+1} = f_2(x,y) &= x_n. \end{aligned} \tag{7.7}$$

Since $x^* = 0$ and $x^* = 1 - 1/\mu$ were the fixed points of the original logistic equation, it follows that $(x,y) = (0,0)$ and $(x,y) = (1-1/\mu, 1-1/\mu)$ are the fixed points of (7.7). We are particularly interested in the stability of the latter fixed point, which corresponds to the more interesting fixed point of the original logistic mapping (7.1).

The Jacobian matrix associated with the right hand side of (7.7) is

$$Jf(x,y) = \begin{bmatrix} \mu - 2\mu x + 2\gamma x - \gamma y - 3\gamma x^2 + 2\gamma xy & -\gamma x + \gamma x^2 \\ 1 & 0 \end{bmatrix}.$$

Evaluating the Jacobian at the fixed point $(1-1/\mu, 1-1/\mu)$ yields

$$Jf(1-1/\mu, 1-1/\mu) = \begin{bmatrix} 2 - \mu + \gamma\left(\frac{1}{\mu} - \frac{1}{\mu^2}\right) & \gamma\left(\frac{1}{\mu^2} - \frac{1}{\mu}\right) \\ 1 & 0 \end{bmatrix}.$$

We are now in a position to use the Jury stability test (Lemma (7.4.6)) to determine whether there exists a range of $\gamma$ values which would make our fixed point stable. Letting $A$ denote our Jacobian matrix, the trace and determinant are given by

$$\mathrm{tr}(A) = 2 - \mu + \gamma\left(\frac{1}{\mu} - \frac{1}{\mu^2}\right) \quad \text{and} \quad \det(A) = \gamma\left(\frac{1}{\mu} - \frac{1}{\mu^2}\right).$$

According to Lemma (7.4.6), there are three conditions that must be satisfied to ensure stability. First, we need $\mathrm{tr}(A) - \det(A) < 1$, which imposes the constraint $\mu > 1$. This actually is not restrictive at all, because there is no need to attempt TDAS for $\mu \leq 1$. Next, we need $\det(A) < 1$. By algebra, this leads to an inequality that $\gamma$ must satisfy, namely

$$\gamma < \frac{\mu^2}{\mu - 1}.$$

The right hand side of this inequality is forced to be positive because of the other constraint that $\mu > 1$. Finally, we require that $\mathrm{tr}(A) + \det(A) > -1$, which by algebra imposes the constraint

$$\gamma > \frac{\mu^2(\mu - 3)}{2(\mu - 1)}.$$

This lower bound on $\gamma$ only becomes interesting when $\mu > 3$, because our fixed point is certainly stable for $1 < \mu < 3$.

In order to determine whether the TDAS scheme may successfully stabilize $(1 - 1/\mu, 1 - 1/\mu)$ when $\mu > 3$, we must determine whether it is possible to simultaneously satisfy both of the above inequalities for $\gamma$; i.e.,

$$\frac{\mu^2(\mu - 3)}{2(\mu - 1)} < \gamma < \frac{\mu^2}{\mu - 1}.$$

Figure 7.3 shows that it is, indeed, possible to satisfy these inequalities even in the regime of $\mu$ values where chaos occurs. The figure shows a plot of the two curves defined by the above inequality on $\gamma$. Notice that the lower boundary is 0 when $\mu = 3$, which makes sense because our fixed point was already stable when $\mu < 3$. As $\mu$ increases, the lower boundary of the "control domain" increases, indicated that larger $\gamma$ would be necessary to control the response. In fact, as $\mu$ approaches 4.0, the figure suggests that we must choose $\gamma$ considerably larger than 2 in order to successfully stabilize our fixed point. The range of $\gamma$ for which control is predicted to succeed becomes narrower as $\mu$ increases. However, even in the regime where chaos exists ($\mu > 3.57$), there are substantial ranges of $\gamma$ for which control may be successful. The figure indicates that if $\mu = 3.65$, which is well within the chaos zone, then using $\gamma = 3.0$ should be enough for TDAS to terminate the chaos. However, below we will point out a flaw in this line of reasoning.

Figure 7.4 illustrates the use of TDAS control to terminate the period-2 response in the discrete logistic equation with $\mu = 3.2$. The first 20 iterates were computed by iterating equation (7.1) (no control), resulting in the alternating pattern. The next 20 iterates were computed from Equation (7.6), using $\gamma = 1.5$, to simulate the effect of TDAS control. After a very brief transient, the iterates converge to the fixed point $1 - 1/\mu = 1 - 1/3.2$. In the absence of control, this fixed point had been unstable.

**Bad news:** Although Figure 7.3 suggests that TDAS should succeed for a wide range of $\gamma$, unfortunately the figure gives a *far* too optimistic prediction of the control domain. There are two notable reasons why this is the case. First, the TDAS method adjusts the parameter $\mu$ based only upon the *two most recent* iterates of the underlying difference equation. In the chaos regime ($\mu > 3.5699$), the iterates of the equation behave so erratically that we should really incorporate more "history" when deciding how $\mu$ should be adjusted. Second, Figure 7.3 was generated based upon a *local* stability analysis. Again, the erratic dynamical behavior for $\mu > 3.5699$ can land us in trouble if we blindly apply the TDAS technique. If $\gamma$ is reasonably large, it is certainly possible for $\gamma(x_n - x_{n-1})$ to be so large that the computed value of $x_{n+1}$ lands *outside* of the interval $[0, 1]$. This causes the iterates to behave *very* badly for the discrete logistic mapping.

**Good news:** There is an improved version of TDAS known as *extended* TDAS, *or* ETDAS, which is better-suited for chaos control (see Socolar et al. [9]). Whereas TDAS modifies $\mu$ by an amount proportional to two preceding iterates, ETDAS incorporates many previous iterates.

**More good news:** In the example of the discrete logistic mapping, we have the luxury of knowing the exact value of the unstable fixed point. Amazingly, neither TDAS nor ETDAS control methods require us to actually know the location of the unstable fixed point(s) in advance. Simply put, these methods can *find* unstable

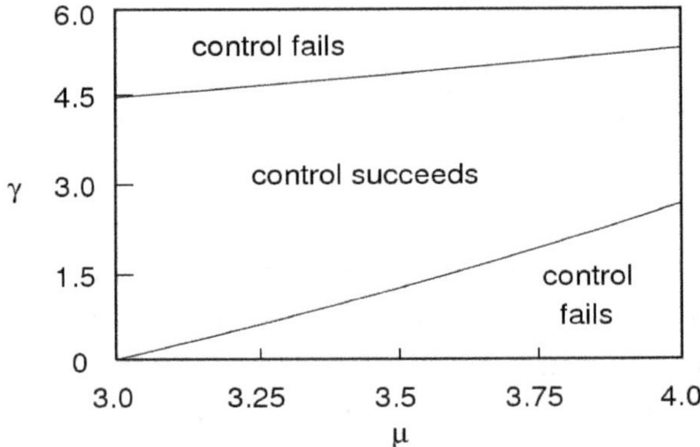


**Figure 7.3.** Predicted domain in which TDAS control succeeds for the discrete logistic equation (7.1).

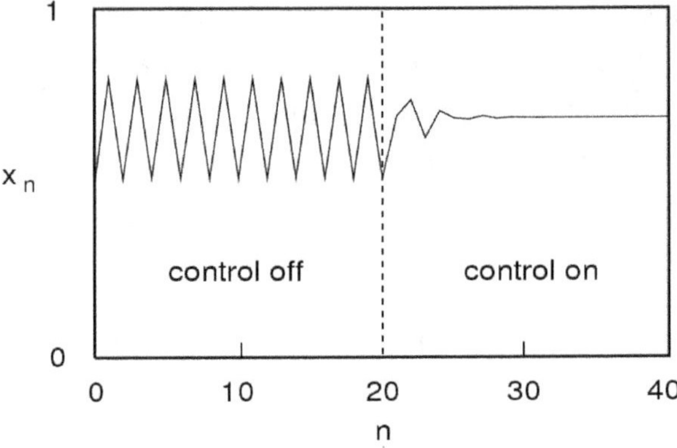


**Figure 7.4.** Illustration of TDAS control in the discrete logistic mapping using $\mu = 3.2$ and $\gamma = 1.5$. The first 20 iterates show the stable period-2 response of the logistic equation (7.1). To simulate the onset of TDAS control, the next 20 iterates are computed using Equation (7.6), which includes control.

fixed points by reversing their stability, eliminating any periodic and/or chaotic behavior in the process.

---

## Exercises

1. Find a closed formula for the solution of $x_{n+1} = 2x_n + 3x_{n-1}$, $x_0 = 1$, $x_1 = 0$.

2. Find a closed formula for the solution of $x_{n+1} = -4x_n - 4x_{n-1}$, $x_0 = 1$, $x_1 = 2$.
3. Find a closed formula for the solution of $x_{n+1} = -4x_{n-1}$, $x_0 = 1$, $x_1 = 2$. Notice that this is a *second-order* difference equation. The coefficient of $x_n$ happens to be 0.

**An alternate approach for solving constant-coefficient equations:** In your first course on ODEs, you probably studied linear, homogeneous constant-coefficient problems such as $y'' + 5y' - 6y = 0$. If you seek exponential solutions of the form $y = e^{\lambda t}$, you obtain a characteristic equation $\lambda^2 + 5\lambda - 6 = 0$, which has distinct, real roots $\lambda = -6$ and $\lambda = 1$. This implies that $y = e^{-6x}$ and $y = e^x$ are solutions of the ODE and, since they are linearly independent, the general solution is $y = C_1e^{-6x} + C_2e^x$. On the other hand, when the characteristic equation has a real root $\lambda$ with multiplicity $k$, then we expect solutions of the form $e^{\lambda t}, te^{\lambda t}, \ldots t^{k-1}e^{\lambda t}$.
In Problems 4 and 5, you will apply the same approach to solve linear, homogeneous, constant-coefficient difference equations.

4. Consider the initial value problem

$$x_{n+1} = -2x_n + 8x_{n-1}, \qquad x_0 = 5, \qquad x_1 = -2.$$

   (a) Suppose we seek solutions of the form $x_n = \lambda^n$, excluding the uninteresting case $\lambda = 0$. Show that $\lambda$ must satisfy a *characteristic equation* $\lambda^2 + 2\lambda - 8 = 0$.
   (b) Note that the roots of this characteristic equation are real and distinct. By drawing an analogy with the above ODE example, write down the general solution of this difference equation.
   (c) Finally, use the initial conditions to find the closed formula for $x_n$. Verify that your solution satisfies the difference equation and both initial conditions.

5. Consider the initial value problem

$$x_{n+1} = 6x_n - 9x_{n-1}, \qquad x_0 = 2, \qquad x_1 = 3.$$

   (a) Suppose we seek solutions of the form $x_n = \lambda^n$, excluding the uninteresting case $\lambda = 0$. Show that $\lambda$ must satisfy a *characteristic equation* $\lambda^2 - 6\lambda + 9 = 0$.

(b) Note that $\lambda = 3$ is a repeated real root of the characteristic equation. By analogy with ODEs, we not only expect solutions of the form $x_n = \lambda^n$, but also solutions of the form $x_n = n\lambda^n$. Use this intuition to write down the general solution of the difference equation.

(c) Finally, use the initial conditions to find the closed formula for $x_n$. Verify that your solution satisfies the difference equation and both initial conditions.

6. Find all fixed points of $x_{n+1} = x_n^3$ and classify their stability.

7. Consider the difference equation $x_{n+1} = f(x_n)$ where $f$ is the piecewise linear function

$$f(x) = \begin{cases} -3x & \text{if } x \le 100 \\ 0 & \text{if } x > 100. \end{cases}$$

(a) Show that $x^* = 0$ is an unstable fixed point.

(b) Despite the instability, show that $x^* = 0$ is *globally attracting*! That is, regardless of the choice of initial condition $x_0$, we have $x_n \to x^*$ as $n \to \infty$. Hint: In fact, the iterates $x_n$ converge to $x^*$ after finitely many iterations.

8. Sometimes, even the Schwarzian derivative test for stability (Theorem 7.3.12) can fail to help us. Consider the difference equation

$$x_{n+1} = \frac{2}{x_n}.$$

(a) Find all fixed points of this equation, and show that they are non-hyperbolic.

(b) Show that Theorem 7.3.12 is inconclusive.

(c) Suppose we choose any non-zero initial condition $x_0$, excluding the fixed points. Show that the solution of the difference equation will always alternate between two distinct values.

(d) Explain why Part (c) allows us to conclude that the fixed points are stable, but not asymptotically stable.

9. Computing square roots by hand. Suppose $k$ is a positive real number and consider the difference equation

$$x_{n+1} = x_n - \frac{x_n^2 - k}{2x_n}.$$

(a) Show that $x^* = \sqrt{k}$ is a fixed point and is locally asymptotically stable.

(b) The result of Part (a) implies that if we start with any initial condition $x_0$ that is reasonably close to $\sqrt{k}$, then the sequence of iterates $x_n$ will converge rapidly to $\sqrt{k}$. Try this with $k = 2$, using an initial guess of $x_0 = 1$. Compute $x_1$, $x_2$, and $x_3$. If you do this correctly, you will find that $x_3$ approximates $\sqrt{2}$ accurate to within $2.2 \times 10^{-6}$.

**10.** Find all fixed points of the system

$$\begin{aligned} x_{n+1} &= 3x_n - x_n y_n \\ y_{n+1} &= -2y_n + x_n y_n \end{aligned}$$

and classify the local stability of each fixed point.

**11.** Consider the difference equation

$$x_{n+1} = \mu(x_n - 1)(2 - x_n),$$

where $\mu$ is a positive parameter.

(a) Show that there are two fixed points if $\mu < 3 - \sqrt{8}$, zero fixed points if $3 - \sqrt{8} < \mu < 3 + \sqrt{8}$, and two fixed points if $\mu > 3 + \sqrt{8}$.

(b) Show that if $\mu = 6$, then the smaller fixed point is unstable and the larger fixed point is locally asymptotically stable.

(c) Show that if $\mu = 7$, then both fixed points are unstable. A period-doubling bifurcation occurs between $\mu = 6$ and $\mu = 7$. If $x^*$ denotes the larger of the two fixed points, then $f'(x^*)$ decreases below $-1$ for some $\mu$ between 6 and 7.

**12.** In this exercise, you will apply the TDAS control algorithm to terminate a period-2 cycle in the discrete logistic equation.

(a) Using Equation (7.1) with $\mu = 3.2$ and initial condition $x_0 = 0.5$, compute $x_1, x_2, \ldots x_{20}$.

(b) To simulate TDAS control, now we will suddenly switch from Equation (7.1) to Equation (7.6). Let $\gamma = 1.0$. Using the value for $x_{20}$ that you computed in Part (a) as an initial condition, generate the first 20 iterates of Equation (7.6). Describe what happens.

(c) Repeat Part (b) with $\gamma = 0.3$. Does TDAS successfully stop the alternation? Explain. (You may wish to refer to Figure 7.3.)

13. In this exercise, you will apply the TDAS control algorithm to terminate a period-4 cycle in the discrete logistic equation.

 (a) Using Equation (7.1) with $\mu = 3.5$ and initial condition $x_0 = 0.5$, compute $x_1, x_2, \ldots x_{20}$.

 (b) To simulate TDAS control, now we will suddenly switch from Equation (7.1) to Equation (7.6). Let $\gamma = 2.0$. Using the value for $x_{20}$ that you computed in Part (a) as an initial condition, generate the first 20 iterates of Equation (7.6). Describe what happens.

# CHAPTER 8

## Introduction to Partial Differential Equations

Many of the natural phenomena that we wish to mathematically model involve several independent variables. For example, the outdoor temperature $T$ depends not only upon time, but also upon spatial location. If $x$ and $y$ denote latitude and longitude and $t$ denotes time, then the function $T(x,y,t)$ describes how temperature varies in space and time. Weather reports usually render this function by using animations in which the variable $t$ increases. Instead of plotting $T(x,y,t)$ as a surface in three dimensions for each fixed $t$, the maps are usually color-coded, with red corresponding to high temperature and blue corresponding to low temperature.

Mathematical models of phenomena incorporating *several* independent variables frequently lead to equations involving partial derivatives. Usually, the independent variables correspond to time and position. Before defining what we mean by a partial differential equation[1], let us establish notation. If $u$ is a quantity that depends upon a single spatial variable (e.g., latitude) as well as time, we will usually write $u = u(x,t)$. Here $x$ denotes the spatial variable and $t$ denotes time. When three spatial dimensions are involved, we will write $u = u(x,y,z,t)$. Instead of using the Leibniz notation for partial derivatives, we use subscripts as follows:

$$\begin{aligned} u_x &= \frac{\partial u}{\partial x} & u_t &= \frac{\partial u}{\partial t} & & \\ u_{xx} &= \frac{\partial^2 u}{\partial x^2} & u_{tt} &= \frac{\partial^2 u}{\partial t^2} & u_{xt} &= (u_x)_t = \frac{\partial}{\partial t}\left(\frac{\partial u}{\partial x}\right) = \frac{\partial^2 u}{\partial t \partial x}. \end{aligned}$$

[1]All of our subsequent presentation is based heavily on the text of Strauss [10].

Unless otherwise stated, we will *always* assume that our functions $u$ are sufficiently smooth to ensure that mixed partial derivatives are equal. That is, $u_{xt} = u_{tx}$ and $u_{xxt} = u_{xtx} = u_{txx}$, and so on.

Roughly speaking, a partial differential equation is any equation involving partial derivatives of some function $u$. With the above notational conventions in mind, we state a more precise definition.

**Definition 8.0.2.** Suppose $u = u(x,t)$ is a function of two variables. An equation of the form

$$F(x, t, u, u_x, u_t, u_{xx}, u_{xt}, u_{tt}, \ldots) = 0$$

is called a *partial differential equation* (PDE).

In this definition, it is understood that the function $F$ has only finitely many arguments. The definition is easily extended to allow for more than two independent variables. The *order* of a PDE is the order of the highest derivative present in the equation. For example, the equation $u_t + u_x = 0$ is a first-order PDE, and the equation $u_{xxt} - (u_x)^8 = 0$ is a third-order PDE. The most general form of a first-order PDE with three independent variables $t$, $x$, and $y$ would be $F(t, x, y, u, u_t, u_x, u_y) = 0$. Here are some well-known examples of PDEs.

**The transport or advection equation:** Let $u = u(x,t)$. Then the equation $u_t + cu_x = 0$ where $c$ is a constant is called the simple *advection equation*. It can be used to model the transport of a pollutant being carried (but not diffusing) in a long, straight river with velocity $c$.

**The heat or diffusion equation:** Let $u = u(x,t)$. Then the equation $u_t = \kappa u_{xx}$ where $\kappa > 0$ is a constant is called the simple *diffusion equation*. It can be used to model the transfer of heat in a long, thin wire or diffusion of a dye in water.

**The wave equation:** Let $u = u(x,t)$. Then the equation $u_{tt} - c^2 u_{xx} = 0$ where $c$ is a constant is called the *wave equation*. It can be used to model the displacement of a plucked guitar string.

**Burgers' equation:** Let $u = u(x,t)$. Then the equation $u_t + uu_x = 0$ is called Burgers' equation. It arises in the study of shock waves.

**Laplace's equation:** Let $u = u(x,y,z)$. Then the equation $u_{xx} + u_{yy} + u_{zz} = 0$ is called Laplace's equation. It models steady-state heat distribution.

**Korteweg-de Vries equation:** Let $u = u(x,t)$. Then the third-order equation $u_t + u_{xxx} - 6uu_x = 0$ is called the Korteweg-de Vries equation. It arises in the context of modeling shallow-water surface waves.

**Definition 8.0.3.** A *solution* of the PDE

$$F(x,t,u,u_x,u_t,u_{xx},u_{xt},u_{tt},\ldots) = 0$$

is any function $u(x,t)$ that satisfies the equation.

This definition, which is easily extended to allow for more than two independent variables, tacitly assumes the existence of partial derivatives of $u$. When we use the word *solution*, we refer to what more advanced textbooks would call a *classical solution*. It is possible to introduce the notion of *weak solutions* of a PDE, which does not require that the partial derivatives of $u$ exist for all $x$ and $t$.

**Example 8.0.4.** Consider the first-order constant-coefficient PDE $\alpha u_x + \beta u_y = 0$. We claim that $u(x,y) = \cos(\beta x - \alpha y)$ is a solution of this PDE. Taking the first-order partial derivatives,

$$\frac{\partial}{\partial x}\cos(\beta x - \alpha y) = -\beta \sin(\beta x - \alpha y) \quad \text{and} \quad \frac{\partial}{\partial y}\cos(\beta x - \alpha y) = \alpha \sin(\beta x - \alpha y).$$

Therefore,

$$\alpha u_x + \beta u_y = -\alpha\beta \sin(\beta x - \alpha y) + \alpha\beta \sin(\beta x - \alpha y) = 0,$$

which proves the claim. You can also show that the function $u(x,y) = (\beta x - \alpha y)^3$ is a solution of the same PDE. This may seem surprising, since the cosine function and the cubic function are *very* different.

**Example 8.0.5.** For $(x,y) \neq (0,0)$, the function $u(x,y) = \ln(x^2 + y^2)$ is a solution of the two-dimensional Laplace equation $u_{xx} + u_{yy} = 0$. To see why, start by taking the first-order partial derivatives

$$u_x = \frac{2x}{x^2 + y^2} \quad \text{and} \quad u_y = \frac{2y}{x^2 + y^2}.$$

Using the quotient rule, the relevant second-order partial derivatives are

$$\begin{aligned} u_{xx} &= \frac{2(x^2+y^2)-(2x)(2x)}{(x^2+y^2)^2} = \frac{-2x^2+2y^2}{(x^2+y^2)^2}, \\ u_{yy} &= \frac{2(x^2+y^2)-(2y)(2y)}{(x^2+y^2)^2} = \frac{2x^2-2y^2}{(x^2+y^2)^2}. \end{aligned}$$

Adding these expressions, we see that $u_{xx}+u_{yy}=0$ as claimed.

In subsequent chapters, we will develop analytical techniques for determining all solutions of certain first and second-order PDEs. First, we introduce several notions that can be used to classify PDEs (e.g., linearity and homogeneity). By simply classifying a PDE, we can often determine whether (i) it is reasonable to expect to solve the PDE by hand, and (ii) if so, which solution techniques are most likely to succeed.

---

## 8.1. Basic Classification of Partial Differential Equations

In ODEs, we learn that linear equations are amenable to analytical techniques, whereas nonlinear ones are usually intractable. The same principle holds for PDEs, and therefore it is useful to determine whether a PDE is linear or nonlinear. A brief review of some notions from linear algebra will assist in our definition of linearity.

**Linearity.** When we think of *functions*, we usually have in mind a rule that assigns *numbers* to numbers—for example, $f : \mathbb{R} \to \mathbb{R}$. An *operator* is essentially a special type of function that acts on a *function* to produce another *function*. One example of an operator is the differentiation operator $T = \frac{\mathrm{d}}{\mathrm{d}x}$. If we feed $T$ a function of the variable $x$, it returns another function. For example, $T(x^3) = 3x^2$, and $T(\mathrm{e}^x) = \mathrm{e}^x$. The (implied) domain of $T$ is the set of all differentiable functions of $x$.

**Definition 8.1.1.** An operator $L$ is called *linear* if $L(\alpha u + \beta v) = \alpha L(u) + \beta L(v)$ for all functions $u$, $v$ and all real scalars $\alpha$, $\beta$. Equivalently, $L$ is a linear operator if both $L(u+v) = L(u) + L(v)$ and $L(cu) = cL(u)$ for all functions $u$ and $v$ and scalars $c$.

**Example 8.1.2.** Consider the set $S$ of all functions $f : \mathbb{R} \to \mathbb{R}$ that are differentiable on the entire real line. The differential operator $L = \frac{\mathrm{d}}{\mathrm{d}x}$ is a linear operator on the set $S$. To see this, suppose $u(x)$ and $v(x)$ are in $S$ and let $\alpha$ and $\beta$ be any real constants. Then

$$L(\alpha u + \beta v) = \frac{\mathrm{d}}{\mathrm{d}x}(\alpha u + \beta v) = \alpha\frac{\mathrm{d}}{\mathrm{d}x}u + \beta\frac{\mathrm{d}}{\mathrm{d}x}v = \alpha L(u) + \beta L(v).$$

**Example 8.1.3.** Consider the set $S$ of all continuous functions on the closed interval $x \in [a, b]$. For functions $f(x) \in S$, define the integral operator

$$I(f(x)) = \int_a^x f(t)\,\mathrm{d}t \qquad (a \le x \le b).$$

For example, if $f(x) = \cos(x)$, then $I(f(x)) = \sin x - \sin a$, and we see that $I$ transforms functions of $x$ into new functions of $x$. We claim that $I$ is a linear operator. Indeed, suppose that $u$ and $v$ are functions in the set $S$, and let $c_1$ and $c_2$ be any real numbers. Then

$$\begin{aligned} I(c_1 u + c_2 v) &= \int_a^x (c_1 u(t) + c_2 v(t))\,\mathrm{d}t = c_1\int_a^x u(t)\,\mathrm{d}t + c_2\int_a^x v(t)\,\mathrm{d}t \\ &= c_1 I(u) + c_2 I(v). \end{aligned}$$

**Example 8.1.4.** Let $S$ be the set of all continuous functions from $\mathbb{R}$ into $\mathbb{R}$. The operator $\Phi$ defined by the rule $\Phi(u) = u^2$ is a nonlinear operator. To see why, suppose $u$ and $v$ are functions from the set $S$. Notice that $\Phi(u + v) = (u + v)^2 = u^2 + 2uv + v^2$, whereas $\Phi(u) + \Phi(v) = u^2 + v^2$. In general, it is not the case that $\Phi(u + v) = \Phi(u) + \Phi(v)$, and therefore $\Phi$ is a nonlinear operator.

**Example 8.1.5.** New linear operators can be formed by taking combinations of other linear operators. For example, consider the set $S$ of all functions $u(x, t)$ that are differentiable with respect to both $x$ and $t$. The operator

$$L = \left(t^2\frac{\partial}{\partial x} + \mathrm{e}^x\frac{\partial}{\partial t}\right)$$

is linear. Given a function $u(x, t)$, the operator $L$ acts on $u$ according to the rule

$$L(u) = \left(t^2\frac{\partial}{\partial x} + \mathrm{e}^x\frac{\partial}{\partial t}\right)u = t^2\frac{\partial u}{\partial x} + \mathrm{e}^x\frac{\partial u}{\partial t}.$$

Now suppose that $u$ and $v$ are functions from the set $S$. Then

$$\begin{aligned} L(u+v) &= \left(t^2\frac{\partial}{\partial x} + e^x\frac{\partial}{\partial t}\right)(u+v) = t^2\frac{\partial}{\partial x}(u+v) + e^x\frac{\partial}{\partial t}(u+v) \\ &= t^2\left(\frac{\partial u}{\partial x} + \frac{\partial v}{\partial x}\right) + e^x\left(\frac{\partial u}{\partial t} + \frac{\partial v}{\partial t}\right). \end{aligned}$$

Comparing the latter expression to

$$\begin{aligned} L(u) + L(v) &= \left(t^2\frac{\partial}{\partial x} + e^x\frac{\partial}{\partial t}\right)u + \left(t^2\frac{\partial}{\partial x} + e^x\frac{\partial}{\partial t}\right)v \\ &= t^2\frac{\partial u}{\partial x} + e^x\frac{\partial u}{\partial t} + t^2\frac{\partial v}{\partial x} + e^x\frac{\partial v}{\partial t}, \end{aligned}$$

we see that $L(u+v) = L(u) + L(v)$. It is also straightforward to show that if $u \in S$ and $c$ is any constant, then $L(cu) = cL(u)$. Therefore, the operator $L$ is linear.

The latter example motivates our definition of linearity for PDEs. In what follows, we restrict ourselves to functions $u = u(x,t)$ of two independent variables, although these concepts are readily extended to more general cases. Moreover, we assume that $L$ is a linear partial differential operator—a linear operator which incorporates partial derivatives with respect to at least one of the independent variables.

**Definition 8.1.6.** A PDE is called *linear* if it can be written in the form $L(u) = f(x,t)$, where $L$ is a linear operator of the sort described in the preceding paragraph. The function $f(x,t)$ can be any function of the two independent variables.

**Example 8.1.7.** Let $\kappa$ be a positive constant. The heat equation

$$\frac{\partial u}{\partial t} = \kappa\frac{\partial^2 u}{\partial x^2}$$

is a linear PDE. Writing the equation in the form $\frac{\partial u}{\partial t} - \kappa\frac{\partial^2 u}{\partial x^2} = 0$ suggests that we define the operator

$$L = \left(\frac{\partial}{\partial t} - \kappa\frac{\partial^2}{\partial x^2}\right)$$

and the function $f(x,t) = 0$. The PDE takes the form $L(u) = f(x,t)$, and you can show that $L$ is a linear operator. It follows that the heat equation is linear.

**Example 8.1.8.** The PDE

$$\sqrt{1+x^2+t^2}\,\frac{\partial^2 u}{\partial x^2} + 3\sin(xt)\,\frac{\partial^2 u}{\partial t^2} - 8\ln(1+x^2+t^4) = 0$$

is linear. If we define the operator

$$L = \left(\sqrt{1+x^2+t^2}\right)\frac{\partial^2}{\partial x^2} + (3\sin xt)\,\frac{\partial^2}{\partial t^2}$$

and the function $f(x,t) = 8\ln(1+x^2+t^4)$, then the PDE takes the form $L(u) = f(x,t)$. To demonstrate that $L$ is a linear operator, use computations similar to those in the preceding example.

**Example 8.1.9.** Burgers' equation $u_t + uu_x = 0$ is a nonlinear PDE. If we define the operator $N$ according to the rule

$$N(u) = \left(\frac{\partial}{\partial t} + u\frac{\partial}{\partial x}\right)u = u_t + uu_x,$$

then Burgers' equation takes the form $N(u) = 0$. To see that the operator $N$ is nonlinear, suppose that $u$ and $v$ are functions. Then

$$\begin{aligned} N(u+v) &= \left(\frac{\partial}{\partial t} + (u+v)\frac{\partial}{\partial x}\right)(u+v) = \frac{\partial}{\partial t}(u+v) + (u+v)\frac{\partial}{\partial x}(u+v) \\ &= \frac{\partial u}{\partial t} + \frac{\partial v}{\partial t} + u\frac{\partial u}{\partial x} + u\frac{\partial v}{\partial x} + v\frac{\partial u}{\partial x} + v\frac{\partial v}{\partial x}. \end{aligned}$$

Comparing the latter expression with

$$N(u) + N(v) = \left(\frac{\partial}{\partial t} + u\frac{\partial}{\partial x}\right)u + \left(\frac{\partial}{\partial t} + v\frac{\partial}{\partial x}\right)v = \frac{\partial u}{\partial t} + u\frac{\partial u}{\partial x} + \frac{\partial v}{\partial t} + v\frac{\partial v}{\partial x},$$

in general it is not the case that $N(u+v) = N(u) + N(v)$. Therefore the PDE is nonlinear.

**Homogeneity.** One further way of classifying linear PDEs is provided by the following definition. As before, we assume that $u = u(x,t)$ is a function of two independent variables and that $L$ is a linear partial differential operator.

**Definition 8.1.10.** A linear PDE $L(u) = f(x,t)$ is called *homogeneous* if $f(x,t) = 0$. Otherwise, the PDE is called *inhomogeneous*.

**Example 8.1.11.** The equation

$$\frac{\partial u}{\partial t} - \frac{\partial^2 u}{\partial x^2} - 8 = 0$$

is inhomogeneous. In order to write this PDE in the form $L(u) = f(x,t)$, we would define

$$L = \left( \frac{\partial}{\partial t} - \frac{\partial^2}{\partial x^2} \right)$$

and $f(x,t) = 8$. Since $f \neq 0$, the PDE is inhomogeneous.

As in ODEs, the combination of linearity and homogeneity is very advantageous, because it allows us to construct new solutions by the "superposition principle".

**Theorem 8.1.12** (Superposition Principle). Suppose $L(u) = 0$ is a linear, homogeneous PDE with particular solutions $u_1, u_2, \ldots u_n$. Then any linear combination

$$c_1 u_1 + c_2 u_2 + \cdots + c_n u_n$$

is also a solution, for any choice of constants $c_1, c_2, \ldots c_n$.

*Proof.* Since $u_1, u_2, \ldots u_n$ are solutions of the PDE, it follows that $L(u_k) = 0$ for each $k = 1, 2, \ldots n$. Now let $u = c_1 u_1 + c_2 u_2 + \cdots + c_n u_n$. Then by linearity of $L$,

$$L(u) = L\left( \sum_{k=1}^{n} c_k u_k \right) = \sum_{k=1}^{n} L(c_k u_k) = \sum_{k=1}^{n} c_k L(u_k) = 0.$$

Therefore, $u$ is also a solution of the PDE. □

**Example 8.1.13.** In a previous example, we noted that $u_1(x,y) = \cos(\beta x - \alpha y)$ and $u_2(x,y) = (\beta x - \alpha y)^3$ are solutions of the PDE $\alpha u_x + \beta u_y = 0$, where $\alpha$ and $\beta$ are constants. Since the PDE is both linear and homogeneous, the Superposition Principle (Theorem 8.1.12) ensures that

$$3u_1 - 5u_2 = 3\cos(\beta x - \alpha y) - 5(\beta x - \alpha y)^3$$

is also a solution of the PDE.

**Second-order constant-coefficient partial differential equations.** Three of the most important examples of PDEs that we will analyze in subsequent chapters are the heat equation $u_t - u_{xx} = 0$, the wave equation $u_{tt} - u_{xx} = 0$, and the Laplace equation $u_{xx} + u_{tt} = 0$. In some sense, every second-order constant-coefficient PDE will behave like one of these three special equations. (This can be shown via appropriate changes of variables; for an explicit example, see Strauss [10].) Since the heat, wave, and Laplace equations have *very* different solutions, classifying second-order PDEs as either "heat-like", "wave-like", or "Laplace-like" can be very illuminating.

Consider the general, second-order, homogeneous, constant-coefficient PDE

$$Au_{xx} + Bu_{xt} + Cu_{tt} + Du_x + Eu_t + Fu = 0, \tag{8.1}$$

where $A$, $B$, $C$, $D$, $E$, and $F$ are constants.

**Definition 8.1.14.** Equation (8.1) is called

☞ *Elliptic* if $B^2 - 4AC < 0$,

☞ *Hyperbolic* if $B^2 - 4AC > 0$,

☞ *Parabolic* if $B^2 - 4AC = 0$ and $A$, $B$, and $C$ are not all zero.

**Example 8.1.15.** Consider the wave equation $u_{xx} - u_{tt} = 0$. Using the notation of Definition 8.1.14, we have $A = 1$, $C = -1$, and $B = D = E = F = 0$. Since $B^2 - 4AC = 4 > 0$, we see that the wave equation is *hyperbolic*. For the heat equation $u_t - u_{xx} = 0$, we have $A = -1$ and $B = C = 0$, from which it follows that $B^2 - 4AC = 0$. Thus, the heat equation is *parabolic*. Finally, the Laplace equation $u_{xx} + u_{tt} = 0$ satisfies $A = C = 1$ and $B = 0$, implying that this equation is *elliptic*.

To reinforce this terminology, it is useful to note the parallels between these PDEs and corresponding algebraic equations. For example, we can associate the wave equation $u_{xx} - u_{tt} = 0$ with the algebraic equation $x^2 - t^2 = 0$, the graph of which is a *hyperbola* in the $xt$-plane. Similarly, the heat equation $u_t - u_{xx} = 0$ can be associated with the algebraic equation $t - x^2 = 0$, the graph of which is a *parabola* in the $xt$-plane. Generally, solutions of parabolic PDEs tend to behave similarly to those of the heat equation. Linear, hyperbolic PDEs tend to be "wave-like", and elliptic PDEs are "Laplace-like" in terms of the behavior of their solutions.

## 8.2. Solutions of Partial Differential Equations

Finding all solutions of a PDE is considerably more challenging than solving and algebraic or ordinary differential equation. Solutions of algebraic equations such as $x^2 - 6x + 5 = 0$ are *numbers*: $x = 1$ and $x = 5$. Solutions of ODEs are *functions* of the independent variable. For example, the general solution of the equation

$$\frac{d^2y}{dx^2} - 6\frac{dy}{dx} + 5y = 0$$

is $y(x) = C_1e^x + C_2e^{5x}$, where $C_1$ and $C_2$ are arbitrary constants. We know that the general solution of a linear $m^{th}$-order ODE will contain $m$ arbitrary constants. What do general solutions of PDEs look like?

**Example 8.2.1.** Find all functions $u(x,t)$ satisfying the PDE $u_x = 0$. *Solution:* We are given the $x$ derivative of $u$, and we need only integrate with respect to $x$ in order to solve for $u$. Our temptation may be to write $u(x,t) =$ constant, but this is incorrect—after all, there is a second independent variable, $t$. Thus, instead of including an arbitrary integration *constant* when we integrate with respect to $x$, we must include an arbitrary *function* of $t$. The general solution of this PDE is $u(x,t) = f(t)$, where $f(t)$ is *any* arbitrary function of $t$.

**Example 8.2.2.** Solve the PDE $u_{xt} = 3$. *Solution:* This second-order PDE can also be solved by integration: since $(u_x)_t = 3$, we integrate with respect to $t$, treating $x$ as constant. Doing so reveals that $u_x = 3t + f(x)$, where $f(x)$ is an *arbitrary* function of $x$. Next, we integrate with respect to $x$, treating $t$ as a constant. We find that $u(x,t) = 3xt + F(x) + g(t)$, where $F(x) = \int f(x)\,dx$ represents an antiderivative of $f$ and $g(t)$ is an arbitrary function of $t$. Since $f(x)$ was arbitrary, so is $F(x)$. It follows that *any* function of the form $u(x,t) = 3xt + F(x) + g(t)$ will satisfy our original PDE, and there are no other solutions.

**Observation.** Whereas general solutions of ODEs contain arbitrary *constants*, the general solutions of PDEs contain arbitrary *functions* of the independent variables. This feature of PDEs opens the door to having *tons* of solutions, and later we will devote considerable effort toward singling out particular (physically relevant) solutions. Doing so will require us to invoke physical intuition regarding the correct number (and types) of initial and boundary conditions to impose.

**Example 8.2.3.** Solve the linear second-order PDE $u_{tt} + u_t = 0$ by finding all functions $u(x,t)$ that satisfy it. *Solution:* If we were not told that there is a second independent variable $x$, we would have no way of distinguishing this PDE from an ODE. Ignoring the variable $x$ for the moment, consider the analogous ODE $u''(t) + u'(t) = 0$. Using techniques from your first course in ODEs, you could solve the associated characteristic equation and find the general solution $u(t) = C_1 + C_2 e^{-t}$, where $C_1$ and $C_2$ are arbitrary constants. Returning to our original PDE, we obtain the general solution by replacing the constants $C_1$ and $C_2$ by arbitrary functions of $x$: that is, $u(x,t) = f(x) + e^{-t}g(x)$ is the general solution of the original PDE.

Another way to attack this problem is to note that we could have integrated once with respect to $t$ to reduce the PDE from second-order to first-order:

$$\int u_{tt} + u_t \, dt = 0.$$

Integration yields $u_t + u - f(x) = 0$, where the minus sign in front of the arbitrary function $f(x)$ is merely included for convenience. The resulting inhomogeneous first-order equation $u_t + u = f(x)$ can be solved by the variation of parameters technique (see Theorem 2.18). Multiplying both sides of this first-order equation by the integrating factor $e^t$, we have $e^t(u_t + u) = e^t f(x)$. Equivalently,

$$\frac{\partial}{\partial t}\left(e^t u\right) = e^t f(x).$$

Integration with respect to $t$ yields

$$e^t u = \int e^t f(x) \, dt = e^t f(x) + g(x),$$

where $g(x)$ is an arbitrary function of $x$. Finally, multiplying both sides by $e^{-t}$ yields the general solution of the original PDE, $u(x,t) = f(x) + e^{-t}g(x)$.

---

## 8.3. Initial Conditions and Boundary Conditions

In our study of ODEs, we learned that it is important to single out specific solutions that are of particular physical relevance. For ODEs, we typically specify initial conditions (or sometimes boundary conditions) in such a way that we may solve for any arbitrary constants that appear in the general solution. The general solutions of PDEs contain arbitrary *functions* of the independent variables, making it more difficult to single out particular solutions. For example, suppose that a pollutant[2] is carried by water moving with constant speed $c$ through a long, "one-dimensional" pipe. Letting $u(x,t)$ represent the concentration of pollutant at position $x$ and time $t$, the advection equation $u_t + cu_x = 0$ can be used to model the spatiotemporal dynamics of $u$. Since we would expect the general solution of this first-order PDE to contain at least one arbitrary function of the variables $x$ and $t$, we should impose at least one auxiliary condition on $u(x,t)$ in order to select a particular solution.

**Initial Conditions.** Perhaps the most intuitive way to prescribe an auxiliary condition is to impose an *initial condition*—i.e., give a formula for $u(x,t)$ at some fixed time $t = t_0$. In our above example of a transported pollutant, the initial condition should describe the concentration of pollutant at *all* spatial locations at time $t_0$. Mathematically, such an initial condition would have the form $u(x,t_0) = \phi(x)$, where $\phi$ is a function only of the spatial variable $x$.

**Example 8.3.1.** Soon, we will know how to show that the general solution of the advection equation $u_t + 8u_x = 0$ on the domain $-\infty < x < \infty$ is given by $u(x,t) = f(x - 8t)$, where $f$ is any differentiable function of a single variable. To verify that any such function satisfies the advection equation, we calculate the first partial derivatives of $u(x,t) = f(x - 8t)$ using the chain rule: $u_t = -8f'(x - 8t)$ and $u_x = f'(x - 8t)$. Hence, $u_t + 8u_x = 0$, as claimed. Now suppose we impose an initial condition at time $t = 0$ by requiring that $u(x,0) = x^2 + \cos x$. Setting $t = 0$ in our general solution $u(x,t) = f(x - 8t)$ implies that $u(x,0) = f(x)$, from which it follows (by the initial condition) that $f(x) = x^2 + \cos x$. Finally, we see that the only function satisfying both the PDE and the initial condition is

$$u(x,t) \;=\; f(x - 8t) \;=\; (x - 8t)^2 + \cos(x - 8t).$$

In the above example, observe that only one auxiliary condition was needed in order to isolate a particular solution. It is also worth noting that (i) the PDE was first-order in both space and time and (ii) the spatial domain $-\infty < x < \infty$ consisted of all real numbers. As we shall see, imposing initial and boundary

[2]Assume that the pollutant is simply carried by the current and does not diffuse into the water. In other words, think of small, plastic beads as opposed to a chemical dye.

conditions requires more care (and physical intuition) when dealing with higher-order PDEs and/or PDEs in which the spatial domain is finite.

**Example 8.3.2.** Consider the wave equation $u_{tt} - 4u_{xx} = 0$ on the domain $-\infty < x < \infty$. Later, we will learn how to show that the general solution of this PDE is given by $u(x,t) = f(x+2t) + g(x-2t)$, where $f$ and $g$ are arbitrary [twice differentiable] functions of a single variable. Suppose that we impose two initial conditions: $u(x,0) = \cos(x)$ and $u_t(x,0) = \sin(x)$. From the general solution of the PDE, we use the chain rule to calculate

$$u_t(x,t) = 2f'(x+2t) - 2g'(x-2t).$$

Setting $t = 0$ in our expressions for $u(x,t)$ and $u_t(x,t)$ allows us to use our initial conditions:

$$\begin{aligned} u(x,0) = f(x) + g(x) &= \cos x \\ u_t(x,0) = 2f'(x) - 2g'(x) &= \sin x. \end{aligned}$$

Taking the derivative of the equation $f(x) + g(x) = \cos x$ yields $f'(x) + g'(x) = -\sin x$ which, combined with the other equation $2f'(x) - 2g'(x) = \sin x$ allows us to algebraically solve for $f'(x)$ and $g'(x)$. Specifically, we find that $f'(x) = -\frac{1}{4}\sin x$ and $g'(x) = -\frac{3}{4}\sin x$. Integrating with respect to $x$, we have $f(x) = \frac{1}{4}\cos x + C_1$ and $g(x) = \frac{3}{4}\cos x + C_2$, where $C_1$ and $C_2$ are constants of integration. However, we may drop these integration constants because (i) they must sum to zero in order for the first initial condition to be satisfied and (ii) the overall solution involves a sum of $f$ and $g$. It follows that the only solution of the wave equation satisfying both initial conditions is

$$u(x,t) = \frac{1}{4}\cos(x+2t) + \frac{3}{4}\cos(x-2t).$$

When we discuss the physical interpretation of the wave equation on the domain $-\infty < x < \infty$, we will see that prescribing two initial conditions $u(x,0)$ and $u_t(x,0)$ is analogous to stating an initial position and velocity.

When the spatial domain is infinite and no boundary conditions are assigned, a PDE together with its initial conditions is sometimes referred to as the *Cauchy problem*, which is basically another way of saying "initial value problem". In the preceding two examples, the first shows a solution of the Cauchy problem for the advection equation, and the second shows a solution of the Cauchy problem for the wave equation.

**Boundary Conditions.** In contrast with the preceding two examples, for practical purposes we are usually interested in solving PDEs over *finite* spatial domains. Depending upon what sorts of phenomena our PDEs are meant to model, we may have special information about the physical state of our system at the boundaries of the spatial domain. When imposing a boundary condition, we must describe the physical state of our system at a *specific* spatial location as time $t$ varies. Contrast this with how initial conditions are prescribed. Boundary conditions "freeze" the spatial variable $x$ and describe the physical state of the system at that location as $t$ varies. Initial conditions "freeze" the time variable $t$ and describe the physical state of the system at all spatial locations $x$.

To give an intuitive illustration of how boundary conditions arise, let us consider the heat equation $u_t - u_{xx} = 0$ in one spatial dimension. This PDE models how heat diffuses in a long, thin insulated[3] wire. Letting $L$ denote the length of the wire, we may assume that our spatial domain is given by the interval $0 \leq x \leq L$. The boundary of our spatial domain consists of two points: $x = 0$ and $x = L$. One way to impose boundary conditions at would be to simply give formulas for $u$ at these two points—i.e., specify $u(0,t)$ and $u(L,t)$ for all time $t \geq 0$. A boundary condition which specifies the exact value of $u$ at the boundary is called a *Dirichlet condition*. For example, if we dip the $x = L$ end of the wire in a bath of boiling water of constant temperature 100 degrees Celsius, this imposes the Dirichlet boundary condition $u(L,t) = 100$ for all $t \geq 0$. If the other end of the wire is placed in contact with a block of ice of temperature 0 degrees Celsius, that would impose another Dirichlet condition $u(0,t) = 0$ for all $t \geq 0$. Dirichlet boundary conditions can also allow us to vary the temperature at the boundary. For example, $u(L,t) = 80 + 20\cos(t)$ would simulate the effects of varying the temperature between 60 and 100 degrees at the $x = L$ end of the wire.

Another way to impose a boundary condition is to describe the temperature *gradient* at the boundary of the spatial domain. Suppose that we insulate the $x = 0$ end of the wire so that no heat can escape (or enter) the wire at the point. Then there can be no spatial gradient of heat at $x = 0$, implying that

[3]The wire's insulation prevents heat from diffusing outward from the wire so that it is only conducted longitudinally within the wire.

$u_x(0,t) = 0$ for all $t \geq 0$. The boundary condition $u_x(0,t) = 0$ is an example of a *Neumann condition*. In general, a Neumann condition is a boundary condition that describes the outward normal derivative of $u$ along the boundary. If our spatial domain is given by an interval $0 \leq x \leq L$ in one dimension, Neumann conditions specify $u_x(0,t)$ and $u_x(L,t)$ for all time $t$. For example, the Neumann condition $u_x(L,t) = t$ would indicate that the spatial temperature gradient at $x = L$ increases linearly in time, with higher temperature outside the wire $(x > L)$ than inside the wire $x < L$. Assuming that heat flows in the direction *opposite* the heat gradient (i.e., from regions of higher temperature towards regions of lower temperature), the steepening gradient would cause heat to flow from right to left (into the wire) at the boundary $x = L$.

In more than one spatial dimension, the boundary of our domain is typically a curve, surface, or hypersurface. For example, Figure 8.1 shows a domain $\Omega$ in two space dimensions. The boundary of $\Omega$, sometimes denoted as $\partial\Omega$, is a closed curve. If $x$ and $y$ are the spatial variables, then a Dirichlet condition would give a formula for the value of $u(x,y,t)$ for all points $(x,y) \in \partial\Omega$ for all time $t$. A Neumann condition would be used to describe the net outward flow across the boundary $\partial\Omega$ for all time $t$. More specifically, let $\nabla u = (u_x, u_y)$ denote the gradient of $u$ with respect to its spatial variables, and let $\mathbf{n}$ denote an outward unit normal vector on $\partial\Omega$ (see figure). Neumann boundary conditions would provide a formula for $\nabla u \bullet \mathbf{n}$ for all points $(x,y) \in \partial\Omega$. Assuming that $\nabla u$ is not the zero vector, we know that $\nabla u$ points in the direction in which $u$ increases most rapidly. The dot product $\nabla u \bullet \mathbf{n}$ indicates whether the gradient vector $\nabla u$ points outward from $\Omega$ (if $\nabla u \bullet \mathbf{n} > 0$), inward (if $\nabla u \bullet \mathbf{n} < 0$), or tangent to the boundary (if $\nabla u \bullet \mathbf{n} = 0$). In the case of the heat/diffusion equation, heat diffuses from regions of high temperature towards regions of lower temperature, flowing in the direction opposite the heat gradient. If $u(x,y,t)$ denotes temperature at position $(x,y) \in \Omega$ at time $t$ and we find that $\nabla u \bullet n > 0$ at all points on $\partial\Omega$, then heat would flow *inward* into our domain $\Omega$. If we insulate the boundary to prevent heat flow across $\partial\Omega$, then we impose the Neumann condition $\nabla u \bullet \mathbf{n} = 0$ at all points on $\partial\Omega$.

**Well-posed problems.** Knowing how to prescribe initial and boundary conditions in such a way that we single out precisely one physically relevant solution of our PDEs is very important. Ideally, we always hope to work with *well-posed* problems—those with exactly one solution and for which small changes in initial

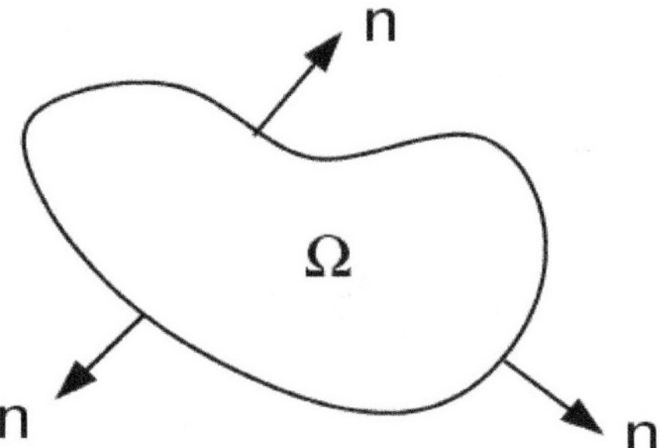


**Figure 8.1.** A spatial domain $\Omega$ in two dimensions. Several outward unit normal vectors $\mathbf{n}$ are shown.

and boundary conditions do not elicit major changes in the behavior of the solution. As we learn more about the physical interpretation of certain PDEs, we will learn how to impose initial and boundary conditions in such a way that our problems are well-posed.

---

## 8.4. Visualizing Solutions of Partial Differential Equations

Visualizing the behavior of solutions for systems of ordinary differential equations is rather straightforward. If $t$ is the independent variable and

$$x_1(t), x_2(t), \ldots x_n(t)$$

are the dependent variables, then one way to graphically render the solution is to plot each dependent variable versus $t$. Now suppose that $u(x,t)$ represents a solution of a PDE with exactly two independent variables, $x$ and $t$. At first glance, you may be tempted to plot $u$ as a function of both $x$ and $t$, but this would require a three-dimensional plot in which $u$ is plotted as a surface. Although this is certainly a valid way of presenting the solution, it is often more convenient to provide two-dimensional plots that illustrate $u$ versus $x$ at several different times $t_1, t_2, \ldots t_n$. This effectively provides us with frames of an "animation", illustrating the spatial distribution of $u$ as we advance forward in time.

**Example 8.4.1.** Consider the PDE $u_t + u_x = 0$ on the spatial domain $-\infty < x < \infty$, and with the initial condition $u(x,0) = \mathrm{e}^{-x^2}$. In the next Chapter, you will learn

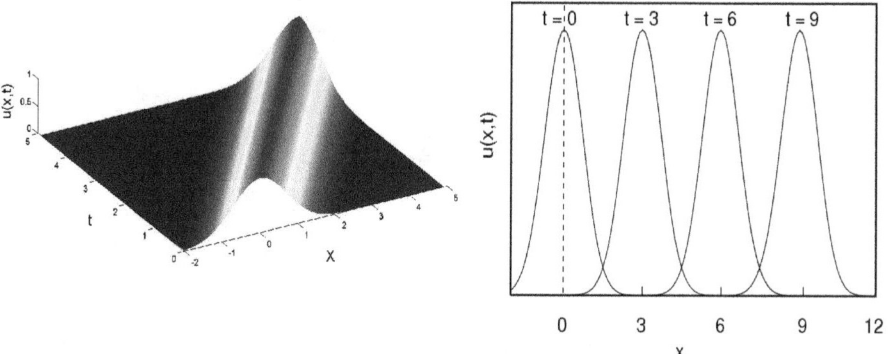


**Figure 8.2.** Left panel: A surface plot of $u(x,t) = \mathrm{e}^{-(x-t)^2}$ versus both $x$ and $t$. Right panel: A plot of $u(x,t^*)$ versus $x$ at four different choices of $t^*$, namely 0, 3, 6, and 9. Note that the initial spatial distribution of $u$ at $t = 0$ is transported from left to right with constant speed 1.0 as $t$ increases.

to show that the solution of this initial value problem is given by

$$u(x,t) = \mathrm{e}^{-(x-t)^2}.$$

Two ways of rendering this solution are illustrated in Figure 8.2. The left panel shows a plot of the function $u(x,t)$ as a surface in three dimensions. The right panel gives four frames of an "animation", showing that our initial distribution of $u$ is transported from left to right with constant speed as $t$ increases. The four traces of $u$ versus $x$ suggest that the speed is 1.0 spatial units per unit time. To further convince ourselves that the propagation speed is 1.0, we can use an analytical approach to track the movement of the peaks. Setting $u = 1$ in our formula for $u(x,t)$ yields $1 = \mathrm{e}^{-(x-t)^2}$, and taking logarithms reveals that $(x-t)^2 = 0$. To calculate the propagation velocity $\frac{\mathrm{d}x}{\mathrm{d}t}$, differentiate the latter expression (implicitly) with respect to $t$, and use algebra to show that $\frac{\mathrm{d}x}{\mathrm{d}t} = 1$.

---

## Exercises

1. Let $S$ denote the set of functions of two variables ($x$ and $y$) whose partial derivatives of all orders exist. Which of the following operators $T$ are linear on the set $S$? Show computations to support your answers.

(a) $T(u) = u_x u_y$

(b) $T(u) = y u_x + x u_y$

(c) $T(u) = 1$

(d) $T(u) = [\arctan(x)] u_{xxx}$

(e) $T(u) = \sqrt{u_{xx} + u_{yy}}$

(f) $T(u) = u_x + u_y + x.$

2. Determine whether or not each of the following PDEs is linear or nonlinear. Show computations to support your answers.

$$u_x + u u_y = 0$$
$$u_{tt} - u_{xx} + u^3 = 0$$
$$u_{tt} + u_{xxxx} = \cos(xt).$$

3. For each of the following equations, state the order and whether it is linear or nonlinear. If an equation is linear, state whether it is homogeneous or inhomogeneous.

(a) $u_t - 4u_{xx} - \ln(1 + x^2) = 0$

(b) $u_{tt} - u_{xx} + xu = 0$

(c) $u_t + u_{xxx} - 6uu_x = 0$

(d) $u_x^2 - u_y^2 = 0$

(e) $u_t + u_{xx} + \frac{u}{x^2} = 0$

(f) $u + u_{xy} + \sqrt{1 + u^2} = 0$

(g) $u_x + \mathrm{e}^{-xy} u_y + 2 = 0$

(h) $u_x u_y - 1 = 0$

4. Let $\kappa > 0$ be a constant. Show that $u(x,t) = A + Bx + 2\kappa Ct + Cx^2$ is a solution of the PDE $u_t = \kappa u_{xx}$ for any choice of constants $A$, $B$, and $C$.

5. Suppose that $f$ and $g$ are arbitrary differentiable functions of a single variable, and that $c$ is a constant. Show that $u(x,t) = f(x + ct) + g(x - ct)$ is a solution of the PDE $u_{tt} - c^2 u_{xx} = 0$.

6. Show that $u(x,y) = \cos(x^2 + y^3)$ is a solution of the PDE $u_{xy} + 6xy^2 u = 0$.

7. Solve the PDE $u_{xt} = 6xt^2$.

8. Find a constant solution of the initial value problem $\frac{dy}{dx} = 3y^{2/3}, \quad y(0) = 0$. Then, use separation of variables to find a non-constant solution. Conclude that this problem is not well-posed since it violates uniqueness.

9. Is the boundary value problem

$$\frac{dy}{dx} + y = 0, \qquad y(0) = 1, \qquad y(2) = \frac{1}{2}$$

a well-posed problem? If so, find the solution. If not, explain why not.

# CHAPTER 9

## Linear, First-Order Partial Differential Equations

In this chapter, we will discuss the first of several special classes of PDEs that can be solve via analytical techniques. In particular, we will investigate linear, first-order PDEs

$$a(x,t)\frac{\partial u}{\partial t} + b(x,t)\frac{\partial u}{\partial x} + f(x,t)u = g(x,t), \tag{9.1}$$

where $u = u(x,t)$ is our dependent variable, and the functions $a$, $b$, $f$, and $g$ are given. Our goal is to develop a systematic method for determining all functions $u$ that satisfy the PDE. A little geometric intuition will help use devise a rather useful technique for solving such equations, and for that reason we will review a few notions from multivariate calculus.

Suppose that $f(x,y)$ is a differentiable function of two variables, and let $\mathbf{v} = (v_1, v_2)$ denote a unit vector in $\mathbb{R}^2$.

**Definition 9.0.2.** The *directional derivative* of $f$ in the direction of the vector $\mathbf{v}$ is given by

$$\lim_{h \to 0} \frac{f(x + v_1 h, y + v_2 h) - f(x,y)}{h},$$

provided that this limit exists.

It is straightforward to prove (see your multivariate calculus text) that this limit is equal to the dot product

$$\nabla f(x,y) \bullet \mathbf{v},$$

thereby providing us with an equivalent (and computationally convenient) definition of the directional derivative. The directional derivative measures the instantaneous rate of change of $f$ in the direction of the vector $\mathbf{v}$. If we choose the unit vector $\mathbf{v} = (1,0)$, which points in the direction of the positive $x$-axis, then the directional derivative is simply

$$\nabla f(x,y) \bullet \mathbf{v} = \left(\frac{\partial f}{\partial x}, \frac{\partial f}{\partial y}\right) \bullet (1,0) = \frac{\partial f}{\partial x},$$

the partial derivative of $f$ with respect to $x$. Similarly, the directional derivative in the direction of $\mathbf{v} = (0,1)$ is given by $\partial f/\partial y$. We now use the notion of directional derivatives to give a quick, clean solution of a special case of Equation (9.1).

**A linear, homogeneous, constant-coefficient equation.** Consider the PDE

$$\alpha u_t + \beta u_x = 0, \tag{9.2}$$

where $\alpha$ and $\beta$ are non-zero constants. Our goal is to determine all functions $u(x,t)$ that satisfy this equation. Observe that since the gradient of $u$ is given by $\nabla u = (u_x, u_t)$, the PDE (9.2) is equivalent to the equation

$$\nabla u(x,t) \bullet (\beta, \alpha) = 0.$$

In other words, the solutions of the PDE are precisely those functions $u(x,t)$ whose directional derivative in the direction of the vector $(\beta, \alpha)$ is 0. Geometrically, this implies that $u(x,t)$ must remain *constant* as we move along any line parallel to the vector $(\beta, \alpha)$. These lines have slope $\alpha/\beta$ in the $xt$ plane, and the general equation for such lines is $t = (\alpha/\beta)x + \text{constant}$. Equivalently, $u(x,t)$ must remain constant along any line of the form $\alpha x - \beta t = C$, where $C$ is an arbitrary constant. Since $u$ remains constant along each such line, $u$ depends only on the value of the constant $C$. In other words, $u$ depends only upon the quantity $\alpha x - \beta t$, but otherwise has no restrictions at all. It follows that the general solution of Equation (9.2) is

$$u(x,t) = f(\alpha x - \beta t), \tag{9.3}$$

where $f$ is any (differentiable) function of a single variable.

**Definition 9.0.3.** In the above example, the "geometric" technique used to solve the PDE is called the *method of characteristics*. The lines $\alpha x - \beta t =$ constant in the above example are called *characteristic curves* associated with the PDE.

By differentiation, it is easy to verify that any function of the form (9.3) certainly satisfies the PDE (9.2). The method of characteristics demonstrated that there are no other solutions of the PDE. Later, we will give more examples illustrating the use of the method of characteristics.

**Example 9.0.4.** To check that $u(x,t) = \cos(\alpha x - \beta t)$ is a solution of the PDE (9.2), we compute the partial derivatives $u_t = \beta \sin(\alpha x - \beta t)$ and $u_x = -\alpha \sin(\alpha x - \beta t)$. From here it is evident that $\alpha u_t + \beta u_x = 0$, as required.

**Example 9.0.5.** Solve the PDE $2u_t + u_x = 0$ with initial condition $u(x,0) = \frac{1}{1+x^2}$, and sketch $u$ versus $x$ for several choices of time $t$. *Solution:* This PDE is a special case of Equation (9.2), so we may quote (9.3) to see that the general solution is $u(x,t) = f(2x - t)$, where $f$ is an arbitrary differentiable function of a single variable. To use the initial condition, we set $t = 0$ to find that

$$f(2x) = \frac{1}{1+x^2}.$$

Dividing $x$ by 2 in the preceding equation leads us to

$$f(x) = \frac{1}{1+(x/2)^2}.$$

Combining the general solution $u(x,t) = f(2x-t)$ with our formula for $f$, the overall solution of the initial value problem is

$$u(x,t) = \frac{1}{1+\left(\frac{2x-t}{2}\right)^2} = \frac{1}{1+\left(x - \frac{t}{2}\right)^2}.$$

To visualize the solution, it is useful to plot $u(x,t)$ versus $x$ for several choices of time $t$. Figure 9.1 shows a plot of $u$ versus $x$ at $t = 0$ (the initial condition) as well as $t = 10$ and $t = 20$. Visual inspection suggests a constant propagation speed of 0.5 units, and straightforward calculus reveals that this is, indeed, the case.

The preceding example gives insight into the dynamics of the constant-coefficient PDE (9.2). Namely, the initial ($t = 0$) spatial distribution of $u$ is

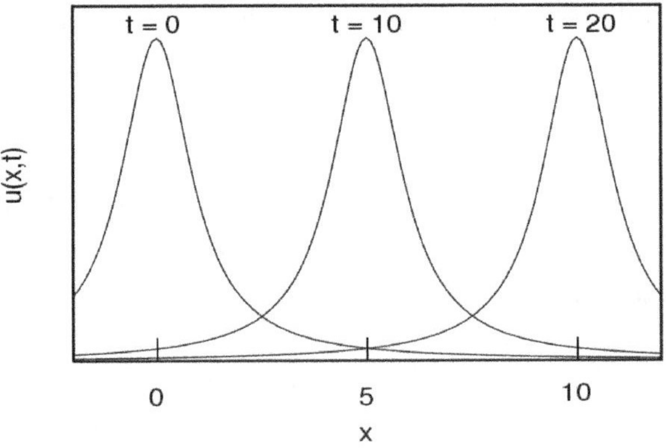


**Figure 9.1.** A plot of $u(x,t)$ versus $x$ for $t=0$, $t=10$, and $t=20$. The three frames of this "animation" suggest that the initial distribution of $u$ is transported from left to right with speed 0.5 units.

*transported* with constant speed while retaining the same spatial profile. We now give an illustration of how this PDE arises naturally in the context of modeling transport.

---

## 9.1. Derivation and Solution of the Transport Equation

Assuming $\alpha \neq 0$, we may divide Equation (9.2) by $\alpha$ to obtain

$$u_t + cu_x = 0, \tag{9.4}$$

where $c = \beta/\alpha$ is a constant. We will refer to this PDE as the *transport equation* or the *advection equation*.

**Derivation.** A physical interpretation of the transport equation can be given as follows. Suppose that a fluid moves with constant speed $c$ in a long, thin pipe. Assume that a pollutant is suspended in the water, and is simply carried by the fluid without diffusing. Moreover, we assume that the pipe can be treated as one-dimensional and that there is no drag/friction between the fluid and the walls of the pipe. If $x$ represents our position along the length of the pipe, we will let $u(x,t)$ denote the concentration (mass per unit length) of pollutant at position $x$ and time $t$. The *mass* $m$ of pollutant in an interval $[0,x]$ at time $t$ is obtained by integrating concentration with respect to the length variable:

$$m = \int_0^x u(s,t)\, ds.$$

If we advance $\Delta t$ units forward in time, the mass that was in the interval $[0, x]$ at time $t$ has moved $c\Delta t$ units down the pipe. In other words, that same mass would be in the interval $[c\Delta t, x + c\Delta t]$ at time $t + \Delta t$, implying that

$$m = \int_{c\Delta t}^{x+c\Delta t} u(s, t + \Delta t)\, ds.$$

Equating our two expressions for mass,

$$\int_0^x u(s,t)\, ds = \int_{c\Delta t}^{x+c\Delta t} u(s, t + \Delta t)\, ds.$$

Differentiating with respect to $x$ and using the Fundamental Theorem of Calculus, we find that

$$u(x,t) = u(x + c\Delta t, t + \Delta t)$$

for all choices of $\Delta t$. Equivalently, $u(x + c\Delta t, t + \Delta t) - u(x,t) = 0$ and, dividing by $\Delta t$, we have

$$\frac{u(x + c\Delta t, t + \Delta t) - u(x,t)}{\Delta t} = 0.$$

Letting $\Delta t \to 0$,

$$\lim_{\Delta t \to 0} \frac{u(x + c\Delta t, t + \Delta t) - u(x,t)}{\Delta t} = 0.$$

Referring to Definition 9.0.2, the preceding equation asserts that the directional derivative in the direction of the vector $(c, 1)$ is zero. Equivalently, $(c, 1) \bullet \nabla u(x,t) = 0$, which is identical to Equation (9.4).

The general solution of the transport equation is readily obtained from Equation (9.3), namely $u(x,t) = g(x - ct)$ where $g$ is an arbitrary differentiable function. The Cauchy problem (i.e., initial value problem on an infinite spatial domain) for the transport equation is given by

$$\begin{aligned} u_t + cu_x &= 0, & (-\infty < x < \infty \text{ and } t > 0) \\ u(x,0) &= f(x), & (-\infty < x < \infty \text{ and } t = 0). \end{aligned}$$

Setting $t = 0$ in the general solution $u(x,t) = g(x - ct)$ reveals that $f(x) = g(x)$. Therefore, the solution of the Cauchy problem for the transport equation is $u(x,t) = f(x - ct)$. Note that at $t = 0$, the spatial concentration profile of the pollutant is described by the graph of $f(x)$. At $t = 1$, we have $u(x,t) = f(x - c)$, implying that the initial distribution of pollutant has been translated horizontally by $c$ units. If $c > 0$, the movement is from left-to-right, and if $c < 0$ then the movement is from right-to-left.

The transport equation can be generalized to higher-dimensional spatial domains. Let $\mathbf{x} = (x_1, x_2, \ldots x_n) \in \mathbb{R}^n$ and suppose that $\mathbf{b} = (b_1, b_2, \ldots b_n)$ is a constant vector in $\mathbb{R}^n$. The transport equation is given by

$$u_t + \mathbf{b} \bullet \nabla u = 0,$$

where the gradient of $u = u(\mathbf{x}; t)$ is taken with respect to the spatial variables only. That is,

$$u_t + b_1 \frac{\partial u}{\partial x_1} + b_2 \frac{\partial u}{\partial x_2} + \cdots + b_n \frac{\partial u}{\partial x_n} = 0.$$

The general solution of this PDE is given by

$$u(\mathbf{x}; t) = u(x_1, x_2, \ldots x_n, t) = g(\mathbf{x} - t\mathbf{b}) = g(x_1 - tb_1, x_2 - tb_2, \ldots, x_n - tb_n),$$

where $g : \mathbb{R}^n \to \mathbb{R}$ is an arbitrary differentiable function. It follows that the solution of the Cauchy problem

$$\begin{aligned} u_t + \mathbf{b} \bullet \nabla u &= 0, && (x \in \mathbb{R}^n \text{ and } t > 0) \\ u(\mathbf{x}, 0) &= f(\mathbf{x}), && (x \in \mathbb{R}^n \text{ and } t = 0), \end{aligned}$$

is given by $u(\mathbf{x}; t) = f(\mathbf{x} - \mathbf{b}t)$ for all $t \geq 0$ and all $\mathbf{x} \in \mathbb{R}^n$.

---

## 9.2. Method of Characteristics: More Examples

When using the method of characteristics, we exploit the fact that solutions of our first-order linear PDEs must remain constant along certain curves in the $xt$-plane. In the case of the transport equation $(c, 1) \bullet \nabla u(x,t) = 0$, the characteristic curves are straight lines parallel to the vector $(c, 1)$. The lines have slope $1/c$ in the $xt$-plane (see Figure 9.2) and solutions $u(x,t)$ must remain constant along these

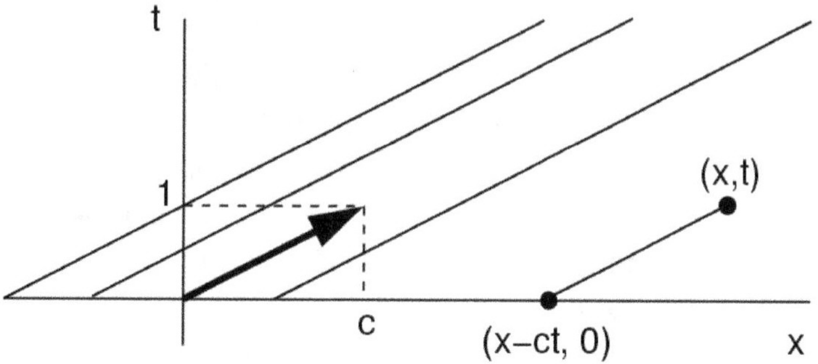


**Figure 9.2.** Characteristic curves for the transport equation $u_t + cu_x = 0$ are lines parallel to the vector $(c, 1)$, shown in bold. Starting from any point $(x, t)$, following the characteristic curve backwards in time to $t = 0$ lets us express $u(x, t)$ in terms of the initial condition.

lines. Now suppose $t > 0$ and let $(x, t)$ be any point in the $xt$-plane. Then to determine $u(x, t)$, we need only trace backwards in time along a characteristic line until we reach $t = 0$. Indeed, since $u(x, t)$ must remain constant as we follow lines of slope $1/c$, it must be the case that $u(x, t) = u(x - ct, 0)$. If we are given an initial condition $u(x, 0) = f(x)$, this implies that $u(x, t) = f(x - ct)$, which we recognize as the solution of the Cauchy problem (i.e., initial value problem) for the transport equation. We now use the method of characteristics to solve a linear, homogeneous, first-order PDE with variable coefficients.

**A partial differential equation with a variable coefficient.** Consider the PDE $u_t + xu_x = 0$, which is equivalent to $\nabla u(x, t) \bullet (x, 1) = 0$. The latter expression says that the directional derivative of $u$ vanishes in the direction of the vector $(x, 1)$. Notice that as $x$ varies, so does the orientation of the vector $(x, 1)$, and sketching this vector field (direction field) helps us visualize the characteristic curves. More exactly, the characteristic curves are the curves in the $xt$-plane that have $(x, 1)$ as their tangent vectors. Since these vectors have slope $1/x$, the characteristic curves satisfy the ordinary differential equation

$$\frac{dt}{dx} = \frac{1}{x}.$$

Assuming $x \neq 0$, we may use separation of variables to solve this ODE:

$$\int 1 \, \mathrm{d}t = \int \frac{1}{x} \, \mathrm{d}x,$$

which implies that $t = \ln|x| + C$, where $C$ is a constant. The family of characteristic curves in the $xt$-plane is shown in Figure (9.3). Since a solution $u(x,t)$ of the PDE must remain constant as we move along a characteristic curve, the value of $u(x,t)$ is completely determined by the value of the constant $C = t - \ln|x|$ if $x \neq 0$. Therefore, any function of the form $u(x,t) = g(t - \ln|x|)$ is a solution of the PDE $u_t + xu_x = 0$ whenever $x \neq 0$. To determine what happens when $x = 0$, notice that the vector $(0,1)$ is always tangent to the characteristic curve when $x = 0$. This implies that the vertical axis $x = 0$ in the $xt$-plane is also a characteristic curve which, in turn, implies that the value of $u(0,t)$ can never change as $t$ varies.

We make several important observations concerning this example:

☞ The characteristic curves completely fill the $xt$-plane without intersecting each other.

☞ Each characteristic curve intersects the horizontal axis $t = 0$. Therefore, if we are provided with an initial condition $u(x,0) = f(x)$, then we can determine the value of $u(x,t)$ by following the characteristic curve through $(x,t)$ until we reach $t = 0$.

☞ Every characteristic curve (excluding $x = 0$) is the graph of an invertible function.

The latter two observations can be used to give a clean solution of the Cauchy problem for this PDE. Suppose that $(x,t)$ is any point in the $xt$-plane and, for convenience, assume $x > 0$. The point $(x,t)$ lies on some characteristic curve, $t = \ln x + C$ where $C$ is a constant. Following this curve to $t = 0$, we find that $0 = \ln x + C$, which implies that $x = e^{-C}$ when $t = 0$. Therefore, since $u(x,t)$ remains constant as we follow a characteristic curve, it must be the case that $u(x,t) = u(e^{-C},0)$. Finally, since $C = t - \ln x$, we have shown that

$$u(x,t) = u\left(e^{\ln x - t},0\right) = u\left(xe^{-t},0\right) = f(xe^{-t}).$$

The case $x < 0$ is handled similarly. Finally, if $x = 0$ we have $u(x,0) = f(0)$, implying that $u$ must remain constant when $x = 0$ (which we already knew).

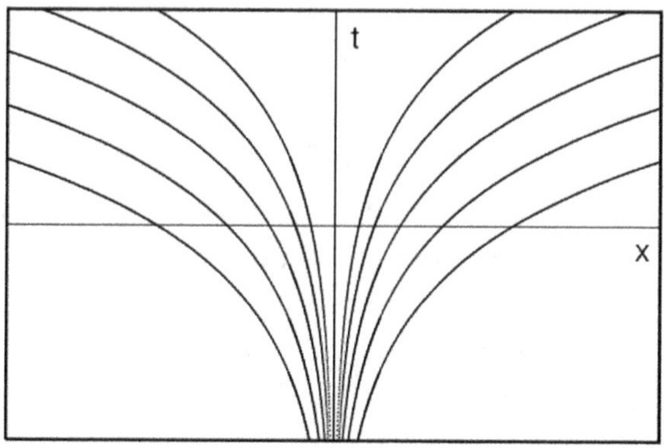


**Figure 9.3.** The family of characteristic curves for the equation $u_t + xu_x = 0$ consists of logarithmic functions $t = \ln|x| + C$ and the vertical axis $x = 0$.

We have shown that the solution of the initial value problem $u_t + xu_x = 0$ with $u(x,0) = f(x)$ is given by $u(x,t) = f(xe^{-t})$.

A graphical interpretation of this solution can be given as follows. As $t$ increases from 0, the exponential function $e^{-t}$ decays from 1 to 0. This has the effect of dilating the initial distribution of $u$, stretching it horizontally as $t$ increases. For example, when $t = \ln 2$, the solution is given by $u(x, \ln 2) = f(x/2)$. The graph of $f(x/2)$ has the same qualitative appearance as the initial condition $f(x)$, but stretched horizontally outward (from $x = 0$) by a factor of 2.

**Example 9.2.1.** Solve the PDE

$$2u_x + 3x^2y^2u_y = 0 \qquad (-\infty < y < \infty)$$

with the auxiliary condition $u(0,y) = e^{-y}$. *Solution:* Writing the PDE in the form $\nabla u(x,y) \bullet (2, 3x^2y^2) = 0$, we see that the characteristic curves must satisfy the ODE

$$\frac{dy}{dx} = \frac{3}{2}x^2y^2.$$

Separation of variables leads us to

$$\int \frac{1}{y^2}\,dy = \frac{3}{2}\int x^2\,dx,$$

and performing the integration reveals that

$$-\frac{1}{y} = \frac{1}{2}x^3 + C,$$

where $C$ is a constant. By algebra, the characteristic curves are given by

$$x = \sqrt[3]{-\frac{2}{y} - 2C}.$$

Since $u(x,y)$ must remain constant along characteristic curves, the value of $u(x,y)$ depends only upon the value of $C = -\frac{1}{y} - \frac{1}{2}x^3$. Therefore, the general solution of the PDE is

$$u(x,y) = f\left(-\frac{1}{y} - \frac{1}{2}x^3\right),$$

where $f$ is an arbitrary differentiable function. Setting $x = 0$ allows us to use the auxiliary condition $u(0,y) = e^{-y}$:

$$f\left(-\frac{1}{y}\right) = e^{-y}.$$

Making the substitution $z = -\frac{1}{y}$ provides the explicit formula

$$f(z) = e^{1/z}.$$

Finally, combining this formula for $f$ with the general solution of the PDE yields the overall solution

$$u(x,y) = \exp\left(-\frac{1}{\frac{1}{y} + \frac{1}{2}x^3}\right),$$

where we have written $\exp(z)$ instead of $e^z$ for notational convenience.

The preceding example merits a few general remarks summarizing the method of characteristics. For $a(x,t)u_t + b(x,t)u_x = 0$, a linear, homogeneous, first-order PDE, the associated characteristic equation is given by

$$\frac{dt}{dx} = \frac{a(x,t)}{b(x,t)} \quad \text{or} \quad \frac{dx}{dt} = \frac{b(x,t)}{a(x,t)}.$$

Ideally, this ODE can be solved by hand, although this is certainly not the case if the coefficient functions $a(x,t)$ and $b(x,t)$ are too complicated. The solutions of the ODE form the family of characteristic curves. Assuming that the characteristic curves do not intersect one another (as in the above examples), the solution $u(x,t)$ is completely determined by which characteristic curve the pair $(x,t)$ lies on. In such cases, it is desirable that all characteristic curves exist for all $t \geq 0$. That way, determining the solution $u(x,t)$ when $t > 0$ is a simple matter of following the characteristic curve through $(x,t)$ backwards in time until we reach $t = 0$, thereby allowing us to invoke any initial conditions. Unfortunately, there are many simple-looking first-order PDEs giving rise to characteristic curves that are far from ideal.

**Example 9.2.2.** The first-order nonlinear PDE $u_t + uu_x = 0$ is known as Burgers' Equation. Equivalently, $\nabla u(x,t) \bullet (u(x,t),1) = 0$, which leads us to the characteristic equation

$$\frac{dx}{dt} = u(x,t).$$

Without knowing the solutions $u(x,t)$ of the PDE, it would appear that solving this ODE is impossible. However, we actually *can* learn quite a bit about the characteristic curves. Suppose that $(x(t),t)$ is a parametrization of a characteristic curve. To see that $u$ remains constant along this curve, we first use the chain rule to calculate

$$\frac{d}{dt}u(x(t),t) = \frac{\partial u}{\partial x}\frac{dx}{dt} + \frac{\partial u}{\partial t}.$$

However, we know that $\frac{dx}{dt} = u(x,t)$ according to the above characteristic ODE. Therefore,

$$\frac{d}{dt}u(x(t),t) = u_t + uu_x = 0,$$

which means that $u(x(t),t)$ is constant. Since $u$ must remain constant along each characteristic curve, the ODE $\frac{dx}{dt} = u(x,t)$ reduces to $\frac{dx}{dt} = \text{constant}$. In other words, we have shown that the characteristic curves are actually *straight lines*. The slope of the characteristic curve through the point $(x,t)$ is equal to the value of $u(x,t)$.

To explore the consequences of these observations, suppose that we impose the initial condition

$$u(x,0) = \phi(x) = \begin{cases} 1 & \text{if } x \leq -1 \\ -x & \text{if } -1 < x < 0 \\ 0 & \text{if } 0 \leq x. \end{cases} \tag{9.5}$$

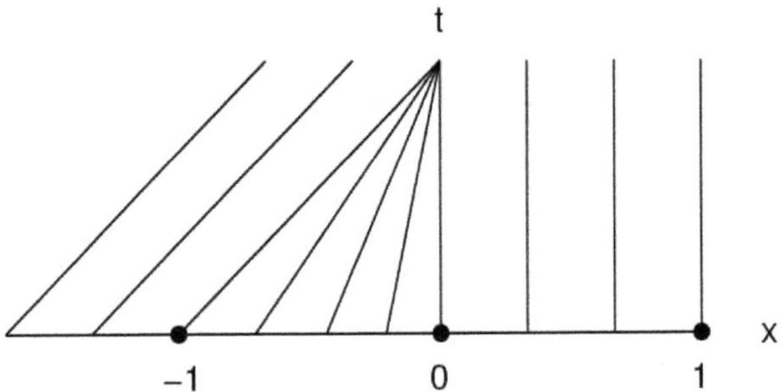


**Figure 9.4.** The family of characteristic curves for the equation $u_t + uu_x = 0$ for the initial condition (9.5).

To sketch the characteristic curves in the $xt$-plane, it helps to rewrite the characteristic ODE as

$$\frac{dt}{dx} = \frac{1}{u(x,t)},$$

with $\frac{dt}{dx}$ undefined if $u(x,t) = 0$. According to our initial condition, the characteristic curves have slope $\frac{dt}{dx} = 1$ for $x \leq -1$, slope $-1/x$ for $-1 < x < 0$, and undefined slope for $x \geq 0$. A sketch of the characteristic curves appears in Figure 9.4. Notice that the lines intersect one another at $t = 1$, a feature of this PDE that has some rather interesting implications. The slopes of the characteristic lines determine the velocity at which our initial distribution of $u(x,t)$ will propagate. For the initial condition (9.5), the propagation velocity is 0 when $x > 0$ but is positive when $x < 0$. Three snapshots of the solution $u(x,t)$ are shown in Figure 9.5. The wave moves from left to right, and the regions where $u$ is highest have the fastest propagation speed. At time $t = 1$, the spatial profile of $u(x,t)$ develops a discontinuous jump at $x = 0$. Examining how the characteristic lines cross each other helps explain the formation of this *shock wave*. The shock forms when the fast moving parts of the wave encounter the "stationary" region $x > 0$. Describing what happens after the shock is formed is a subject more suitable for an advanced PDE course.

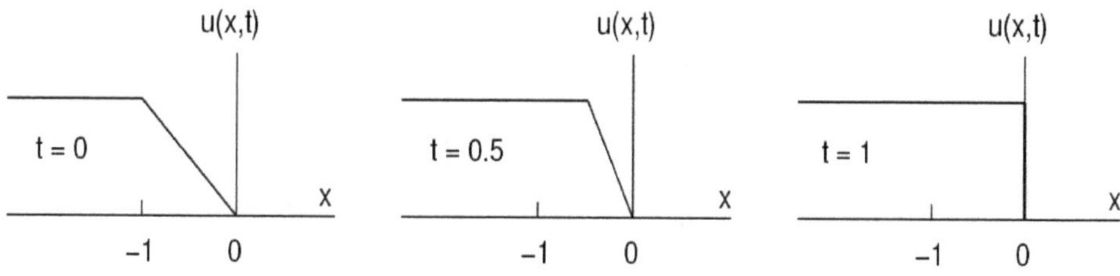


**Figure 9.5.** Solution of $u_t + uu_x = 0$ with initial condition (9.5) at times $t = 0$ (the initial condition), $t = 1/2$ and $t = 1$. A shock has formed at $t = 1$.

---

## Exercises

1. Solve the Cauchy problem

$$\begin{aligned} 2u_t + 5u_x &= 0 \\ u(x,0) &= \frac{1}{1+x^2} \end{aligned}$$

and sketch $u(x,t)$ versus $x$ when $t = 0$, $t = 5$ and $t = 10$.

2. Use the substitution $v = u_y$ to find the general solution of $2u_y + u_{xy} = 0$.
3. Solve the PDE $(e^x + e^{-x})u_t + 2u_x = 0$, and sketch a few of the characteristic curves.
4. Let $g$ be a function of one variable and let $G$ denote an antiderivative of $g$. Use the method of characteristics to show that the general solution of $g(x)u_t + u_x = 0$ is given by $u(x,t) = f(t - G(x))$, where $f$ is an arbitrary differentiable function.
5. Let $g$ and $h$ be a functions of one variable and let $G$ and $H$ denote antiderivatives of $g$ and $h$, respectively. Use the method of characteristics to show that the general solution of $g(x)u_t + h(t)u_x = 0$ is given by $u(x,t) = f(H(t) - G(x))$, where $f$ is an arbitrary differentiable function.
6. Solve the PDE $u_t + 2tx^2u_x = 0$ with initial condition $u(x,0) = \sin x$.
7. Solve the initial value problem

$$\begin{aligned} (1+t^2)u_t + u_x &= 0 \qquad (-\infty < x < \infty), \\ u(x,0) &= e^{-x^2}. \end{aligned}$$

Then, sketch three different characteristic curves. Finally, plot $u(x,t)$ versus $x$ for three different choices of $t$.

8. These questions concern the initial value problem for the inhomogeneous transport equation on the whole real line:

$$\begin{aligned} u_t + cu_x &= g(x,t) \qquad (-\infty < x < \infty), \\ u(x,0) &= f(x). \end{aligned}$$

It is possible to show that the solution of this initial value problem is

$$u(x,t) = f(x-ct) + \int_0^t g\left(x + c(s-t), s\right)\, ds. \tag{9.6}$$

Note that the inhomogeneity $g(x,t)$ gives rise to an integral which was not present when we solved the homogeneous transport equation.

(a) Using Equation (9.6), show that the solution of the initial value problem

$$\begin{aligned} u_t + u_x &= x \cos t, \qquad (-\infty < x < \infty) \\ u(x,0) &= \frac{1}{1+x^2} \end{aligned}$$

is given by

$$u(x,t) = \frac{1}{1+(x-t)^2} + x \sin t + \cos t - 1.$$

(b) Now, using direct substitution, check that $u(x,t)$ really *does* satisfy the initial value problem in Part (a).

# CHAPTER 10

# The Heat and Wave Equations on an Unbounded Domain

At first glance, the heat equation $u_t - \kappa u_{xx} = 0$ and the wave equation $u_{tt} - c^2 u_{xx} = 0$ appear very similar. Since $\kappa$ and $c^2$ are always assumed to be positive constants, the only apparent distinction is between the $u_t$ in the heat equation and the $u_{tt}$ in the wave equation. As we shall see, this makes a *profound* difference in the behavior of the solutions of these two equations. We begin this Chapter with a derivation of these two PDEs from basic physical principles. Then, we will solve both PDEs on the unbounded one-dimensional spatial domain $-\infty < x < \infty$.

## 10.1. Derivation of the Heat and Wave Equations

The heat equation arises in the context of modeling diffusive processes. For example, suppose heat is distributed within a long, thin wire that can be treated as one-dimensional. Further suppose that the wire is insulated so that heat is only transferred within the wire (as opposed to radially outward). If $T(x,t)$ denotes the temperature at position $x$ and time $t$, then $T$ satisfies the heat equation.

**The Heat Equation.** For the purposes of deriving the heat equation, we actually have in mind another phenomenon that is modeled by the same PDE. Namely, suppose that a dye (such as food coloring) diffuses in a motionless liquid that is confined to a long, thin "one-dimensional" pipe. Letting $u(x,t)$ denote the concentration of dye at position $x$ and time $t$, our goal is to derive an equation that captures the dynamics of $u$. Generally, dye diffuses from regions of higher concentration to regions of lower concentration, and the relevant physical law that governs diffusion is

**Fick's Law of Diffusion.** The rate of motion of dye is proportional to the concentration gradient.

Fick's Law will let us track how the total mass of dye in a given region of pipe, say $[x_0, x_1]$, changes over time. Mass is obtained by integrating concentration; i.e., the mass of dye in the region $[x_0, x_1]$ at time $t$ is given by

$$M(t) = \int_{x_0}^{x_1} u(x,t)\,\mathrm{d}x.$$

Differentiating with respect to $t$ yields

$$\frac{\mathrm{d}M}{\mathrm{d}t} = \int_{x_0}^{x_1} u_t(x,t)\,\mathrm{d}x.$$

Since $\frac{\mathrm{d}M}{\mathrm{d}t}$ measures the rate of change of mass, it is equal to the difference between the rate of flow of dye into the interval $[x_0, x_1]$ and the rate of flow of dye out of the interval $[x_0, x_1]$. By Fick's Law, the rate of flow of the dye is proportional to the (spatial) concentration gradient $u_x$. Letting $\kappa > 0$ denote the proportionality constant, the rate of inward flow at the left endpoint $x_0$ is given by $-\kappa u_x(x_0, t)$. To explain why the negative sign is included, suppose $u_x(x_0, t) > 0$. Then the concentration is higher for $x > x_0$ than for $x < x_0$, which implies that the flow at $x_0$ would be from right-to-left (see Figure 10.1). By contrast, if $u_x(x_0, t) < 0$, then the concentration is higher for $x < x_0$ than for $x > x_0$, thereby resulting in a flow from left-to-right at $x_0$. Similar reasoning shows that the rate of inward flow at the right endpoint $x_1$ is given by $\kappa u_x(x_1, t)$. Therefore,

$$\frac{\mathrm{d}M}{\mathrm{d}t} = \kappa u_x(x_1, t) - \kappa u_x(x_0, t),$$

and equating this to our original expression for $\frac{\mathrm{d}M}{\mathrm{d}t}$ yields

$$\int_{x_0}^{x_1} u_t(x,t)\,\mathrm{d}x = \kappa u_x(x_1, t) - \kappa u_x(x_0, t).$$

Taking the derivative of both sides with respect to $x_1$, we see that

$$u_t(x_1, t) = \kappa u_{xx}(x_1, t).$$

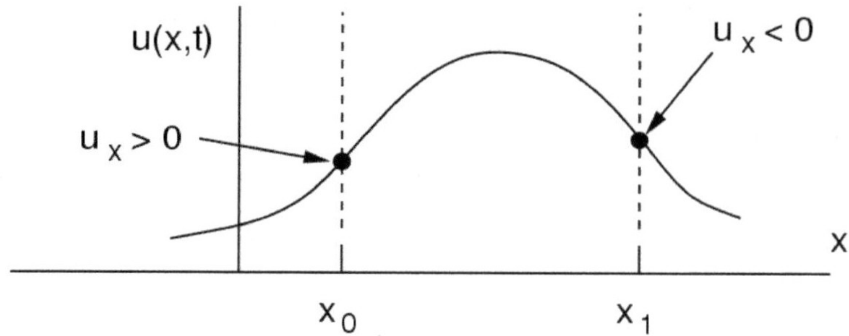


**Figure 10.1.** Equation (10.1) describes diffusive processes in which diffusion is proportional to the concentration gradient $u_x$. In this figure, $u_x > 0$ at the left endpoint $x_0$, and diffusion should cause a net outward flow (from right to left) at $x_0$. Similarly, $u_x < 0$ at the right endpoint $x_1$. The flow should be from left to right at $x_1$, i.e., from higher concentration towards lower concentration. Note that the direction of motion has the "opposite sign" of the gradient $u_x$.

Since $x_1$ was arbitrary, we have derived the *heat equation* or *diffusion equation*

$$u_t = \kappa u_{xx}. \tag{10.1}$$

**The Wave Equation.** We now use physical principles to derive the wave equation $u_{tt} - c^2 u_{xx} = 0$. Whereas the heat equation was derived from Fick's Law, the wave equation can be derived from Newton's second law of motion, $F = ma$.

Suppose that we pluck a flexible, elastic guitar string of uniform density. Our goal is to completely describe the displacement $u(x,t)$ from the equilibrium position of the string (see Figure 10.2). We will make several assumptions concerning the dynamics of the vibrating string. First, we assume that all vibrations are *transverse* (i.e., up-and-down in Figure 10.2) as opposed to longitudinal (i.e., side-to-side in the figure). Second, we assume that the vibrations are "small" in the sense that both $u(x,t)$ and $u_x(x,t)$ have low magnitudes. Finally, we assume that the string has constant density $\rho$ and that air resistance is negligible.

Let $\mathbf{T}(x,t)$ denote the tension (force) vector at position $x$ and time $t$. Tension $\mathbf{T}(x,t)$ is directed tangent to the string, and therefore has the same direction as the vector $(1, u_x(x,t))$ (see Figure 10.2). We will examine the force and acceleration on a region $[x_0, x_1]$ of the guitar string. The idea is to write Newton's Law

$\mathbf{F} = m\mathbf{a}$ in terms of its transverse and longitudinal components where, in this case, the force vector $\mathbf{F}$ is tension. Mathematically,

$$\mathbf{T}(x,t) \;=\; \mathbf{T}_{\text{long}}(x,t) + \mathbf{T}_{\text{trans}}(x,t).$$

Since $\mathbf{T}(x,t)$ is parallel to the vector $(1, u_x(x,t))$, we can derive relationships between the lengths of transverse and longitudinal components of the vectors by using a "similar triangles" argument (see Figure 10.3). Letting $\|\mathbf{T}\|$ denote the length of the tension vector, we see that

$$\frac{\|\mathbf{T}_{\text{long}}\|}{\|\mathbf{T}\|} \;=\; \frac{1}{\sqrt{1+u_x^2}},$$

from which it follows that the magnitude of the longitudinal force is given by

$$\|\mathbf{T}_{\text{long}}\| \;=\; \frac{\|\mathbf{T}\|}{\sqrt{1+u_x^2}}.$$

Similarly,

$$\frac{\|\mathbf{T}_{\text{trans}}\|}{\|\mathbf{T}\|} \;=\; \frac{u_x}{\sqrt{1+u_x^2}},$$

from which it follows that the magnitude of the transverse force is given by

$$\|\mathbf{T}_{\text{trans}}\| \;=\; \frac{\|\mathbf{T}\| u_x}{\sqrt{1+u_x^2}}. \tag{10.2}$$

By assumption, the longitudinal acceleration of the region $[x_0, x_1]$ is 0, and therefore

$$\left.\frac{\|\mathbf{T}(x,t)\|}{\sqrt{1+u_x^2}}\right|_{x_0}^{x_1} \;=\; 0. \tag{10.3}$$

Since $u(x,t)$ represents transverse displacement, the transverse acceleration is $u_{tt}(x,t)$, the second derivative of $u$ with respect to $t$. If we partition the interval $[x_0, x_1]$ into small segments of width $\Delta x$, then the mass of each segment is $\rho\Delta x$. Letting $x_j^*$ denote the right endpoint of the $j^{th}$ segment, the transverse acceleration of the $j^{th}$ segment is approximately $u_{tt}(x_j^*, t)$. Transverse force on the $j^{th}$ segment is given by Newton's law

$$\mathbf{F}_{\text{trans}} = m\mathbf{a}_{\text{trans}} \approx \rho u_{tt}(x_j^*, t)\Delta x.$$

Summing over all segments yields a Riemann sum for the transverse force exerted on the whole interval $[x_0, x_1]$, namely

$$\mathbf{F}_{\text{trans}} \approx \sum_j \rho u_{tt}(x_j^*, t)\Delta x.$$

As $\Delta x \to 0$, this Riemann sum converges to an integral

$$\mathbf{F}_{\text{trans}} = \int_{x_0}^{x_1} \rho u_{tt}(x,t)\, \mathrm{d}x.$$

Since the transverse force is due to tension, this expression for transverse force must match our earlier expression (10.2):

$$\left.\frac{\|\mathbf{T}(x,t)\| u_x}{\sqrt{1+u_x^2}}\right|_{x_0}^{x_1} = \int_{x_0}^{x_1} \rho u_{tt}(x,t)\, \mathrm{d}x. \tag{10.4}$$

Since we have assumed that the transverse vibrations are small, we know that $|u_x| \approx 0$. Thus, we may approximate $\sqrt{1+u_x^2} \approx 1$ with a high degree of accuracy. This simplification reduces the longitudinal equation (10.3) to

$$\left.\frac{\|\mathbf{T}(x,t)\|}{\sqrt{1}}\right|_{x_0}^{x_1} = 0,$$

which means that $\|\mathbf{T}(x_1,t)\| = \|\mathbf{T}(x_0,t)\|$. In other words, tension has approximately constant magnitude along the entire vibrating string, and so $\mathbf{T}$ is actually independent of $x$. It is also reasonable to assume that the tension is independent of time $t$, and from now on we will let $T$ denote the (constant) tension force. With these simplifications, the equation (10.4) for transverse force reduces to

$$\left. Tu_x \right|_{x_0}^{x_1} = \int_{x_0}^{x_1} \rho u_{tt}(x,t)\, \mathrm{d}x.$$

Differentiating with respect to $x_1$ and applying the Fundamental Theorem of Calculus, we are left with $Tu_{xx} = \rho u_{tt}$. Introducing the constant $c = \sqrt{T/\rho}$, we have derived the *wave equation*

$$u_{tt} - c^2 u_{xx} = 0. \tag{10.5}$$

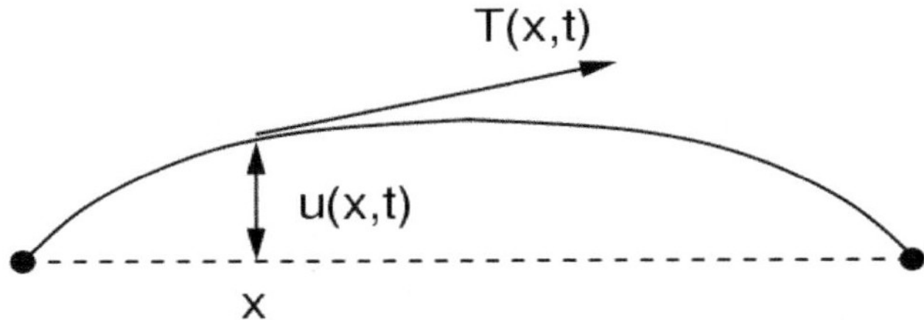


**Figure 10.2.** The displacement $u(x,t)$ of a guitar string (solid curve) from its resting position (dashed line) at a fixed time $t$. The tension vector $\mathbf{T}(x,t)$ is tangent to the curve.

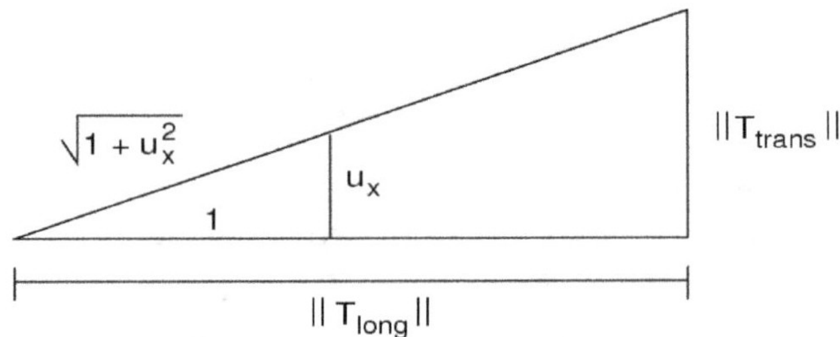


**Figure 10.3.** Longitudinal and transverse components of the tension vector $\mathbf{T}$. The hypotenuse of the larger triangle has length $\|\mathbf{T}\|$, the magnitude of the tension vector.

---

## 10.2. Cauchy Problem for the Wave Equation

In this section we will solve the wave equation (10.5) on the unbounded domain $-\infty < x < \infty$. Before imposing initial conditions, we will derive the general solution of the equation.

**General solution.** Starting from the wave equation

$$\left(\frac{\partial^2}{\partial t^2} - c^2\frac{\partial^2}{\partial x^2}\right) u = 0,$$

the idea is to factor the operator as

$$\left(\frac{\partial}{\partial t} - c\frac{\partial}{\partial x}\right)\left(\frac{\partial}{\partial t} + c\frac{\partial}{\partial x}\right) u = 0. \tag{10.6}$$

Note that separately, the operators

$$\left(\frac{\partial}{\partial t} - c\frac{\partial}{\partial x}\right) \quad \text{and} \quad \left(\frac{\partial}{\partial t} + c\frac{\partial}{\partial x}\right)$$

remind us of the transport equation. Since we know that the general solution of the transport equation

$$\left(\frac{\partial}{\partial t} + c\frac{\partial}{\partial x}\right) u = 0$$

is given by $u(x,t) = g(x - ct)$ where $g$ is an arbitrary differentiable function, we claim that the general solution of the wave equation (10.6) is given by

$$u(x,t) = f(x + ct) + g(x - ct), \tag{10.7}$$

where $f$ and $g$ are arbitrary differentiable functions. To prove this claim, we write the second-order PDE (10.6) as a system of two first-order PDEs by introducing the new variable

$$w = \left(\frac{\partial}{\partial t} + c\frac{\partial}{\partial x}\right) u = u_t + cu_x.$$

The resulting system is

$$w_t - cw_x = 0 \quad \text{and} \quad u_t + cu_x = w.$$

The solution of the homogeneous equation for $w$ is $w(x,t) = h(x + ct)$, where $h$ is an arbitrary differentiable function. Substituting this formula for $w$ into the other equation yields

$$u_t + cu_x = h(x + ct).$$

How should we go about solving this inhomogeneous PDE? As with ODEs, one option is to produce one *particular* solution of the inhomogeneous equation and add it to the *general* solution of the associated homogeneous PDE. The homogeneous problem $u_t + cu_x = 0$ has general solution $u(x,t) = g(x - ct)$, where $g$ is an arbitrary differentiable function. To produce a particular solution of the inhomogeneous equation, it is natural to guess that if $H$ is an antiderivative for $h$, then $H(x + ct)$ will satisfy the PDE. This is almost the case—suppose that $u(x,t) = H(x + ct)$. Then

$$u_t = cH'(x + ct) = ch(x + ct) \quad \text{and} \quad u_x = H'(x + ct) = h(x + ct).$$

Therefore, $u_t + cu_x = 2ch(x + ct)$, and we see that our guess was merely off by a factor of $2c$. This is easily remedied by re-scaling our guess; specifically, define

$$f(\xi) = \frac{1}{2c}H(\xi) = \frac{1}{2c}\int h(\xi)\,d\xi,$$

a scalar multiple of the antiderivative of $h$. It is straightforward to check that $u(x,t) = f(x+ct)$ actually *is* a particular solution of $u_t + cu_x = h(x+ct)$. Adding this particular solution to the general solution of the homogeneous equation $u_t + cu_x = 0$ establishes that (10.7) really is the general solution of the wave equation.

Notice that the general solution of the wave equation contains two arbitrary functions, which is no coincidence since the wave equation is second-order in the variable $t$. If $c > 0$, then the function $f(x+ct)$ corresponds to a wave moving right-to-left with speed $c$, and the function $g(x-ct)$ corresponds to a wave moving left-to-right with speed $c$. Whereas the transport equation $u_t + cu_x = 0$ gives rise to one family of characteristic curves ($x - ct =$ constant), the wave equation gives rise to *two* families of characteristic curves: $x \pm ct =$ constant. In other words, the wave equation transmits waves in both directions, whereas the transport equation can only transmit a wave in one direction.

**The Cauchy Problem.** Recall that a *Cauchy problem* for a PDE is essentially an initial value problem on an unbounded domain (i.e., boundary conditions are not needed). Since the wave equation models the displacement of a vibrating guitar string from its resting position, intuitively we expect that two initial conditions will be needed: initial displacement and initial velocity. Indeed, the Cauchy problem for the wave equation is given by

$$\begin{aligned} u_{tt} - c^2 u_{xx} &= 0 && (-\infty < x < \infty), \\ u(x,0) &= \phi(x) && (-\infty < x < \infty), \\ u_t(x,0) &= \psi(x) && (-\infty < x < \infty), \end{aligned} \tag{10.8}$$

where $\phi(x)$ and $\psi(x)$ represent the initial displacement and velocity of the string, respectively. Since we know the general solution of the wave equation is given by

$$u(x,t) = f(x+ct) + g(x-ct),$$

we must incorporate the initial conditions to solve for $f$ and $g$ in terms of $\phi$ and $\psi$. Setting $t = 0$ yields

$$u(x,0) = \phi(x) = f(x) + g(x).$$

To use the initial condition for velocity, differentiate with respect to $t$ to obtain

$$u_t(x,t) = cf'(x+ct) - cg'(x-ct),$$

and then set $t = 0$:

$$u_t(x,0) = \psi(x) = cf'(x) - cg'(x).$$

Integrating $\psi = cf' - cg'$ over the interval $[0,x]$,

$$\frac{1}{c}\int_0^x \psi(s)\,ds \;=\; \int_0^x f'(s) - g'(s)\,ds \;=\; f(x) - g(x) - f(0) + g(0).$$

Combined with our earlier equation $\phi = f + g$, we obtain the system

$$f(x) + g(x) = \phi(x) \tag{10.9}$$

$$f(x) - g(x) = f(0) - g(0) \;+\; \frac{1}{c}\int_0^x \psi(s)\,ds. \tag{10.10}$$

Adding (10.9) and (10.10) and solving for $f(x)$, we have

$$f(x) \;=\; \frac{1}{2}\phi(x) \;+\; \frac{1}{2c}\int_0^x \psi(s)\,ds \;+\; \frac{f(0) - g(0)}{2}, \tag{10.11}$$

and subtracting (10.10) from (10.9) yields

$$g(x) \;=\; \frac{1}{2}\phi(x) \;-\; \frac{1}{2c}\int_0^x \psi(s)\,ds \;-\; \frac{f(0) - g(0)}{2}. \tag{10.12}$$

Substituting the expressions (10.11) and (10.12) into the general solution (10.7) of the wave equation,

$$\begin{aligned} u(x,t) = f(x+ct) \;+\; g(x-ct) \;&=\; \frac{1}{2}\phi(x+ct) \;+\; \frac{1}{2c}\int_0^{x+ct} \psi(s)\,ds \\ &\quad + \frac{1}{2}\phi(x-ct) \;-\; \frac{1}{2c}\int_0^{x-ct} \psi(s)\,ds. \end{aligned} \tag{10.13}$$

(Notice that all of the stray constants in (10.11) and (10.12) have now dropped out.) Combining the two integrals in (10.13), we have shown that

$$u(x,t) = \frac{1}{2}\left[\phi(x+ct)+\phi(x-ct)\right] + \frac{1}{2c}\int_{x-ct}^{x+ct} \psi(s)\,ds \qquad (10.14)$$

is the solution of the Cauchy problem for the wave equation. We will refer to it as *D'Alembert's formula* in honor of the mathematician who first discovered it.

By the very construction of D'Alembert's formula (10.14), it is evident that it represents the *unique* solution of the Cauchy problem for the wave equation, and that small changes in $\phi$ and $\psi$ do not cause major changes in the behavior of solutions. In this sense, the Cauchy problem for the wave equation is *well-posed*—there is precisely one solution and it is not hyper-sensitive to small changes in the initial conditions.

Inspection of D'Alembert's formula (10.14) provides some insight regarding how solutions of the wave equation should behave. If the initial velocity is zero, then the integral term vanishes and we are left with $u(x,t) = \frac{1}{2}[\phi(x+ct)+\phi(x-ct)]$. Effectively, this says that our initial displacement $\phi$ will be "split" into two waves traveling in opposite directions with speed $|c|$ and with half the amplitude of the initial wave profile. The next example reinforces this observation.

**Example 10.2.1.** Solve the Cauchy problem

$$\begin{aligned} u_{tt} - u_{xx} &= 0 && (-\infty < x < \infty), \\ u(x,0) &= \phi(x) && (-\infty < x < \infty), \\ u_t(x,0) &= 0 && (-\infty < x < \infty), \end{aligned}$$

where the initial displacement $\phi(x)$ is given by $\phi(x) = 1$ if $-1 \le x \le 1$ and $\phi(x) = 0$ otherwise.

**Solution:** Although it may seem odd that the initial displacement is discontinuous, nothing prevents us from quoting D'Alembert's formula. In this case, the wave speed is $c = 1$ and the initial velocity is $\psi = 0$. According to (10.14), the solution of this initial value problem is $u(x,t) = \frac{1}{2}[\phi(x+t)+\phi(x-t)]$, where $\phi$ is the given initial displacement. By definition of $\phi$, we have

$$\phi(x+t) = \begin{cases} 1 \text{ if } -1 \le x+t \le 1 \\ 0 \text{ otherwise} \end{cases}$$

and

$$\phi(x-t) = \begin{cases} 1 \text{ if } -1 \leq x-t \leq 1 \\ 0 \text{ otherwise.} \end{cases}$$

It is instructive to sketch the domains $-1 \leq x+t \leq 1$ and $-1 \leq x-t \leq 1$ in the $xt$ plane (see Figure 10.4). The Figure shows the regions in which $\phi(x+t) = 1$, $\phi(x-t) = 1$, both are 1, or both are 0.

**Example 10.2.2.** Solve the Cauchy problem

$$\begin{aligned} u_{tt} - 9u_{xx} &= 0 & (-\infty < x < \infty), \\ u(x,0) &= e^{-x^2} & (-\infty < x < \infty), \\ u_t(x,0) &= xe^{-x^2} & (-\infty < x < \infty). \end{aligned}$$

**Solution:** In this case, the speed is $c = 3$. Quoting D'Alembert's formula, the solution is given by

$$u(x,t) = \frac{1}{2}\left[e^{-(x+3t)^2} + e^{-(x-3t)^2}\right] + \frac{1}{6}\int_{x-3t}^{x+3t} se^{-s^2}\,ds.$$

A substitution $v = -s^2$, $-\frac{1}{2}dv = sds$ facilitates the integration:

$$u(x,t) = \frac{1}{2}\left[e^{-(x+3t)^2} + e^{-(x-3t)^2}\right] + \frac{1}{6}\int_{-(x-3t)^2}^{-(x+3t)^2} -\frac{1}{2}e^{v}\,dv.$$

Evaluating the integral, the overall solution is given by

$$\begin{aligned} u(x,t) &= \frac{1}{2}\left[e^{-(x+3t)^2} + e^{-(x-3t)^2}\right] - \frac{1}{12}\left[e^{-(x+3t)^2} - e^{-(x-3t)^2}\right] \\ &= \frac{5}{12}e^{-(x+3t)^2} + \frac{7}{12}e^{-(x-3t)^2}. \end{aligned}$$

Notice that in this example, the non-zero initial velocity breaks the "symmetry" of the two waves that propagate outward. In some sense, we see that $\frac{7}{12}$ of our initial displacement profile travels to the right with speed 3, while only $\frac{5}{12}$ travels to the left with speed 3. This may seem a bit surprising considering that both the initial displacement and velocity have even and odd symmetries, respectively.

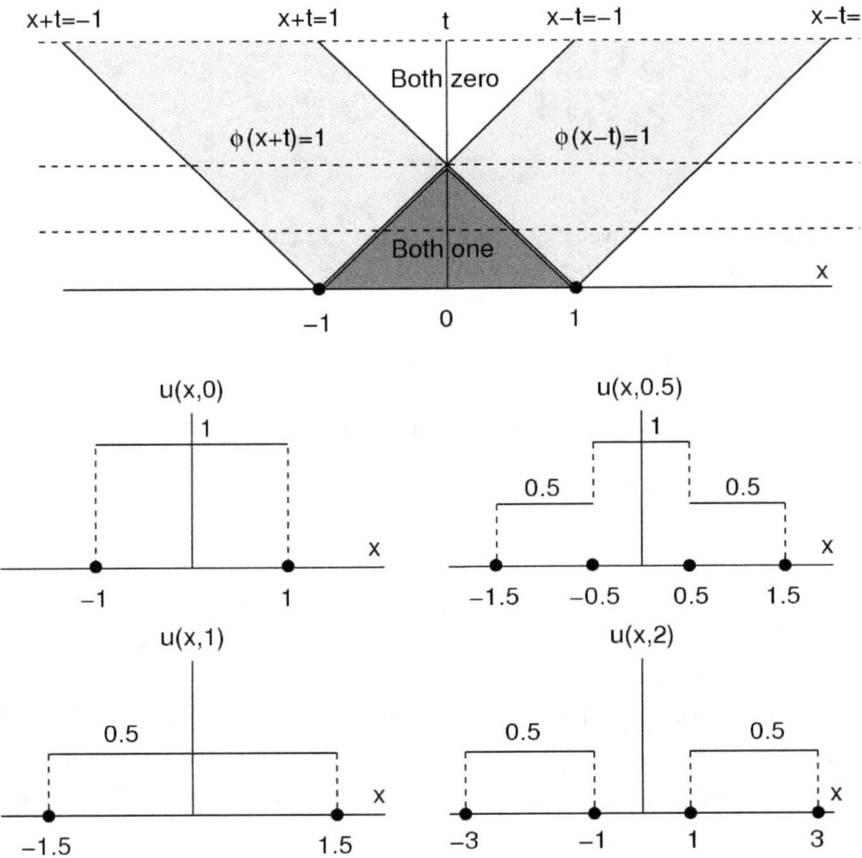


**Figure 10.4.** Wave propagation with a discontinuous initial displacement $\phi(x)$ (see example in text). Upper figure: Regions in which either $\phi(x+t) = 1$, $\phi(x-t) = 1$, both are 1, or both are 0. The dashed horizontal lines, included for reference, correspond to $t = 0.5$, 1 and 2. Lower figure: Four frames of an "animation" of the solution. The initial condition $u(x,0)$ is discontinuous, with jumps occurring at $x = \pm 1$. At time $t = 1/2$, we see that the initial displacement is beginning to split into two waves of half amplitude propagating outward. At $t = 1$, the two waves have almost separated from each other, and by $t = 2$ we have two completely separate waves. The two waves are identical to the initial displacement but have half the amplitude.

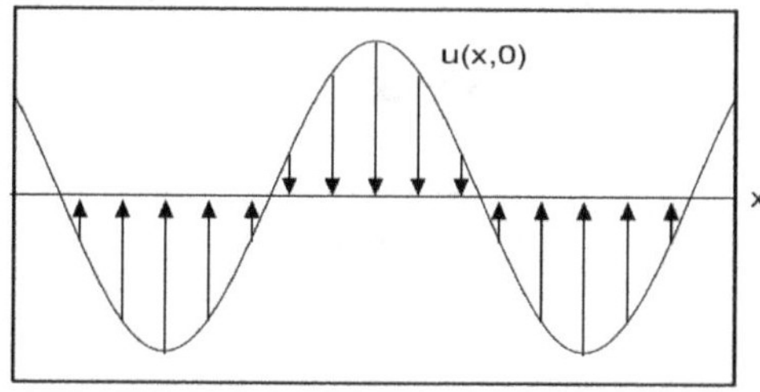


**Figure 10.5.** Initial velocity $u_t(x,0) = -\cos x$ (indicated by arrows) acts opposite the initial displacement $u(x,0) = \cos x$.

**Example 10.2.3.** Solve the Cauchy problem

$$\begin{aligned} u_{tt} - c^2 u_{xx} &= 0 && (-\infty < x < \infty), \\ u(x,0) &= \cos x && (-\infty < x < \infty), \\ u_t(x,0) &= -\cos x && (-\infty < x < \infty). \end{aligned}$$

*Solution:* Before quoting D'Alembert's formula, examine the initial conditions carefully. The initial velocity acts completely "opposite" the initial displacement as illustrated in Figure 10.5. Intuitively, this would seem to suggest that the solution of the Cauchy problem will be a *standing wave*: the wave will oscillate up and down, with motion perpendicular to the $x$-axis. Moreover, the string should remain stationary at all the points where $u(x,0) = 0$. By D'Alembert's formula (10.14), the solution of the Cauchy problem is

$$\begin{aligned} u(x,t) &= \frac{1}{2}\left[\cos(x+ct) + \cos(x-ct)\right] - \frac{1}{2c}\int_{x-ct}^{x+ct} \cos(s)\, ds \\ &= \frac{1}{2}\left[\cos(x+ct) + \cos(x-ct)\right] - \frac{1}{2c}\left[\sin(x+ct) - \sin(x-ct)\right]. \end{aligned} \tag{10.15}$$

Although this is the correct solution of our Cauchy problem, visualizing how the solution behaves is not easy unless we write it in a simpler form. The double-angle identities from trigonometry

$$\begin{aligned} \sin(\alpha+\beta) &= \sin\alpha\cos\beta \;+\; \cos\alpha\sin\beta \\ \cos(\alpha+\beta) &= \cos\alpha\cos\beta \;-\; \sin\alpha\sin\beta \end{aligned}$$

combined with the symmetry properties $\sin(-\alpha) = -\sin\alpha$ and $\cos(-\alpha) = \cos(\alpha)$ will greatly simplify (10.15). The first term in (10.15) is

$$\begin{aligned}
&\frac{1}{2}\left[\cos(x+ct)+\cos(x-ct)\right] \\
&\quad = \frac{1}{2}\left[\cos(x)\cos(ct)-\sin(x)\sin(ct)+\cos(x)\cos(-ct)-\sin(x)\sin(-ct)\right] \\
&\quad = \frac{1}{2}\left[\cos(x)\cos(ct)-\sin(x)\sin(ct)+\cos(x)\cos(ct)+\sin(x)\sin(ct)\right] \\
&\quad = \cos(x)\cos(ct).
\end{aligned}$$

In the same way, we can simplify the other term in (10.15):

$$-\frac{1}{2c}\left[\sin(x+ct)-\sin(x-ct)\right] = -\frac{1}{c}\cos(x)\sin(ct).$$

Combining our two simplifications, our overall solution is now written as

$$u(x,t) = \cos(x)\cos(ct)-\frac{1}{c}\cos(x)\sin(ct) = \cos(x)\left[\cos(ct)-\frac{1}{c}\sin(ct)\right].$$

With $u(x,t)$ expressed in this form, solutions are much easier to visualize. Notice that the $\cos(x)$ factor corresponds to the initial displacement and depends only on $x$. The time-dependent factor $\cos(ct) - \frac{1}{c}\sin(ct)$ experiences periodic oscillations as $t$ varies and, in doing so, modifies the amplitude of the displacement. If we use a computer to create an animation of $u$ versus $x$ as $t$ increases, we would see a standing wave. At points where $\cos(x) = 0$, the string would appear to remain stationary, whereas at all other points the string would continually oscillate up and down.

**Domains of Dependence and Influence.** D'Alembert's solution of the wave equation $u_{tt} - c^2u_{xx} = 0$ shows that the initial displacement profile $\phi(x)$ effectively splits into two waves traveling in opposite directions with speed $c$. Certainly the initial velocity $\psi(x)$ may influence the propagating waves, *but no part of the waves can travel with speed faster than* $c$. There are two important consequences of this remark. First, suppose that we select a point $x_0$ from within our spatial domain $-\infty < x < \infty$. The characteristic curves through the point $(x_0, 0)$ are the lines $x = x_0 \pm ct$, as illustrated in the upper panel of Figure 10.6. The initial conditions $\phi(x_0), \psi(x_0)$ at the point $x_0$ cannot influence the behavior

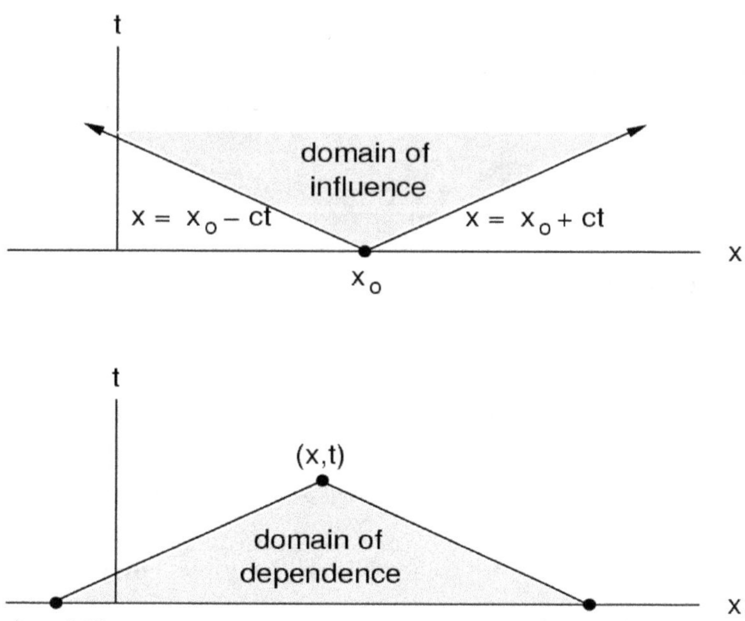


**Figure 10.6.** Upper panel: The domain of influence for $x_0$. The initial condition at $x_0$ cannot affect how $u(x,t)$ behaves outside the shaded region because waves cannot travel faster than speed $c$. Lower panel: The domain of dependence for $u(x,t)$. If $t > 0$, the value of $u(x,t)$ can only depend upon the initial conditions within the closed interval $[x - ct, x + ct]$.

of the solution outside the shaded region in the $xt$-plane, because the waves cannot travel faster than speed $c$. For that reason, the shaded region is called the *domain of influence* of $x_0$. Likewise, if we select a point $(x,t)$, it is natural to ask for the range of $x$-values for which the initial conditions could have any impact on the value of $u(x,t)$. The lower panel of Figure 10.6 illustrates the *domain of dependence* for $u(x,t)$—i.e., the region in the $xt$-plane which could influence the value of $u(x,t)$. In particular, because waves can travel no faster than speed $c$, the value of $u(x,t)$ can only depend upon the behavior of the initial conditions within the interval $[x - ct, x + ct]$ as shown in the Figure.

## 10.3. Cauchy Problem for the Heat Equation

We now turn our attention to the Cauchy problem for the heat equation:

$$u_t - \kappa u_{xx} = 0 \qquad (-\infty < x < \infty) \tag{10.16}$$

$$u(x,0) = \phi(x) \qquad (-\infty < x < \infty), \tag{10.17}$$

where $\kappa > 0$ is a constant. In contrast with the Cauchy problem for the wave equation, we have specified only one initial condition. Recalling the physical intuition that guided our derivation of the heat equation, this is not surprising. Indeed, we need only know the initial heat (or concentration) profile of $u$ versus $x$, and then Fick's Law will govern how the distribution of heat (or concentration) will evolve over time.

Perhaps the most elegant way to attack the Cauchy problem for the heat equation is by means of Fourier transforms. However, we do not presume that the reader is acquainted with certain techniques from complex analysis that are useful in understanding how the transform procedure works. Instead, we will follow the text of Strauss [10], exploiting certain invariance properties of the heat equation in order to construct the solution of (10.16) and (10.17). The idea is to solve that system for a very special choice of the initial condition $\phi(x)$. Then, we will use the solution of that special problem to construct the general solution of the Cauchy problem. There are five properties of the heat equation and its solutions that we will invoke:

**Property 1: Translation invariance.** If $u(x,t)$ is a solution of (10.16), then for any fixed number $x_0$, the function $u(x - x_0, t)$ is also a solution. Indeed, direct differentiation yields $u_t(x - x_0, t) = \kappa u_{xx}(x - x_0, t)$.

**Property 2: Derivatives of solutions.** If $u(x,t)$ satisfies the heat equation (10.16) then the partial derivatives of $u$ also satisfy the heat equation. For example, to see that $u_t$ is also a solution, let $v = u_t$. Then

$$v_t = (u_t)_t = (\kappa u_{xx})_t = (\kappa u_t)_{xx} = \kappa v_{xx}.$$

Thus, $v = u_t$ is also a solution.

**Property 3: Linearity.** Since the heat equation is linear, any finite linear combination of solutions of the heat equation (10.16) is also a solution.

**Property 4: Integrals and Convolutions.** Suppose $S(x,t)$ is a solution of the heat equation (10.16). Then by translation invariance (Property 1), so is $S(x-y,t)$ for any fixed choice of $y$. If $g$ is any other function, we define the *convolution* of $S$ with $g$ as

$$(S \star g)(x,t) = \int_{-\infty}^{\infty} S(x-y,t)g(y)\,dy,$$

provided that this improper integral converges. We claim that, regardless of $g$, the convolution $(S \star g)(x,t)$ is also a solution of the heat equation. To check this, we need to show that $(S \star g)_t - \kappa(S \star g)_{xx} = 0$. Taking the derivatives of the convolution,

$$\begin{aligned}
&(S \star g)_t - \kappa(S \star g)_{xx} \\
&\quad = \frac{\partial}{\partial t}\int_{-\infty}^{\infty} S(x-y,t)g(y)\,dy - \kappa\frac{\partial^2}{\partial x^2}\int_{-\infty}^{\infty} S(x-y,t)g(y)\,dy \\
&\quad = \int_{-\infty}^{\infty} S_t(x-y,t)g(y)\,dy - \int_{-\infty}^{\infty} \kappa S_{xx}(x-y,t)g(y)\,dy \\
&\quad = \int_{-\infty}^{\infty} [S_t(x-y,t) - \kappa S_{xx}(x-y,t)]g(y)\,dy = 0,
\end{aligned}$$

as claimed.

**Remark.** In the above chain of calculations, we are only justified in moving the derivatives under the integral sign provided that all of the functions involved are reasonably well-behaved. Certainly, all of the integrals need to converge. For precise conditions under which interchanging derivatives and integrals is justified, consult a textbook on mathematical analysis.

**Property 5: Dilation.** Suppose $a > 0$ is a constant. If $u(x,t)$ is a solution of the heat equation, then the *dilated* function $v(x,t) = u(\sqrt{a}\,x, at)$ is also a solution. To prove this, we calculate

$$\begin{aligned}
\frac{\partial}{\partial t}v(x,t) &= \frac{\partial}{\partial t}u(\sqrt{a}\,x,at) = a\,u_t(\sqrt{a}\,x,at) \\
\frac{\partial}{\partial x}v(x,t) &= \frac{\partial}{\partial x}u(\sqrt{a}\,x,at) = \sqrt{a}\,u_x(\sqrt{a}\,x,at)
\end{aligned}$$

and

$$\frac{\partial^2}{\partial x^2} v(x,t) = a\, u_{xx}(\sqrt{a}\, x, at).$$

It follows that

$$\begin{aligned} v_t(x,t) - \kappa v_{xx}(x,t) &= a\, u_t(\sqrt{a}\, x, at) - \kappa a\, u_{xx}(\sqrt{a}\, x, at) \\ &= a\left[u_t(\sqrt{a}\, x, at) - \kappa\, u_{xx}(\sqrt{a}\, x, at)\right] = 0, \end{aligned}$$

which establishes that $v$ is also a solution of the heat equation.

We now use these five Properties of the heat equation to build the general solution of the Cauchy problem (10.16)–(10.17). Following Strauss [10], we first solve the heat equation with a special initial condition:

$$Q_t - \kappa Q_{xx} = 0 \qquad (-\infty < x < \infty), \tag{10.18}$$

$$Q(x,0) = \phi(x) = \begin{cases} 1 & \text{if } x > 0 \\ 0 & \text{if } x \leq 0 \end{cases} \qquad (-\infty < x < \infty). \tag{10.19}$$

This seemingly strange initial condition $\phi(x)$ is very convenient because it is *dilation invariant*: for any constant $a > 0$, the graph of $\phi(ax)$ is identical to that of $\phi(x)$. Moreover, we also know (Property 5 above) that any solution $Q(x,t)$ of the heat equation is unaffected by the dilation $x \mapsto \sqrt{a}x$ and $t \mapsto at$. These observations have a nice implication: notice that the quantity $x/\sqrt{t}$ is unaffected by these dilations because

$$\frac{x}{\sqrt{t}} \mapsto \frac{\sqrt{a}x}{\sqrt{at}} = \frac{x}{\sqrt{t}}.$$

Consequently, solutions $Q(x,t)$ of (10.18)–(10.19) depend only on the quantity $x/\sqrt{t}$. Therefore, we will seek solutions of the form

$$Q(x,t) = g\left(\frac{x}{\sqrt{4\kappa t}}\right), \tag{10.20}$$

where the constant $4\kappa$ included for later convenience. (The following computations could be carried out without including the $4\kappa$, but some equations would look a bit messy.)

Now that we are seeking special solutions of the form (10.20) and $g$ is a function of *one* variable $p = x/\sqrt{4\kappa t}$, we can reduce the heat equation to an *ordinary* differential equation. Substituting $Q(x,t) = g(p)$ into (10.18), the chain rule yields

$$Q_t = \frac{dg}{dp}\frac{\partial p}{\partial t} = \frac{\partial p}{\partial t} g'(p).$$

By calculus and algebra,

$$\frac{\partial p}{\partial t} = \frac{\partial}{\partial t}\frac{x}{\sqrt{4\kappa t}} = -\frac{1}{2t}\frac{x}{\sqrt{4\kappa t}} = -\frac{p}{2t}$$

and, combining this with the previous equation, we have

$$Q_t = -\frac{1}{2t} p g'(p).$$

Similarly, taking derivatives with respect to $x$ reveals that

$$Q_x = \frac{dg}{dp}\frac{\partial p}{\partial x} = \frac{1}{\sqrt{4\kappa t}} g'(p)$$

and

$$Q_{xx} = \frac{dQ_x}{dp}\frac{\partial p}{\partial x} = \left[\frac{1}{\sqrt{4\kappa t}} g''(p)\right]\frac{1}{\sqrt{4\kappa t}} = \frac{1}{4\kappa t} g''(p).$$

Combining these new expressions for $Q_t$ and $Q_{xx}$ with the heat equation (10.18), we find that

$$-\frac{1}{2t} p g'(p) - \frac{\kappa}{4\kappa t} g''(p) = 0$$

for all $t > 0$. By algebra, this is equivalent to the second-order variable-coefficient ODE

$$g''(p) + 2p\,g'(p) = 0.$$

For emphasis, notice that we have effectively reduced the heat equation to a second-order ODE by seeking solutions of a special type (10.20).

To solve the ODE $g''(p) + 2pg'(p) = 0$, we first use algebra to write the equation as

$$\frac{g''(p)}{g'(p)} = -2p.$$

The sneaky observation that

$$\frac{g''(p)}{g'(p)} = \frac{d}{dp}\ln[g'(p)]$$

makes it easier to solve this ODE. Indeed, integrating both sides of

$$\frac{d}{dp}\ln[g'(p)] = -2p$$

with respect to $p$ reveals that

$$\ln[g'(p)] = -p^2 + C,$$

where C is a constant of integration. Exponentiating both sides, we have

$$g'(p) = e^{-p^2+C} = e^C e^{-p^2} = C_1 e^{-p^2},$$

where $C_1 = e^C$ is a constant. Integrating both sides,

$$g(p) = C_2 + \int_0^p C_1 e^{-r^2}\, dr,$$

where $C_2 = g(0)$ is another constant (see Equation (2.16)). Thus far, we have shown that the solution of the special problem (10.18)–(10.19) has the form

$$Q(x,t) = C_1 \int_0^{x/\sqrt{4\kappa t}} e^{-r^2}\, dr + C_2. \tag{10.21}$$

Note that this formula only makes sense for $t > 0$. As soon as we figure out the values of the constants $C_1$ and $C_2$, we will have solved the initial value problem (10.18)–(10.19). Determining the values of these constants will require a technical lemma.

**Lemma 10.3.1.**

$$\int_{-\infty}^{\infty} e^{-x^2}\, dx = \sqrt{\pi} \quad \text{and} \quad \int_0^{\infty} e^{-x^2}\, dx = \int_{-\infty}^{0} e^{-x^2}\, dx = \frac{\sqrt{\pi}}{2}.$$

*Proof.* Let $I$ denote the value of the integral over the domain $(-\infty, \infty)$. Then

$$I^2 = \left(\int_{-\infty}^{\infty} e^{-x^2}\, dx\right)\left(\int_{-\infty}^{\infty} e^{-y^2}\, dy\right) = \int_{-\infty}^{\infty}\int_{-\infty}^{\infty} e^{-(x^2+y^2)}\, dx\, dy.$$

Converting to polar coordinates,

$$I^2 = \int_0^{2\pi}\int_0^{\infty} re^{-r^2}\, dr\, d\theta.$$

The substitution $\psi = r^2$ simplifies the interior integral:

$$I^2 = \int_0^{2\pi}\int_0^{\infty} \frac{1}{2}e^{-\psi}\, d\psi\, d\theta = \int_0^{2\pi} \frac{1}{2}\, d\theta = \pi.$$

Therefore, $I = \sqrt{\pi}$, as claimed. The second statement in the Lemma follows from the fact that the integrand has even symmetry. $\square$

With Lemma 10.3.1 in mind, we return to formula (10.21). If $x > 0$, then taking the limit $t \to 0^+$ yields

$$1 = \lim_{t\to 0^+} Q(x,t) = C_1 \int_0^{\infty} e^{-r^2}\, dr + C_2.$$

The leftmost equality is a consequence of our initial condition and the fact that we are temporarily assuming $x > 0$. Using Lemma 10.3.1, this statement simplifies to

$$1 = \frac{\sqrt{\pi}}{2}C_1 + C_2.$$

Now suppose that $x < 0$, the region in which the initial condition is identically 0. Taking the limit $t \to 0^+$ as before, we have

$$0 = \lim_{t\to 0^+} Q(x,t) = C_1 \int_0^{-\infty} e^{-r^2}\, dr + C_2 = -C_1 \int_{-\infty}^{0} e^{-r^2}\, dr + C_2.$$

Recognizing the integral from the Lemma, this equation reduces to

$$0 = -\frac{\sqrt{\pi}}{2}C_1 + C_2.$$

Now that we have a system of two equations for $C_1$ and $C_2$, namely

$$1 = \frac{\sqrt{\pi}}{2}C_1 + C_2 \quad \text{and} \quad 0 = -\frac{\sqrt{\pi}}{2}C_1 + C_2,$$

routine algebra tells us that $C_1 = 1/\sqrt{\pi}$ and $C_2 = 1/2$. Finally, we have shown that the solution of the special Cauchy problem (10.18)–(10.19) is given by

$$Q(x,t) = \frac{1}{2} + \frac{1}{\sqrt{\pi}}\int_0^{x/\sqrt{4\kappa t}} e^{-r^2}\, dr. \tag{10.22}$$

Unfortunately, this integral cannot be simplified, because the function $e^{-r^2}$ does not have an antiderivative that is expressible in terms of elementary functions.

Now that we have solved the Cauchy problem for the heat equation with a special choice of initial conditions (10.19), we can combine our findings with the five Properties to solve the general Cauchy problem (10.16)–(10.17).

All of our above work shows that the function $Q(x,t)$ given by (10.22) is a solution of the heat equation $u_t - \kappa u_{xx} = 0$. By Property 2 (see above) of solutions of the heat equation, we know that the derivative

$$S(x,t) = \frac{\partial Q}{\partial x} = \frac{1}{\sqrt{4\pi\kappa t}}e^{-x^2/4\kappa t}$$

is also a solution of the heat equation. By Property 1 (translation invariance), it follows that $S(x-y,t)$ is also a solution for each fixed $y$. By Property 4, the convolution of $S(x,t)$ with the initial condition $\phi(x)$ is also a solution. That is,

$$(S \star \phi)(x,t) = \int_{-\infty}^{\infty} S(x-y,t)\phi(y)\, dy$$

is also a solution of the heat equation. In fact, we shall soon see that this is the *unique* solution of the Cauchy problem for the heat equation. For emphasis, the solution of (10.16)–(10.17) is given by

$$u(x,t) = (S \star \phi)(x,t) = \frac{1}{\sqrt{4\pi\kappa t}}\int_{-\infty}^{\infty} e^{-(x-y)^2/4\kappa t}\,\phi(y)\, dy. \tag{10.23}$$

Notice that (10.23) only represents the solution for $t > 0$. Checking that this formula obeys the initial condition $u(x,0) = \phi(x)$ requires us to prove that $\lim_{t\to 0^+} u(x,t) = \phi(x)$, a rather technical and tedious calculation.

Solving the Cauchy problem for the heat equation required substantial effort and, unfortunately, the form of the solution (10.23) is not as "illuminating" as what we might have hoped for. There are very few choices of initial conditions $\phi$ for which it is possible to explicitly evaluate the integral in (10.23) by hand. Certainly performing the integration with respect to the dummy variable $y$ should return a function of both $x$ and $t$, but understanding the behavior of solutions requires some thought (see below).

We make several additional remarks about the solution (10.23) before interpreting that formula. First, observe that (10.23) makes no sense if $t \leq 0$. (Keep in mind that the diffusion constant $\kappa$ was assumed to be positive.) Next, we mention that $S(x,t)$ has a special name:

**Definition 10.3.2.** The function

$$S(x,t) = \frac{1}{\sqrt{4\pi\kappa t}} e^{-x^2/4\kappa t} \qquad (t > 0) \tag{10.24}$$

is called the (one-dimensional) *heat kernel*.

The heat kernel was defined as $S(x,t) = Q_x(x,t)$, where $Q$ was the solution of the Cauchy problem with a special initial condition: a piecewise constant function with a single jump discontinuity at $x = 0$. Then, we claimed that the solution of the general Cauchy problem for the heat equation is obtained by taking the convolution of the heat kernel $S(x,t)$ with the initial condition $\phi(x)$. Understanding why this mysterious process successfully solved the Cauchy problem is a topic that is more suitable for an advanced (more theoretical) course in PDEs.

Now let us give a qualitative description of the behavior of the solution (10.23) by analyzing the effects of the two factors that appear in the heat kernel (10.24). For $t > 0$, the graph of the Gaussian function

$$e^{-x^2/4\kappa t}$$

is a "bell curve". As $t$ increases, the graph is dilated (stretched outward). The other factor in the heat kernel, namely

$$\frac{1}{\sqrt{4\pi\kappa t}}$$

modulates the *amplitude* of the Gaussian curves. The amplitude blows up to $\infty$ as $t \to 0^+$ and approaches 0 as $t \to \infty$. Combining the effects of the two factors in the heat kernel, we see that $S(x,t)$ is "tall and thin" for small positive $t$ and "short and broad" for large positive $t$. This is illustrated graphically in Figure 10.7. Note that if $y$ is a fixed number, then the graph of $S(x-y,t)$ is nothing more than a horizontally shifted version of what appears in Figure 10.7 (the peaks would be at $x = y$ instead of at $x = 0$). As a final remark about the heat kernel, we mention that

$$\int_{-\infty}^{\infty} S(x,t)\,dx = 1$$

for all $t > 0$ (see Exercises), from which it follows that for any fixed $y$,

$$\int_{-\infty}^{\infty} S(x-y,t)\,dx = 1$$

as well.

With our new understanding of the behavior of the heat kernel $S(x,t)$, we are in a position to interpret the solution of the Cauchy problem given by formula (10.23). In the convolution

$$u(x,t) = (S \star \phi)(x,t) = \int_{-\infty}^{\infty} S(x-y,t)\phi(y)\,dy,$$

the $S(x-y,t)$ factor essentially gives a "weighted average" of the function $\phi$, exaggerating the points near $y = x$. As $t$ increases, $S(x-y,t)$ becomes shorter and broader, but still has its peak at $y = x$. The convolution $(S \star \phi)(x,t)$ has the effect of "diffusing" $\phi$ outward at each point in the spatial domain.

**Example 10.3.3.** The only initial condition for which we have explicitly calculated the solution of the Cauchy problem for the heat equation is given by (10.19). Using the Heaviside (unit step) function as the initial condition, we found that the solution is given by

$$Q(x,t) = \frac{1}{2} + \frac{1}{\sqrt{\pi}} \int_0^{x/\sqrt{4\kappa t}} e^{-r^2}\,dr.$$

Although the integrand does not have an elementary antiderivative, we could use a computer to estimate the values of this integral for various choices of $x$ and $t$. More systematically, we could pick a specific $t > 0$ and then have a computer approximate the integral for various choices of $x$. Results of such a procedure are

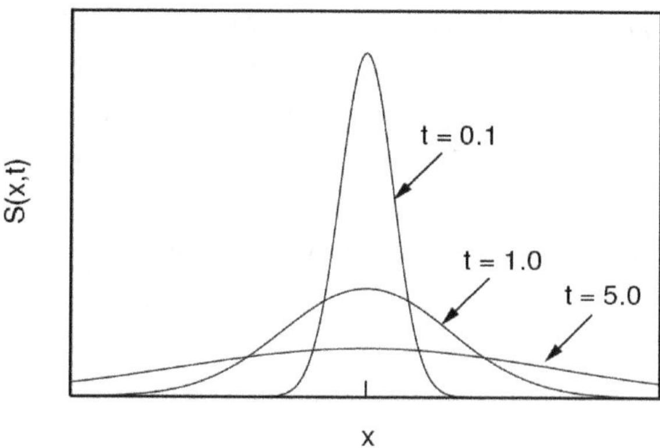


**Figure 10.7.** Plot of the heat kernel $S(x,t)$ versus $x$ at three specific times $t = 0.1$, $t = 1.0$, and $t = 5.0$. The diffusion coefficient was chosen to be $\kappa = 1$. Notice that as $t$ increases, the graph of $S(x,t)$ versus $x$ becomes shorter and broader.

shown in Figure 10.8, which shows plots of the initial condition $Q(x,0)$ as well as the functions $Q(x,1)$ and $Q(x,2)$ versus $x$. Notice that jump discontinuity that appeared in the initial condition is smoothed out. Moreover, the spatial profile of $Q(x,t)$ is stretched/diffused outward as $t$ increases, as described in the paragraph preceding this example. In fact, despite the fact that the initial condition was discontinuous, the function $Q(x,t)$ is *smooth* (differentiable infinitely many times) for all $t > 0$. No matter how "ugly" the initial condition is, solutions of the heat equation become instantly smooth. This is one of that many features of the heat equation that distinguishes it from the wave equation.

**Propagation speed.** Another feature (or liability) of the diffusion/heat equation is that "information" propagates with infinite speed. To illustrate what we mean by this, consider the Cauchy problem

$$u_t - \kappa u_{xx} = 0$$

$$u(x,0) = \phi(x) = \begin{cases} 1 & \text{if } -1 \leq x \leq 1 \\ 0 & \text{otherwise} \end{cases}$$

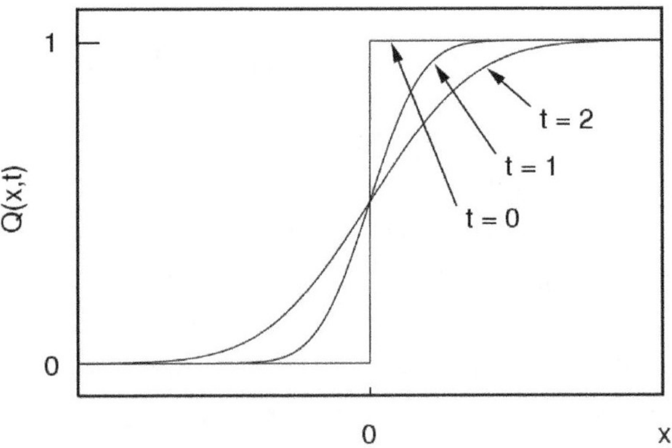


**Figure 10.8.** Plot of $Q(x,t)$ versus $x$ at three specific times $t = 0$, $t = 1$, and $t = 2$.

on the usual domain $-\infty < x < \infty$ and $t \geq 0$. Quoting formula (10.23), the solution is given by

$$u(x,t) = \frac{1}{\sqrt{4\pi\kappa t}} \int_{-\infty}^{\infty} e^{-(x-y)^2/4\kappa t} \phi(y)\, dy = \frac{1}{\sqrt{4\pi\kappa t}} \int_{-1}^{1} e^{-(x-y)^2/4\kappa t}\, dy.$$

Inspecting this integral, it is evident that for *all* real $x$, the value of $u(x,t)$ is *positive* when $t > 0$ because the exponential function in the integrand is always positive. Of course, if $|x|$ is huge, then the value of $u(x,t)$ is barely positive. The important thing to notice here is that although the initial condition is non-zero only in the region $x \in [-1,1]$, this initial "disturbance" has an immediate (albeit small) effect on the *entire* real line. The infinite propagation speed associated with the heat equation is not physically realistic, but the effect is small enough that this PDE is still a useful model of diffusive processes.

The preceding examples highlight some of the important distinctions between solutions of the wave and heat equations. The Cauchy problem for the heat equation is *not* well-posed for $t \leq 0$, whereas D'Alembert's solution of the wave equation makes sense for all real $t$. If we use discontinuous initial data for the wave equation, the discontinuities are preserved as $t$ increases. By contrast, discontinuities are instantly smoothed by the heat equation—solutions are infinitely differentiable for all $t > 0$. Finally, the issue of finite versus infinite propagation speed provides yet another fundamental distinction between waves and diffusions.

## 10.4. Well-Posedness and the Heat Equation

In the previous section, we found a solution (10.23) of the Cauchy problem for the heat equation. In order for the Cauchy problem to be well-posed, existence of a solution is not enough. We also need to know that (i) the solution is unique and (ii) small changes in the initial condition $\phi(x)$ should not cause major changes in how the solution behaves. We will use two different methods to prove that initial/boundary value problems for the heat equation are well-posed. The first method, known as the *energy method*, is more versatile because it can be used to prove well-posedness of the Cauchy, Dirichlet, and Neumann problems for the heat equation as well as other PDEs. The second method exploits a special property of the heat equation known as the Maximum Principle, which offers additional insight into the heat equation's dynamics.

**Energy Method.** Consider the Cauchy problem

$$u_t - \kappa u_{xx} = 0$$
$$u(x,0) = \phi(x)$$

on the usual domain $-\infty < x < \infty$ and $t \geq 0$. In an above example, we argued that the solution of this problem is smooth (infinitely differentiable) for $t > 0$. We now prove that there is a *unique* smooth solution provided that $\phi(x) \to 0$ as $x \to \pm\infty$. The assumption that $\phi$ vanishes as we approach $\pm\infty$ is actually not very restrictive at all.

To prove uniqueness, suppose that $u(x,t)$ and $v(x,t)$ are solutions of the above Cauchy problem. We must show that $u = v$ or, equivalently, that $w = u - v$ is zero. Since both $u_t = \kappa u_{xx}$ and $v_t = \kappa v_{xx}$, by linearity it follows that

$$w_t = (u-v)_t = u_t - v_t = \kappa u_{xx} - \kappa v_{xx} = \kappa(u-v)_{xx} = \kappa w_{xx}.$$

Therefore, $w$ is also a solution of the heat equation. As for the initial conditions, since both $u(x,0) = \phi(x)$ and $v(x,0) = \phi(x)$, we have $w(x,0) = u(x,0) - v(x,0) = 0$. In summary, $w$ satisfies the Cauchy problem

$$w_t = \kappa w_{xx} \quad \text{and} \quad w(x,0) = 0.$$

Intuitively, it seems apparent that $w(x,t) = 0$ for all real $x$ and $t > 0$. To prove this rigorously, we begin by multiplying both sides of $w_t = \kappa w_{xx}$ by $w$ to get

$$ww_t = \kappa w w_{xx}.$$

Next, we make the sneaky observation that the left-hand side can be re-written as follows:

$$\frac{1}{2}\left(w^2\right)_t = \kappa w w_{xx}.$$

Integrating both sides over the entire spatial domain yields

$$\int_{-\infty}^{\infty} \frac{1}{2}\left(w^2\right)_t \, dx = \kappa \int_{-\infty}^{\infty} w w_{xx} \, dx$$

and, re-writing the left-hand side,

$$\frac{1}{2}\frac{d}{dt}\int_{-\infty}^{\infty} w^2 \, dx = \kappa \int_{-\infty}^{\infty} w w_{xx} \, dx.$$

Integrating the right-hand side by parts,

$$\frac{1}{2}\frac{d}{dt}\int_{-\infty}^{\infty} w^2 \, dx = \kappa \left[ \; w w_x \Big|_{-\infty}^{\infty} - \int_{-\infty}^{\infty} w_x^2 \, dx \right].$$

Because of our assumption that $\phi(x) \to 0$ as $x \to \pm\infty$, we know that $w \to 0$ as $x \to \pm\infty$. Therefore, the boundary term that appeared when integrating by parts is actually zero:

$$\frac{1}{2}\frac{d}{dt}\int_{-\infty}^{\infty} w^2 \, dx = -\kappa \int_{-\infty}^{\infty} w_x^2 \, dx.$$

The important thing to notice here is that the integrand on the right-hand side is *non-negative*: $w_x^2 \geq 0$. Thus, we have an inequality

$$\frac{d}{dt}\int_{-\infty}^{\infty} w^2 \, dx \leq 0.$$

In words, this means that the area under the graph of $w^2$ (a non-negative function) cannot increase over time. Hence,

$$\int_{-\infty}^{\infty} w^2(x,t) \, dx \leq \int_{-\infty}^{\infty} w^2(x,0) \, dx$$

for all $t > 0$. However, we know (from the initial condition) that $w^2(x,0) = 0$, which implies that

$$\int_{-\infty}^{\infty} w^2(x,t)\,dx \;\leq\; 0$$

for all $t > 0$. On the other hand, the integral could never be negative because $w^2(x,t) \geq 0$ for all $x$ and $t$. We have shown that

$$\int_{-\infty}^{\infty} w^2(x,t)\,dx \;=\; 0,$$

which could only be true if $w^2 = 0$ throughout the entire domain $-\infty < x < \infty$. Consequently, $w = 0$ so that $u = v$, and we see that the Cauchy problem has a unique solution as claimed.

We can use almost identical computations to prove that the solution is not greatly affected by small changes in the initial condition $\phi$. To see this, let us consider two Cauchy problems

$$u_t - \kappa u_{xx} \;=\; 0 \qquad u(x,0) \;=\; \phi_1(x)$$

and

$$v_t - \kappa v_{xx} \;=\; 0 \qquad v(x,0) \;=\; \phi_2(x).$$

Both $u$ and $v$ satisfy the heat equation but with different initial conditions. To measure the impact of altering the initial conditions, we will compare the gap between $u(x,t)$ and $v(x,t)$ with the gap between the initial conditions $\phi_1$ and $\phi_2$. Letting $w = u - v$ as before, $w$ satisfies the heat equation. However, this time the initial condition on $w$ is non-zero, namely

$$w(x,0) \;=\; u(x,0) - v(x,0) \;=\; \phi_1(x) - \phi_2(x).$$

Retracing the same steps we performed above, we know that

$$\int_{-\infty}^{\infty} w^2(x,t)\,dx \;\leq\; \int_{-\infty}^{\infty} w^2(x,0)\,dx.$$

Equivalently,

$$\int_{-\infty}^{\infty} [u(x,t) - v(x,t)]^2\,dx \;\leq\; \int_{-\infty}^{\infty} [\phi_1(x) - \phi_2(x)]^2\,dx$$

for all time $t \geq 0$. The integral on the right hand side provides a measure of the "gap" between $\phi_1$ and $\phi_2$. Anytime the graphs $\phi_1$ and $\phi_2$ deviate from each other, a positive contribution is made to the integral. Likewise, the integral on the left-hand side of the inequality gives us a measure of the "gap" between $u(x,t)$ and $v(x,t)$ at time $t$. The inequality assures us that the separation between $u$ and $v$ can *never* increase over time. We conclude that small changes in the initial conditions cannot cause major differences between how solutions behave, and hence we have shown that the Cauchy problem for the heat equation is well-posed. We remark that there are other (more natural) ways to measure the separation between two functions, rather than integrating the square of their difference. The above measure of "separation" between $\phi_1(x)$ and $\phi_2(x)$ is the square of the $L^2$ *distance* between those functions, a concept that we will explore in greater depth in subsequent chapters.

**Dirichlet Problem.** The energy method also works to establish well-posedness of the Dirichlet problem for the heat equation:

$$\begin{aligned} u_t &= \kappa u_{xx}, && (0 < x < L) \\ u(x,0) &= \phi(x), && (0 < x < L) \\ u(0,t) &= g(t), && (t > 0) \\ u(L,t) &= h(t), && (t > 0). \end{aligned}$$

Existence of a solution will be established in a later chapter, once we develop the separation of variables technique for PDEs. Assuming that this Dirichlet problem has a solution, we can show that it is unique and is not sensitive to small changes in the initial condition or boundary conditions. Proceeding as above, suppose that $u$ and $v$ are solutions of the Dirichlet problem. Defining $w = u - v$, we see that $w$ satisfies the simpler problem

$$\begin{aligned} w_t &= \kappa w_{xx}, && (0 < x < L) \\ w(x,0) &= 0, && (0 < x < L) \\ w(0,t) &= 0, && (t > 0) \\ w(L,t) &= 0, && (t > 0). \end{aligned}$$

The only difference in the procedure we used for the Cauchy problem is that we integrate the equation

$$\frac{1}{2}\left(w^2\right)_t = \kappa w w_{xx}$$

over the *finite* spatial domain $[0, L]$:

$$\int_0^L \frac{1}{2}\left(w^2\right)_t \, dx = \kappa \int_0^L w w_{xx} \, dx.$$

Integrating the right-hand side by parts,

$$\frac{1}{2}\frac{d}{dt}\int_0^L w^2 \, dx = \kappa \left[ \left. w w_x \right|_0^L - \int_0^L w_x^2 \, dx \right].$$

This time, the boundary term $(ww_x)$ vanishes because the boundary conditions tell us that $w = 0$ at both $x = 0$ and $x = L$. The remainder of the calculation is precisely the same as before, and we conclude that the Dirichlet problem for the heat equation has a unique solution. Notice that unlike the proof of well-posedness for the Cauchy problem, we did not require any special assumptions regarding the initial or boundary conditions. Proving that the solution of the Dirichlet problem is not sensitive to small changes in the initial (or boundary) conditions is essentially identical to the corresponding proof for the Cauchy problem. Again, the only difference is that all integrals are evaluated over the finite interval $[0, L]$ as opposed to the infinite domain $(-\infty, \infty)$. These calculations, as well as the proof of well-posedness for the Neumann problem, are left as exercises.

**The Maximum Principle.** We now use a special property of the heat equation to provide a much quicker proof that the Dirichlet problem is well-posed. In the process, we will gain even more intuition regarding how solutions of the heat equation can behave.

**Theorem 10.4.1** (Maximum Principle). Suppose that $u(x, t)$ satisfies the heat equation on the interval $0 \leq x \leq L$ and for $t \geq 0$. Then the maximum value of $u(x, t)$ occurs either initially (i.e., when $t = 0$) or on the boundary (i.e., $x = 0$ or $x = L$).

The intuitive explanation of the Maximum Principle as follows. If the maximum temperature within a one-dimensional wire occurred somewhere in the middle of the wire, then it would have to cool off unless heat were supplied at the boundary.

The idea behind the proof of the Maximum Principle involves some notions from first-semester calculus. Suppose indirectly that $u$ has a maximum at some point $(x,t)$ in the *interior* of the domain. Then we would expect the first derivatives $u_t$ and $u_x$ to be zero at that point. Moreover, the second derivative $u_{xx}$ should be at most zero. (After all, if $u_{xx}(x,t) > 0$, then a plot of $u$ versus $x$ would reveal a local minimum, not a maximum.) It cannot be the case that $u_{xx} < 0$, because this would imply that $0 = u_t = \kappa u_{xx} < 0$, a contradiction. We have ruled out the possibility that $u$ could have a maximum at a point $(x,t)$ on the interior of the domain if $u_{xx} < 0$. The proof that no interior maximum could occur with $u_{xx} = 0$ requires more care, and the details appear in Strauss [10].

The same intuition that motivated the Maximum Principle gives rise to a similar result:

**Theorem 10.4.2** (Minimum Principle.). Suppose that $u(x,t)$ satisfies the heat equation on the interval $0 \leq x \leq L$ and for $t \geq 0$. Then the minimum value of $u(x,t)$ occurs either initially (i.e., when $t = 0$) or on the boundary (i.e., $x = 0$ or $x = L$).

*Proof.* If $u(x,t)$ satisfies $u_t = \kappa u_{xx}$ then so does $-u$, since the heat equation is linear and homogeneous. Applying the maximum principle to $-u$, we conclude that the maximum value of $-u$ must occur either initially ($t = 0$) or at one of the boundaries ($x = 0$ or $x = L$). Certainly $u$ attains its minimum value wherever $-u$ attains its maximum value. □

The Maximum and Minimum Principles can be used to provide a quick proof that the Dirichlet problem

$$
\begin{aligned}
u_t &= \kappa u_{xx}, && (0 < x < L) \\
u(x,0) &= \phi(x), && (0 < x < L) \\
u(0,t) &= g(t), && (t > 0) \\
u(L,t) &= h(t), && (t > 0)
\end{aligned}
$$

has a unique solution. As before, suppose that $u$ and $v$ are both solutions and define $w = u - v$. Then $w$ satisfies the simpler Dirichlet problem

$$\begin{aligned} w_t &= \kappa w_{xx}, && (0 < x < L) \\ w(x,0) &= 0, && (0 < x < L) \\ w(0,t) &= 0, && (t > 0) \\ w(L,t) &= 0, && (t > 0). \end{aligned}$$

Notice that $w$ is exactly 0 initially and at both boundaries. By the Maximum and Minimum principles, the maximum and minimum values of $w$ are both 0. This means that $w(x,t) = 0$ for all $(x,t)$, implying that $u = v$.

Testing for stability with respect to changes in the initial (or boundary) conditions is also straightforward using the Maximum/Minimum Principles. Suppose that $u$ and $v$ represent solutions to the Dirichlet problems

$$\begin{aligned} u_t &= \kappa u_{xx}, && (0 < x < L) \\ u(x,0) &= \phi_1(x), && (0 < x < L) \\ u(0,t) &= g(t), && (t > 0) \\ u(L,t) &= h(t), && (t > 0) \end{aligned}$$

and

$$\begin{aligned} v_t &= \kappa v_{xx}, && (0 < x < L) \\ v(x,0) &= \phi_2(x), && (0 < x < L) \\ v(0,t) &= g(t), && (t > 0) \\ v(L,t) &= h(t), && (t > 0). \end{aligned}$$

The only difference between the two Dirichlet problems is in the initial condition. Defining $w = u - v$ as usual, we see that $w$ satisfies a new Dirichlet problem:

$$\begin{aligned} w_t &= \kappa w_{xx}, && (0 < x < L) \\ w(x,0) &= \phi_1(x) - \phi_2(x), && (0 < x < L) \\ w(0,t) &= 0, && (t > 0) \\ w(L,t) &= 0, && (t > 0). \end{aligned}$$

We now apply the Maximum/Minimum Principles to $w$. Since $w = 0$ on the boundaries, we have

$$\max_{0 \le x \le L,\ 0 \le t} |w(x,t)| \le \max_{0 \le x \le L} |\phi_1(x) - \phi_2(x)|.$$

Equivalently,

$$\max_{0 \le x \le L,\ 0 \le t} |u(x,t) - v(x,t)| \le \max_{0 \le x \le L} |\phi_1(x) - \phi_2(x)|.$$

This estimate shows that the Dirichlet problem is stable to small changes in the initial condition. Specifically, it tells us that the maximum gap between the solutions $u$ and $v$ of the two problems is at most as large as the maximum separation between the two different initial conditions. Hence, if we start from two different initial conditions $\phi_1(x)$ and $\phi_2(x)$ that are "close", then the solutions $u(x,t)$ and $v(x,t)$ will remain "close" for all time $t$. The same argument reveals that the Dirichlet problem is not sensitive to small changes in the boundary conditions at $x = 0$ and $x = L$.

We have now provided two different proofs that the Dirichlet problem for the heat equation is well-posed: one using the energy method, and one using the maximum principle. The proofs introduced two very different notions of "separation", or distance, between two functions $\phi_1(x)$ and $\phi_2(x)$. When we proved stability via the energy method, we used the integral

$$\int_0^L [\phi_1(x) - \phi_2(x)]^2 \, dx$$

to measure the distance between the two functions. By contrast, the Maximum Principle used a different measure of distance, namely

$$\max_{0 \le x \le L} |\phi_1(x) - \phi_2(x)|.$$

Both of these are perfectly reasonable ways of measuring the "distance" between two functions, and our choice of which metric to use often depends upon the context. Later, during our study of Fourier series, we will explore these different notions of distance in greater detail.

## 10.5. Inhomogeneous Equations and Duhamel's Principle

Suppose that $L$ is a linear differential operator, and consider the linear, inhomogeneous differential equation $L(u) = f$. In your introductory course on differential equations, you learned that the general solution of $L(u) = f$ has the form $u = u_h + u_p$, where $u_h$ is the general solution of the homogeneous problem $L(u) = 0$ and $u_p$ is a particular solution of the inhomogeneous equation $L(u) = f$. In this section, we will solve the Cauchy problems for the inhomogeneous heat, wave, and transport equations by using *Duhamel's Principle*, a generalization of the variation of parameters technique from ordinary differential equations. For this reason, we motivate Duhamel's Principle by performing an alternate derivation of the variation of parameters formula (2.18).

Consider the *ordinary* differential equation

$$\begin{aligned} \frac{du}{dt} &= Au + f(t) \\ u(0) &= u_0, \end{aligned} \tag{10.25}$$

where $A$ is a constant and $f$ is continuously differentiable (i.e., $f$ has a continuous derivative). By linearity, we can split this initial value problem into two simpler problems, writing $u = v + w$ where

$$\begin{aligned} \frac{dv}{dt} &= Av \\ v(0) &= u_0, \end{aligned} \tag{10.26}$$

and

$$\begin{aligned} \frac{dw}{dt} &= Aw + f(t) \\ w(0) &= 0. \end{aligned} \tag{10.27}$$

The solution of the homogeneous problem for $v$ is $v(t) = e^{tA}u_0$. When we solve the inhomogeneous problem (10.27), we will invoke a technical Lemma:

**Lemma 10.5.1.** Suppose that $f(x,t)$ and $\partial f/\partial t$ are continuous and that $\beta(t)$ is differentiable. Then

$$\frac{d}{dt}\int_0^{\beta(t)} f(x,t)\,dx = \beta'(t)f(\beta(t),t) + \int_0^{\beta(t)} \frac{\partial f}{\partial t}(x,t)\,dx.$$

*Proof.* Use the chain rule and the fundamental theorem of calculus. □

Notice that Lemma 10.5.1 is not merely a re-statement of the fundamental theorem of calculus, because the variable $t$ appears both in the integrand *and* in the upper limit of integration.

For our purposes, a special case of Lemma 10.5.1 in which $\beta(t) = t$ will be especially useful:

**Corollary 10.5.2.** Suppose that $f(x,t;s)$ and $\partial f/\partial t$ are continuous functions of the variables $x$ and $t$ as well as the parameter $s$. Then

$$\frac{\mathrm{d}}{\mathrm{d}t}\int_0^t f(x,t-s;s)\,\mathrm{d}s \;=\; f(x,0;s) + \int_0^t \frac{\partial f}{\partial t}(x,t-s;s)\,\mathrm{d}s.$$

We will use Corollary 10.5.2 to show that the solution of (10.27) can be constructed by integrating the solution of a homogeneous initial value problem, namely

$$\begin{aligned} \frac{\mathrm{d}W}{\mathrm{d}t} &= AW \\ W(0;s) &= f(s), \end{aligned} \tag{10.28}$$

where $s$ is a *parameter*. Notice that the inhomogeneous term in (10.27) has been transferred to the initial condition in (10.28).

**Duhamel's Principle:** We claim that if $W(t;s)$ is a solution of (10.28), then

$$w(t) \;=\; \int_0^t W(t-s;s)\,\mathrm{d}s \tag{10.29}$$

is a solution of (10.27). To see why, suppose that $w(t)$ is defined as in (10.29). Then by Corollary (10.5.2),

$$\begin{aligned} w'(t) &= \frac{\mathrm{d}}{\mathrm{d}t}\int_0^t W(t-s;s)\,\mathrm{d}s \;=\; W(0;t) + \int_0^t W_t(t-s;s)\,\mathrm{d}s \\ &= f(t) + A\int_0^t W(t-s;s)\,\mathrm{d}s \;=\; f(t) + Aw(t), \end{aligned}$$

as claimed.

Since the solution of (10.28) is given by $W(t;s) = \mathrm{e}^{tA} f(s)$, the above claim tells us that

$$w(t) \;=\; \int_0^t W(t-s;s)\,\mathrm{d}s \;=\; \int_0^t \mathrm{e}^{(t-s)A} f(s)\,\mathrm{d}s$$

is a solution of (10.27). Finally, the general solution of our original problem (10.25) is given by

$$u(t) \;=\; v(t) + w(t) \;=\; \mathrm{e}^{tA} u_0 \;+\; \int_0^t \mathrm{e}^{(t-s)A} f(s)\, \mathrm{d}s, \tag{10.30}$$

which we recognize as the familiar variation of parameters formula.

The same idea can be used to solve inhomogeneous PDEs, which we now demonstrate via three examples.

**Inhomogeneous heat equation.** Recall that the Cauchy problem for the [homogeneous] heat equation models heat transfer within a long, thin wire. The wire is insulated to prevent heat from radiating outward, and the diffusion of heat is not influenced by any sources/sinks of heat. Now consider the more general diffusion problem in which we *are* allowed to supply/remove heat from the wire, and let $f(x,t)$ describe the heat injected at position $x$ and time $t \geq 0$. (We will assume that $f$ is a continuously differentiable function.) Mathematically, we can model this scenario with the Cauchy problem

$$\begin{aligned} u_t &= \kappa u_{xx} \;+\; f(x,t) \\ u(x,0) &= \phi(x) \end{aligned} \tag{10.31}$$

where, as usual, $\phi(x)$ denotes the initial temperature distribution within the wire.

We will solve the inhomogeneous Cauchy problem (10.31) via the same procedure used to solve the ODE (10.25). First, write $u = v + w$ where $v$ and $w$ are solutions of the simpler problems

$$\begin{aligned} v_t &= \kappa v_{xx} \\ v(x,0) &= \phi(x) \end{aligned} \tag{10.32}$$

and

$$\begin{aligned} w_t &= \kappa w_{xx} \;+\; f(x,t) \\ w(x,0) &= 0. \end{aligned} \tag{10.33}$$

Because $v$ satisfies the homogeneous Cauchy problem for the heat equation, we immediately conclude that

$$v(x,t) = (S \star \phi)(x,t) = \int_{-\infty}^{\infty} S(x-y,t)\,\phi(y)\,dy$$
$$= \frac{1}{\sqrt{4\pi\kappa t}} \int_{-\infty}^{\infty} e^{-(x-y)^2/4\kappa t}\,\phi(y)\,dy,$$

where $S$ denotes the heat kernel (10.24). To solve (10.33), we mimic the preceding example.

**Duhamel's Principle:** We claim that if $W(x,t;s)$ is a solution of the *homogeneous* problem

$$\begin{aligned} W_t &= \kappa W_{xx} \\ W(x,0;s) &= f(x,s), \end{aligned} \tag{10.34}$$

where $s$ is treated as a parameter, then

$$w(x,t) = \int_0^t W(x,t-s;s)\,ds \tag{10.35}$$

is a solution of (10.33). To prove the claim, suppose that $w(x,t)$ is given by (10.35). By computing the relevant partial derivatives of $w(x,t)$, we must show that $w$ satisfies the inhomogeneous heat equation. According to Corollary 10.5.2,

$$w_t(x,t) = W(x,0;t) + \int_0^t W_t(x,t-s;s)\,ds = f(x,t) + \int_0^t W_t(x,t-s;s)\,ds,$$

and by direct differentiation,

$$w_{xx}(x,t) = \int_0^t W_{xx}(x,t-s;s)\,ds.$$

Therefore,

$$\begin{aligned} w_t - \kappa w_{xx} &= f(x,t) + \int_0^t W_t(x,t-s;s) - \kappa W_{xx}(x,t-s;s)\,ds \\ &= f(x,t) + \int_0^t [W_t - \kappa W_{xx}](x,t-s;s)\,ds = f(x,t), \end{aligned}$$

as claimed.

Now since the solution of the homogeneous problem (10.34) is given by

$$W(x,t;s) = (S \star f)(x,t;s) = \int_{-\infty}^{\infty} S(x-y,t) f(y,s)\,dy,$$

it follows from Duhamel's Principle that

$$w(x,t) = \int_0^t W(x, t-s; s)\, ds = \int_0^t \int_{-\infty}^{\infty} S(x-y, t-s) f(y,s)\, dy\, ds$$

is a solution of (10.33). Finally, recalling that $u = v + w$, we conclude that the solution of the Cauchy problem (10.31) for the inhomogeneous heat equation is given by

$$u(x,t) = \int_{-\infty}^{\infty} S(x-y,t)\phi(y)\, dy + \int_0^t \int_{-\infty}^{\infty} S(x-y,t-s) f(y,s)\, dy\, ds \quad (10.36)$$

for $t > 0$. From the definition of the heat kernel $S(x,t)$, formula (10.36) is equivalent to

$$\begin{aligned} u(x,t) = \frac{1}{\sqrt{4\pi\kappa t}} & \int_{-\infty}^{\infty} e^{-(x-y)^2/4\kappa t} \phi(y)\, dy \\ & + \int_0^t \int_{-\infty}^{\infty} \frac{1}{\sqrt{4\pi\kappa(t-s)}} e^{-(x-y)^2/4\kappa(t-s)} f(y,s)\, dy\, ds. \end{aligned}$$

The solution (10.36) is rather unwieldy, as there are very few choices of initial heat profiles $\phi(x)$ and heat source terms $f(x,t)$ for which these integrals can be evaluated explicitly. Hence, one typically uses computer software packages to numerically approximate solutions of inhomogeneous PDEs such as this one.

**Inhomogeneous wave equation.** In the preceding examples, we used Duhamel's Principle to solve inhomogeneous equations whose homogeneous counterparts had already been solved. In effect, Duhamel's Principle tells us that the cumulative effect of a source term $f(t)$ or $f(x,t)$ is measured by taking an appropriate integral involving (i) a solution of the underlying homogeneous problem and (ii) the function $f$ itself. We apply the same concept to solve the Cauchy problem for the inhomogeneous wave equation:

$$\begin{aligned} u_{tt} &= c^2 u_{xx} + f(x,t) \\ u(x,0) &= \phi(x) \\ u_t(x,0) &= \psi(x). \end{aligned} \quad (10.37)$$

Here, $\phi(x)$ and $\psi(x)$ represent the initial displacement and velocity (respectively) of a vibrating string, and $f(x,t)$ describes an external force being applied to the string. We will assume that $f(x,t)$ is a continuously differentiable function. Splitting (10.37) into two simpler problems, we write $u = v + w$ where

$$\begin{aligned} v_{tt} &= c^2 v_{xx} \\ v(x,0) &= \phi(x) \\ v_t(x,0) &= \psi(x). \end{aligned} \tag{10.38}$$

and

$$\begin{aligned} w_{tt} &= c^2 w_{xx} + f(x,t) \\ w(x,0) &= 0 \\ w_t(x,0) &= 0. \end{aligned} \tag{10.39}$$

Because $v$ satisfies the homogeneous Cauchy problem for the wave equation, we quote D'Alembert's formula to obtain

$$v(x,t) = \frac{1}{2}\left[\phi(x+ct) + \phi(x-ct)\right] + \frac{1}{2c}\int_{x-ct}^{x+ct} \psi(s)\, ds.$$

To solve (10.39), we proceed as before.

**Duhamel's Principle:** We claim that if $W(x,t;s)$ is a solution of the *homogeneous* problem

$$\begin{aligned} W_{tt} &= c^2 W_{xx} \\ W(x,0;s) &= 0 \\ W_t(x,0;s) &= f(x,s), \end{aligned} \tag{10.40}$$

where $s$ is treated as a parameter, then

$$w(x,t) = \int_0^t W(x,t-s;s)\, ds \tag{10.41}$$

is a solution of (10.39). To prove the claim, suppose that $w(x,t)$ is given by (10.41). By computing the relevant partial derivatives of $w(x,t)$, we must show that $w$ satisfies the inhomogeneous wave equation. According to Corollary 10.5.2,

$$w_t(x,t) = W(x,0;t) + \int_0^t W_t(x,t-s;s)\, ds = \int_0^t W_t(x,t-s;s)\, ds$$

and, differentiating with respect to $t$ a second time,

$$w_{tt}(x,t) \;=\; W_t(x,0;t) + \int_0^t W_{tt}(x,t-s;s)\,ds \;=\; f(x,t) + \int_0^t W_{tt}(x,t-s;s)\,ds.$$

Derivatives with respect to $x$ can be computed directly:

$$w_{xx}(x,t) \;=\; \int_0^t W_{xx}(x,t-s;s)\,ds.$$

Therefore,

$$\begin{aligned} w_{tt} - c^2 w_{xx} &= f(x,t) + \int_0^t W_{tt}(x,t-s;s) - c^2 W_{xx}(x,t-s;s)\,ds \\ &= f(x,t) + \int_0^t [W_{tt} - c^2 W_{xx}](x,t-s;s)\,ds \;=\; f(x,t), \end{aligned}$$

as claimed.

From D'Alembert's formula, the solution of the homogeneous problem (10.40) is given by

$$W(x,t;s) \;=\; \frac{1}{2c}\int_{x-ct}^{x+ct} f(\eta,s)\,d\eta,$$

from which it follows that

$$w(x,t) \;=\; \int_0^t W(x,t-s;s)\,ds \;=\; \frac{1}{2c}\int_0^t \int_{x-c(t-s)}^{x+c(t-s)} f(\eta,s)\,d\eta\,ds$$

is a solution of (10.39). Finally, recalling that $u = v + w$, we conclude that the solution of the Cauchy problem (10.37) for the inhomogeneous wave equation is given by

$$\begin{aligned} u(x,t) = \frac{1}{2}\left[\phi(x+ct) + \phi(x-ct)\right] &+ \frac{1}{2c}\int_{x-ct}^{x+ct} \psi(s)\,ds \\ &+ \frac{1}{2c}\int_0^t \int_{x-c(t-s)}^{x+c(t-s)} f(\eta,s)\,d\eta\,ds. \end{aligned} \tag{10.42}$$

The iterated integral appearing in (10.42) is not nearly as problematic as the one that appeared in the solution of the inhomogeneous heat equation (see exercises).

**Inhomogeneous transport equation.** In a previous chapter, we showed that the solution of the Cauchy problem

$$\begin{aligned} u_t + cu_x &= 0 \\ u(x,0) &= f(x) \end{aligned} \tag{10.43}$$

for the simple transport equation is given by $u(x,t) = f(x - ct)$. Recall that $u(x,t)$ can be interpreted as the concentration of a pollutant being carried in a river whose current has constant velocity $c$. The pollutant does not diffuse (it is merely suspended in the water) and the total mass of pollutant is conserved. The inhomogeneous transport equation

$$\begin{aligned} u_t + cu_x &= g(x,t) \\ u(x,0) &= f(x), \end{aligned} \tag{10.44}$$

allows for the possibility that we may add/remove pollutant from the stream, as modeled by the source term $g(x,t)$. Solving (10.44) follows the same procedure that we used to solve the inhomogeneous heat and wave equations. As an exercise, use Duhamel's Principle to show that the solution of (10.44) is given by

$$u(x,t) = f(x - ct) + \int_0^t g(x - c(t-s), s)\, \mathrm{d}s. \tag{10.45}$$

---

## Exercises

1. Solve the PDE $u_{tt} - 5u_{xt} - 36u_{xx} = 0$. To do so, factor the operator on $u$ as

$$\left(\frac{\partial}{\partial t} - 9\frac{\partial}{\partial x}\right)\left(\frac{\partial}{\partial t} + 4\frac{\partial}{\partial x}\right) u = 0,$$

   and mimic what we did when solving the wave equation.

2. Solve the PDE $u_{tt} - (\alpha + \beta)u_{xt} + \alpha\beta u_{xx} = 0$, where $\alpha$ and $\beta$ are non-zero constants. (Hint: See previous exercise.)

3. Let $c \neq 0$ be a constant. Solve the Cauchy problem

$$\begin{aligned} u_{tt} - c^2 u_{xx} &= 0 \\ u(x,0) &= \mathrm{e}^{-x} \end{aligned}$$

$$u_t(x,0) = \cos x.$$

4. Solve the Cauchy problem

$$\begin{aligned} u_{tt} - 9u_{xx} &= 0 \\ u(x,0) &= \arctan(x) \\ u_t(x,0) &= 2x\ln(1+x^2). \end{aligned}$$

5. A function $f(x)$ is called *even* if it satisfies $f(-x) = f(x)$ for all real $x$. For example, $\cos(x)$ is even because $\cos(-x) = \cos(x)$ for all $x$. Show that if the initial conditions $\phi(x)$ and $\psi(x)$ for the wave equation are both even, then the solution $u(x,t)$ is even (in the $x$ variable) for all time $t$. That is, show that $u(-x,t) = u(x,t)$.
6. Suppose that $u(x,t)$ satisfies the wave equation $u_{tt} = c^2u_{xx}$. Let $\alpha$ be any non-zero constant and define the dilated function $v(x,t) = u(\alpha x, \alpha t)$. Show that $v$ also satisfies the wave equation.
7. Solve the Cauchy problem

$$\begin{aligned} u_{tt} - c^2u_{xx} &= 0 \qquad (-\infty < x < \infty), \\ u(x,0) &= \sin x, \\ u_t(x,0) &= -\sin x. \end{aligned}$$

Express the solution as a product of two functions, one of which depends on $x$ and one of which depends on $t$. To do so, you may find the following trigonometric identities useful:

$$\sin(\alpha+\beta) = \sin\alpha\cos\beta + \cos\alpha\sin\beta, \qquad \cos(\alpha+\beta) = \cos\alpha\cos\beta - \sin\alpha\sin\beta.$$

8. Consider the Cauchy problem

$$\begin{aligned} u_t &= \kappa u_{xx} \qquad (-\infty < x < \infty), \\ u(x,0) &= \phi(x), \end{aligned}$$

where $\phi(x)$ is an *even* function. Show that the solution $u$ will remain even in $x$ for all $t > 0$. In other words, show that $u(-x,t) = u(x,t)$ for all $t > 0$.

9. The function $u(x,t) = 10 - x + 4t + 2x^2$ is a solution of the heat equation $u_t = u_{xx}$. Find the locations of its maximum and minimum values on the domain $0 \leq x \leq 2$ and $0 \leq t \leq 8$.

10. Show that $u(x,t) = e^{\alpha x + \beta t}$ satisfies the heat equation $u_t = \kappa u_{xx}$ if and only if $\beta = \kappa \alpha^2$. In the special case $\alpha = -2$ and $\kappa = 1$, note that $e^{-2x+4t}$ is a solution of $u_t = u_{xx}$. Find the maximum and minimum values of $e^{-2x+4t}$ over the domain $0 \leq x \leq 2$ and $0 \leq t \leq 3$. Do the Maximum/Minimum Principles hold?

11. Unlike the heat equation, the wave equation $u_{tt} = u_{xx}$ does not obey a Maximum Principle. Consider, for example, the function $u(x,t) = \sin(x + t) + \cos(x - t)$ on the domain $0 \leq x \leq \frac{\pi}{2}$ and $0 \leq t \leq \frac{\pi}{4}$. Show that $u(x,t)$ satisfies the wave equation and find the maximum value of $u(x,t)$ over the given domain. Then, explain why you can conclude that the wave equation does not obey the Maximum Principle.

12. Suppose that $u(x,t)$ and $v(x,t)$ are solutions of the heat equation on the domain $0 \leq x \leq L$ and $t \geq 0$. Suppose that $u \leq v$ both initially ($t = 0$) and at both boundaries ($x = 0$ and $x = L$). Use the Maximum/Minimum Principle to prove that $u(x,t) \leq v(x,t)$ for all $(x,t)$. Hint: Let $w = v - u$.

13. According to Lemma 10.3.1, we know that

$$\int_{-\infty}^{\infty} e^{-r^2}\, dr = \sqrt{\pi}.$$

By making the substitution $r = x/\sqrt{4\kappa t}$, show that

$$\int_{-\infty}^{\infty} S(x,t)\, dx = 1 \qquad (t > 0),$$

where $S(x,t)$ denotes the heat kernel.

14. Consider the Dirichlet problem

$$\begin{aligned} u_t &= \kappa u_{xx} && (0 < x < L \text{ and } t > 0) \\ u(x,0) &= \phi(x) && (0 < x < L) \\ u(0,t) &= g(t) \\ u(L,t) &= h(t). \end{aligned}$$

Show that the solution $u(x,t)$ is not sensitive to small changes in the right boundary condition $h(t)$.

**15.** Let $r > 0$ be a constant. The reaction-diffusion equation

$$\begin{aligned} u_t - \kappa u_{xx} + ru &= 0 \qquad\qquad (-\infty < x < \infty), \\ u(x,0) &= \phi(x), \end{aligned}$$

can be solved by making a substitution that converts it to the heat equation. (a) Let $v(x,t) = e^{rt}u(x,t)$. Show that $v$ satisfies the heat equation $v_t = \kappa v_{xx}$, and obeys the same initial condition as $u$. (b) After quoting (10.23) to write down a formula for $v(x,t)$, find the solution $u(x,t)$ of the original reaction-diffusion problem.

**16.** Let $c > 0$ be a constant. The advection-diffusion equation

$$\begin{aligned} u_t + cu_x &= \kappa u_{xx} \qquad\qquad (-\infty < x < \infty), \\ u(x,0) &= \phi(x) \end{aligned}$$

models concentration of a pollutant that diffuses while being transported with constant speed $c$. In the absence of diffusion (i.e., $\kappa = 0$), the PDE reduces to the simple transport equation which was solved in a previous chapter. From our understanding of the transport equation, the reaction-diffusion equation should simplify if we switch to a moving coordinate system. (a) Make the substitution $v(x,t) = u(x+ct,t)$ and show that $v$ satisfies the heat equation. (b) Set up and solve a Cauchy problem for $v(x,t)$. (c) Use your formula from Part (b) to write down the solution of the advection-diffusion equation.

**17.** Consider the Neumann problem for the heat equation:

$$\begin{aligned} u_t &= \kappa u_{xx} \qquad (0 < x < L) \\ u(x,0) &= \phi(x) \qquad (0 < x < L) \\ u_x(0,t) &= g(t) \\ u_x(L,t) &= h(t). \end{aligned}$$

The purpose of this problem is to show that, assuming a solution exists, it must be unique. (i) Explain why we cannot use the Maximum/Minimum

Principle to prove uniqueness as we did for the Dirichlet problem. (ii) Use the energy method to prove that the solution really is unique.

**18.** The purpose of this exercise is to use the energy method to prove that if the Dirichlet problem

$$\begin{aligned} u_{tt} &= c^2 u_{xx} && (0 < x < L) \\ u(x,0) &= \phi(x) && (0 < x < L) \\ u_t(x,0) &= \psi(x) && (0 < x < L) \\ u(0,t) &= g(t) \\ u(L,t) &= h(t) \end{aligned}$$

for the wave equation has a solution, then the solution must be unique.

(a) Suppose that $u$ and $v$ are solutions of this Dirichlet problem. Show that $w = u - v$ satisfies the wave equation with $w = w_t = 0$ initially and $w = 0$ on the boundaries.

(b) Define the energy function

$$E(t) = \frac{1}{2}\int_0^L w_t^2 + c^2 w_x^2 \, dx.$$

Assuming that you are justified in differentiating under the integral sign, show that

$$E'(t) = \int_0^L w_t w_{tt} + c^2 w_x w_{xt} \, dx = \int_0^L w_t w_{tt} \, dx + c^2 \int_0^L w_x w_{xt} \, dx.$$

(c) Integrate $w_x w_{xt}$ by parts and show that

$$E'(t) = \int_0^L w_t \left[ w_{tt} - c^2 w_{xx} \right] dx = 0.$$

(d) Use the result from Part (c) to explain why $E(t) = 0$ for all $t$. Then, explain why it follows that $w(x,t) = 0$ for $0 \le x \le L$ and all $t$. Finally, conclude that $u(x,t) = v(x,t)$, which means that the solution of the Dirichlet problem is unique.

**19.** Use Duhamel's Principle to derive formula (10.45), the solution of the inhomogeneous transport equation (10.44).

**20.** Solve the Cauchy problem

$$u_t + 8u_x = x \sin t$$
$$u(x,0) = \frac{1}{1+x^2}.$$

**21.** Solve the Cauchy problem

$$u_{tt} = c^2 u_{xx} + xt^2$$
$$u(x,0) = e^{-x^2}$$
$$u_t(x,0) = 0.$$

How does the source term $xt^2$ affect the behavior of the solution? That is, how does your solution compare with that of the associated homogeneous problem?

**22.** Solve the Cauchy problem

$$u_{tt} = 4u_{xx} + e^x$$
$$u(x,0) = 0$$
$$u_t(x,0) = 0$$

and sketch the solution $u(x,1)$ versus $x$. Note: If the source term $e^x$ were absent, the solution would be $u(x,t) = 0$. Physically, this would correspond to a stationary string in its equilibrium position. What effect does the source term have on the displacement of the string?

# CHAPTER 11

## Initial-Boundary Value Problems

The infinite spatial domains considered in the previous chapter give insight regarding the behavior of waves and diffusions. However, since such domains are not physically realistic, we need to develop new techniques for solving PDEs on bounded domains. As a first step towards solving the heat and wave equations over finite spatial domains (such as the interval $0 \le x \le L$ in one space dimension), we will solve these equations on "semi-infinite" domains whose boundaries consist of one point.

### 11.1. Heat and Wave Equations on a Half-Line

**Heat Equation: Dirichlet Problem.** We begin by solving the homogeneous Dirichlet problem for the heat equation on the interval $0 \le x < \infty$; that is,

$$u_t = \kappa u_{xx} \qquad (0 < x < \infty) \tag{11.1}$$

$$u(x,0) = \phi(x) \qquad (0 < x < \infty) \tag{11.2}$$

$$u(0,t) = 0 \qquad (t \ge 0). \tag{11.3}$$

The homogeneous boundary condition is quite important for the solution technique that follows. In the context of heat transfer within a "one-dimensional" wire, this Dirichlet boundary condition is analogous to immersing the $x = 0$ end of the wire in a bath of ice water with temperature zero degrees Celsius.

We will solve the homogeneous Dirichlet problem (11.1)–(11.3) using a *reflection method,* temporarily extending our spatial domain to the entire real line and solving a Cauchy problem instead. By quoting the formula for the solution

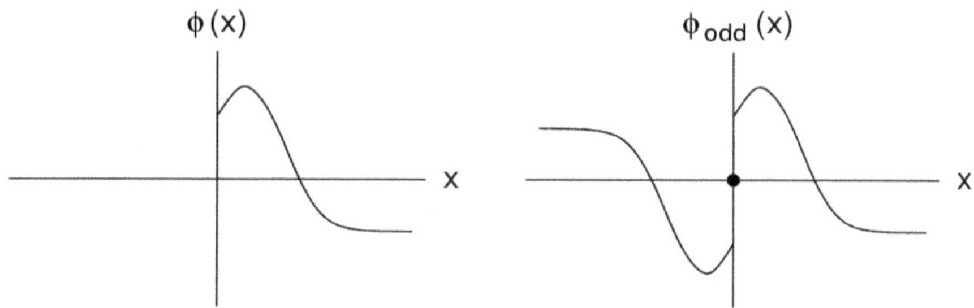


**Figure 11.1.** Illustration of the odd extension of a function $\phi(x)$.

of the Cauchy problem in the preceding chapter, we will obtain the solution of (11.1)–(11.3) by restricting ourselves to the original spatial domain.

First, recall that a function $f(x)$ of a single variable is called *odd* if it has the property that $f(-x) = -f(x)$ for all real $x$. Examples of odd functions include $\sin(x)$ and $x^3$. If $f$ is an odd function, notice that $f(0) = f(-0) = -f(0)$, which implies that $f(0) = 0$. Now, referring to the initial condition (11.2) above, we define the *odd extension* of $\phi(x)$ as

$$\phi_{\text{odd}}(x) = \begin{cases} \phi(x) & \text{if } x > 0 \\ -\phi(-x) & \text{if } x < 0 \\ 0 & \text{if } x = 0. \end{cases}$$

By construction, $\phi_{\text{odd}}$ is an odd function and is defined for all real $x$ (see Figure 11.1). Now consider the *Cauchy* problem

$$\begin{aligned} v_t &= \kappa v_{xx} & (-\infty < x < \infty) \\ v(x,0) &= \phi_{\text{odd}}(x) & (-\infty < x < \infty). \end{aligned}$$

From the previous chapter, we know that the solution is given by the convolution of the heat kernel $S(x,t)$ with the initial condition:

$$v(x,t) = (S \star \phi_{\text{odd}})(x,t) = \int_{-\infty}^{\infty} S(x-y,t)\phi_{\text{odd}}(y)\,\mathrm{d}y.$$

We claim that the restriction of $v(x,t)$ to the domain $x \geq 0$ is the solution of the Dirichlet problem (11.1)–(11.3). To see why, we need to verify that all three conditions of our Dirichlet problem are satisfied. Certainly $v(x,t)$ satisfies the same PDE as $u(x,t)$ on the domain $x > 0$. The initial conditions also match on that domain, because $v(x,0) = \phi(x) = u(x,0)$ whenever $x > 0$. Checking the boundary condition requires a bit more care. As an exercise, you should verify that since the initial condition for $v(x,t)$ is odd, then the solution $v(x,t)$ will remain odd for all $t > 0$. That is, $v(-x,t) = -v(x,t)$ for all $t \geq 0$. By our earlier remarks on odd functions, this implies that $v(0,t) = 0$ for all $t \geq 0$. It follows that $v$ automatically obeys the homogeneous Dirichlet boundary condition that we imposed on $u$. Since $u(x,t)$ and $v(x,t)$ satisfy the heat equation with the same initial and boundary conditions on the domain $x \geq 0$ and $t \geq 0$, we conclude (by uniqueness) that $u(x,t) = v(x,t)$ on that domain.

Now that we have proved that the restriction of $v(x,t)$ to the domain $x \geq 0$ is the solution of the Dirichlet problem (11.1)–(11.3), we can give an explicit formula. The piecewise definition of $\phi_{\text{odd}}$ suggests that we split the region of integration as

$$v(x,t) = \int_0^\infty S(x-y,t)\phi_{\text{odd}}(y)\,dy + \int_{-\infty}^0 S(x-y,t)\phi_{\text{odd}}(y)\,dy.$$

Using the definition of $\phi_{\text{odd}}$,

$$v(x,t) = \int_0^\infty S(x-y,t)\phi(y)\,dy - \int_{-\infty}^0 S(x-y,t)\phi(-y)\,dy,$$

and substituting $w = -y$ in the second integral yields

$$v(x,t) = \int_0^\infty S(x-y,t)\phi(y)\,dy + \int_\infty^0 S(x+w,t)\phi(w)\,dw.$$

Reversing the limits of integration in the second integral, we have

$$v(x,t) = \int_0^\infty S(x-y,t)\phi(y)\,dy - \int_0^\infty S(x+w,t)\phi(w)\,dw.$$

Since $w$ is simply a dummy variable of integration, we may revert to using $y$ instead. Combining the two integrals,

$$v(x,t) = \int_0^\infty [S(x-y,t) - S(x+y,t)]\phi(y)\,dy.$$

Finally, writing out the heat kernel $S$ explicitly, we have shown that the solution of the Dirichlet problem (11.1)–(11.3) is given by

$$u(x,t) = \frac{1}{\sqrt{4\pi\kappa t}} \int_0^\infty \left[ e^{-(x-y)^2/4\kappa t} - e^{-(x+y)^2/4\kappa t} \right] \phi(y)\, dy \qquad (11.4)$$

for $t > 0$. Note that the integral is taken over the entire spatial domain.

In deriving (11.4), it was very important that the boundary condition at $x = 0$ was *homogeneous*. Otherwise, the solution $v(x,t)$ of the auxiliary Cauchy problem would not have automatically satisfied the boundary condition (11.3) for $u(x,t)$. Fortunately, given a more general Dirichlet condition $u(0,t) = g(t)$, it is still possible to use the odd reflection technique. The idea is to make a special substitution $w(x,t) = u(x,t) - g(t)$ that converts the inhomogeneous boundary condition into a homogeneous one. It turns out that $w$ then satisfies an inhomogeneous heat equation which can be solved via yet another substitution (see exercises).

**Example 11.1.1.** To understand the effect of the homogeneous Dirichlet boundary condition, let us compare the solution (10.23) of the Cauchy problem for the heat equation with formula (11.4) for a special choice of initial condition. Namely, if we use the constant heat distribution $\phi(x) = 1$, then the solution of the Cauchy problem would be

$$u(x,t) = (S \star \phi)(x,t) = \int_{-\infty}^{\infty} S(x-y,t)\, dy,$$

where $S$ denotes the heat kernel. From a past exercise, we know that the integral of the heat kernel over the entire real line is precisely 1 for all $t > 0$. Therefore, the solution of the Cauchy problem with this initial condition is $u(x,t) = 1$ for all real $x$ and $t \geq 0$. This makes perfect sense, because we would expect a perfectly uniform initial heat distribution to remain uniform for all time $t$.

Now suppose we restrict ourselves to a half-line and impose a homogeneous Dirichlet condition at $x = 0$:

$$\begin{aligned} u_t &= \kappa u_{xx} && (0 < x < \infty) \\ u(x,0) &= 1 && (0 < x < \infty) \\ u(0,t) &= 0 && (t \geq 0). \end{aligned}$$

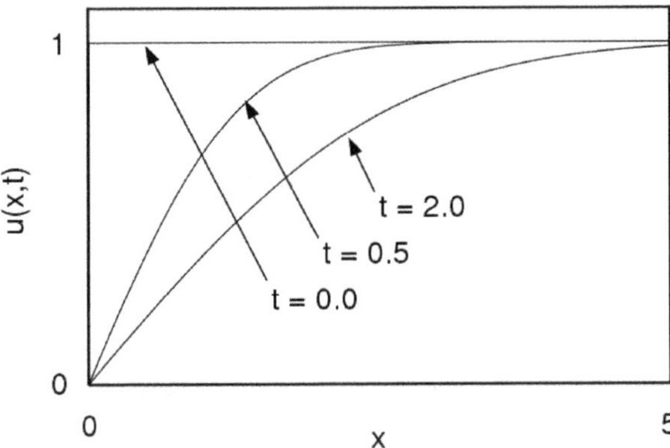


**Figure 11.2.** Solution of (11.1)–(11.3) with constant initial heat distribution $\phi(x) = 1$.

By formula (11.4), the solution is

$$u(x,t) = \int_0^\infty S(x-y,t) - S(x+y,t)\,dy. \tag{11.5}$$

Unlike the solution of the Cauchy problem, this integral is *not* identically equal to 1 for all $x \geq 0$ and $t > 0$. In fact, the integral (11.5) cannot be evaluated explicitly (although it can be written in terms of the standard error function). A graph of the function $u(x,t)$ from (11.5) is shown in Figure 11.2. The "cooling" effect of the boundary condition is felt over a progressively wider region near the boundary.

**Heat Equation, Neumann Problem.** The homogeneous Neumann problem for the heat equation on a half-line is given by

$$u_t = \kappa u_{xx} \qquad (0 < x < \infty) \tag{11.6}$$

$$u(x,0) = \phi(x) \qquad (0 < x < \infty) \tag{11.7}$$

$$u_x(0,t) = 0 \qquad (t \geq 0). \tag{11.8}$$

The boundary condition (11.8) states that the spatial gradient of $u$ is 0 at $x = 0$. Hence, there is no heat flux across the boundary, which is analogous to insulating the $x = 0$ end of the wire to prevent "leakage" of heat. Physically, this is very different from the homogeneous Dirichlet condition that we considered earlier.

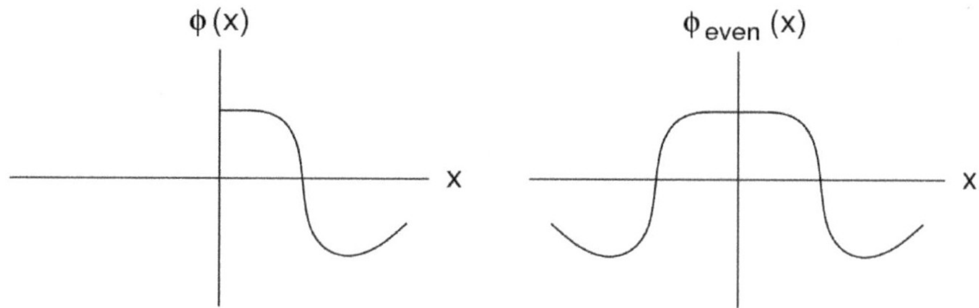


**Figure 11.3.** Illustration of the even extension of a function $\phi(x)$.

Solving (11.6)–(11.8) is accomplished by a similar reflection method as the one we developed for the Dirichlet problem. However, using an odd extension of $\phi(x)$ will not necessarily preserve the boundary condition at $x = 0$. Instead, we will introduce the *even extension* of $\phi(x)$, which is defined as

$$\phi_{\text{even}}(x) = \begin{cases} \phi(x) & \text{if } x \geq 0 \\ \phi(-x) & \text{if } x < 0. \end{cases}$$

This concept is illustrated in Figure 11.3. Notice that if $\phi(x)$ has a right-hand derivative of 0 at $x = 0$, then the even extension is differentiable at $x = 0$ and satisfies $\phi'_{\text{even}}(0) = 0$. To solve our Neumann problem, follow a similar procedure as before: (1) Solve the Cauchy problem using $\phi_{\text{even}}$ as the initial condition. (2) Argue that the solution of the Cauchy problem must remain even for all $t > 0$ since the initial condition was even. (3) Show that restricting the solution of the Cauchy problem to the domain $x \geq 0$ yields the solution of the Neumann problem. By following this procedure, you will find that the solution of (11.6)–(11.8) is given by

$$\begin{aligned} u(x,t) &= \int_0^\infty [S(x-y,t) + S(x+y,t)]\, \phi(y)\, dy \\ &= \frac{1}{\sqrt{4\pi\kappa t}} \int_0^\infty \left[ e^{-(x-y)^2/4\kappa t} + e^{-(x+y)^2/4\kappa t} \right] \phi(y)\, dy \end{aligned} \tag{11.9}$$

Notice that the only difference between the solutions (11.4) (Dirichlet problem) and (11.9) (Neumann problem) is the sign that appears in the integrand.

Again, we remark that the method of even reflection relied upon the fact that the Neumann boundary condition (11.8) was homogeneous. Given the more general Neumann condition $u_x(0,t) = g(t)$, we can make a substitution $w(x,t) = u(x,t) - xg(t)$. Then $w(x,t)$ will satisfy the homogeneous Neumann condition $w_x(0,t) = 0$. Although this substitution improved the boundary condition, it has the unfortunate side-effect that $w$ satisfies an *inhomogeneous* heat equation. You will learn how to solve the inhomogeneous Neumann problem as an exercise.

**Wave Equation, Dirichlet Problem.** The homogeneous Dirichlet problem for the wave equation on a half-line is given by

$$u_{tt} = c^2 u_{xx} \qquad (0 < x < \infty) \tag{11.10}$$

$$u(x,0) = \phi(x) \qquad (0 < x < \infty) \tag{11.11}$$

$$u_t(x,0) = \psi(x) \qquad (0 < x < \infty) \tag{11.12}$$

$$u(0,t) = 0 \qquad (t \geq 0). \tag{11.13}$$

Recall that the two initial conditions correspond to the initial displacement and velocity of a vibrating string. The Dirichlet condition requires that the boundary of the string remain stationary in the equilibrium position.

As with the homogeneous Dirichlet problem for the heat equation, we will use the method of odd extension. Define

$$\phi_{\text{odd}}(x) = \begin{cases} \phi(x) & \text{if } x > 0 \\ -\phi(-x) & \text{if } x < 0 \\ 0 & \text{if } x = 0 \end{cases} \quad \text{and} \quad \psi_{\text{odd}}(x) = \begin{cases} \psi(x) & \text{if } x > 0 \\ -\psi(-x) & \text{if } x < 0 \\ 0 & \text{if } x = 0 \end{cases}$$

and consider the Cauchy problem

$$\begin{aligned} v_{tt} &= c^2 v_{xx} && (-\infty < x < \infty) \\ v(x,0) &= \phi_{\text{odd}}(x) && (-\infty < x < \infty) \\ v_t(x,0) &= \psi_{\text{odd}}(x) && (-\infty < x < \infty). \end{aligned}$$

Once again, we claim that the restriction of $v(x,t)$ to the domain $x \geq 0$ is the solution of the Dirichlet problem (11.10)–(11.13). Certainly $u$ and $v$ satisfy the same PDE and initial conditions on the domain $x > 0$, so it remains to check that $v$ automatically satisfies the boundary condition (11.13) on $u$. However, we know that $v$ will remain odd for all $t$ because the initial conditions $\phi_{\text{odd}}$ and $\psi_{\text{odd}}$ are odd functions (exercise). This implies that $v(0,t) = 0$ for all $t$, which is precisely what we need at the boundary.

The solution of the Cauchy problem is provided by D'Alembert's formula:

$$v(x,t) = \frac{1}{2}\left[\phi_{\text{odd}}(x+ct) + \phi_{\text{odd}}(x-ct)\right] + \frac{1}{2c}\int_{x-ct}^{x+ct} \psi_{\text{odd}}(s)\,\text{ds}. \qquad (11.14)$$

In order to recover the solution of (11.10)–(11.13), we must restrict this solution to $x \geq 0$ and express the solution in terms of $\phi$ and $\psi$ (rather than $\phi_{\text{odd}}$ and $\psi_{\text{odd}}$). We must exercise caution—even if $x > 0$ and $t > 0$, it is possible that $x - ct < 0$. This would mean that the integral term in (11.14) is referencing points that lie outside the spatial domain, a situation that we need to avoid. We will proceed by considering several cases.

**Case 1.** Suppose that $x - ct > 0$ and $x + ct > 0$. Then the domain of integration in formula (11.14) lies entirely within the spatial domain of our Dirichlet problem. Moreover, we know that $\phi_{\text{odd}}(x) = \phi(x)$ and $\psi_{\text{odd}}(x) = \psi(x)$ for $x > 0$. Hence, the solution of the Dirichlet problem is given by

$$u(x,t) = \frac{1}{2}\left[\phi(x+ct) + \phi(x-ct)\right] + \frac{1}{2c}\int_{x-ct}^{x+ct} \psi(s)\,\text{ds}. \qquad (11.15)$$

Notice that, in this case, formula (11.15) is identical to D'Alembert's formula for the solution of the Cauchy problem. This is explained easily if we recall the notion of the domain of dependence for $u(x,t)$. Since we have assumed that both $x - ct$ and $x + ct$ are *positive*, the boundary $x = 0$ has no influence on $u(x,t)$. This is a consequence of the finite propagation speed $c$, as illustrated in the left panel of Figure 11.4. Assuming $t > 0$ as illustrated in the Figure, the value of $u(x,t)$ is only impacted by the initial conditions within the interval $[x - ct, x + ct]$. Since the interval avoids the boundary in this case, the boundary condition cannot possibly influence the behavior of $u$ at the point $(x,t)$.

**Case 2.** Suppose that $x - ct < 0$ and $x + ct > 0$. This time, formula (11.14) references points that lie outside the spatial domain $(x \geq 0)$ of our Dirichlet problem. Writing the solution in terms of $\phi$ and $\psi$ requires more care in this case. Since $x - ct < 0$, we can write $\phi_{\text{odd}}(x - ct) = \phi_{\text{odd}}(-(ct - x)) = -\phi(ct - x)$. Next, we must write the integral term in (11.14) in such a way that the interval of integration does not include negative values. The idea is to split the integral

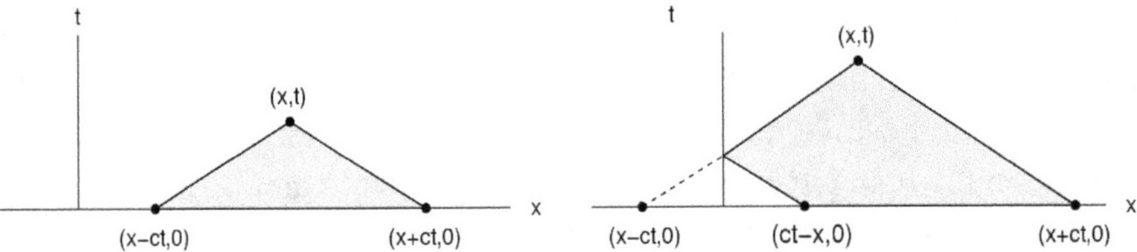


**Figure 11.4.** Examples of domains of dependence for $u(x,t)$, the solution of the homogeneous Dirichlet problem (11.10)–(11.13). Left panel: Domain of dependence of $(x,t)$ assuming that $t > 0$ and both $x \pm ct > 0$. Right panel: Domain of dependence of $(x,t)$ assuming that $t > 0$ and $x - ct < 0 < x + ct$.

as follows:

$$\frac{1}{2c}\int_{x-ct}^{x+ct} \psi_{\text{odd}}(s)\,\mathrm{d}s \;=\; \frac{1}{2c}\int_{x-ct}^{0} \psi_{\text{odd}}(s)\,\mathrm{d}s \;+\; \frac{1}{2c}\int_{0}^{x+ct} \psi_{\text{odd}}(s)\,\mathrm{d}s.$$

Referring to the definition of $\psi_{\text{odd}}$, we can rewrite this equation as

$$\frac{1}{2c}\int_{x-ct}^{x+ct} \psi_{\text{odd}}(s)\,\mathrm{d}s \;=\; \frac{1}{2c}\int_{x-ct}^{0} -\psi(-s)\,\mathrm{d}s \;+\; \frac{1}{2c}\int_{0}^{x+ct} \psi(s)\,\mathrm{d}s.$$

Substituting $w = -s$ gives

$$\frac{1}{2c}\int_{x-ct}^{x+ct} \psi_{\text{odd}}(s)\,\mathrm{d}s \;=\; \frac{1}{2c}\int_{ct-x}^{0} \psi(w)\,\mathrm{d}w \;+\; \frac{1}{2c}\int_{0}^{x+ct} \psi(s)\,\mathrm{d}s,$$

and since $s$ and $w$ are merely dummy variables of integration, we can now combine the integrals as

$$\frac{1}{2c}\int_{x-ct}^{x+ct} \psi_{\text{odd}}(s)\,\mathrm{d}s \;=\; \frac{1}{2c}\int_{ct-x}^{x+ct} \psi(s)\,\mathrm{d}s.$$

In summary, we have now shown that if $x - ct < 0 < x + ct$, then the solution of the Dirichlet problem (11.10)–(11.13) is

$$u(x,t) \;=\; \frac{1}{2}\left[\phi(x+ct) - \phi(ct-x)\right] \;+\; \frac{1}{2c}\int_{ct-x}^{x+ct} \psi(s)\,\mathrm{d}s. \tag{11.16}$$

Formula (11.16) differs from D'Alembert's formula. Interaction with the boundary forced us to "reflect" the negative quantity $x - ct$, expressing the solution in terms of the positive quantity $ct - x$. This reflection off the boundary is illustrated in the right panel of Figure 11.4. The shaded region shows the domain of dependence for a point $(x, t)$ with $t > 0$.

It is certainly possible to consider other possibilities, such as $x \pm ct < 0$. However, given that our spatial domain is $x \geq 0$ and $c > 0$, such situations could only occur for negative time $t$. Although it is perfectly reasonable to solve the wave equation for negative $t$ (contrast this with the heat equation), we will leave the details of the $x \pm ct < 0$ case to the reader.

**Wave Equation, Neumann Problem.** The homogeneous Neumann problem for the wave equation on a half-line is given by

$$\begin{aligned} u_{tt} &= c^2 u_{xx} && (0 < x < \infty) \\ u(x,0) &= \phi(x) && (0 < x < \infty) \\ u_t(x,0) &= \psi(x) && (0 < x < \infty) \\ u_x(0,t) &= 0 && (t \geq 0). \end{aligned}$$

Physically, the boundary condition is a bit more difficult to interpret than a homogeneous Dirichlet condition. Instead of holding the $x = 0$ end of our vibrating string completely stationary, it is free to move transversely. Moreover, there is no tension at that end of the string. To solve this Neumann problem, define the *even* extensions of $\phi$ and $\psi$ and convert it to a Cauchy problem as before. Writing the solution involves several cases according to the signs of $x + ct$ and $x - ct$.

---

## 11.2. Separation of Variables

Henceforth, we will solve PDEs on *finite* spatial domains. Since we generally work in one spatial dimension, we have in mind a finite interval $0 \leq x \leq L$, where $L$ is a positive constant. The boundary of the domain consists of two points ($x = 0$ and $x = L$), and we shall impose boundary conditions at both of these points. In the context of the wave equation, $L$ is the length of our vibrating string. For the heat equation, $L$ would represent the length of the wire within which heat diffuses longitudinally.

The inclusion of a second boundary point prevents us from recycling the reflection methods introduced in the previous section—those methods only apply to semi-infinite domains with a single boundary point. Instead, we will develop a completely different approach known as *separation of variables*. To motivate this technique, let us recall a specific example discussed in a previous chapter. After deriving D'Alembert's formula, we solved the Cauchy problem $u_{tt} - c^2 u_{xx} = 0$ with initial displacement $\phi(x) = \cos(x)$ and initial velocity $\psi(x) = -\cos(x)$. By creative use of trigonometric identities, we were able to express the solution in the form

$$u(x,t) = \cos(x)\left[\cos(ct) - \frac{1}{c}\sin(ct)\right].$$

The solution is written as a product of two functions: a function of $x$ only and a function of $t$ only. The fact that we were able to "separate" the spatial and temporal parts of the solution made it considerably easier to visualize the dynamics.

The idea of the separation of variables technique is to seek special separated solutions of the form $u(x,t) = X(x)T(t)$, where $X$ and $T$ are functions only of the spatial and temporal variables, respectively. By finding enough separated solutions of a linear homogeneous initial-boundary value problem, we will be able to construct the general solution via superposition.

**11.2.1 Wave Equation, Dirichlet Problem.** We illustrate the method of separation of variables by solving the homogeneous Dirichlet problem

$$u_{tt} = c^2 u_{xx} \qquad (0 < x < L) \qquad (11.17)$$

$$u(x,0) = \phi(x) \qquad (0 < x < L) \qquad (11.18)$$

$$u_t(x,0) = \psi(x) \qquad (0 < x < L) \qquad (11.19)$$

$$u(0,t) = 0 \qquad (t \geq 0) \qquad (11.20)$$

$$u(L,t) = 0 \qquad (t \geq 0). \qquad (11.21)$$

Physically, the two Dirichlet boundary conditions indicate that the ends of our vibrating string must remain stationary for all $t$, which is precisely what we would envision for a vibrating guitar string.

**Step 1: Separate the variables.** The first step in attacking this problem is to seek non-trivial[1] separated solutions $u(x,t) = X(x)T(t)$ which satisfy the PDE (11.17) and both boundary conditions (11.20)–(11.21). Substituting $u(x,t) = X(x)T(t)$ into the wave equation (11.17), we obtain

$$X(x)T''(t) = c^2X''(x)T(t).$$

We are now able to use primes to denote differentiation, since $X$ and $T$ are each functions of one variable. By algebra, we may separate the time and space-dependent parts of this equation:

$$\frac{X''}{X} = \frac{T''}{c^2T}.$$

Notice that the left-hand side is a function of $x$ only, whereas the right-hand side is a function of $t$ only. The only way for such functions to be identically equal is if both functions are equal to a common constant. Denoting this constant by $-\lambda$, where the negative sign is included for later convenience, we have

$$\frac{X''}{X} = \frac{T''}{c^2T} = -\lambda.$$

Equivalently, we have obtained a pair of two second-order, linear constant-coefficient ODEs:

$$X'' + \lambda X = 0 \quad \text{and} \quad T'' + \lambda c^2 T = 0. \tag{11.22}$$

The fact that we are now dealing with ODEs echoes a recurring theme: seeking special solutions of PDEs often reduces the PDE to an ODE.

**Step 2: Boundary conditions.** The boundary conditions (11.20)–(11.21) will impose special requirements on the $X$ equation in (11.22). Combining $u(x,t) = X(x)T(t)$ with the fact that $u(0,t) = 0$ for all time $t$, it follows that $X(0)T(t) = 0$ for all time $t$. There are two ways this could happen: either $X(0) = 0$ or $T(t) = 0$

[1]That is, we wish to exclude the constant solution $u(x,t) = 0$.

for all $t$. The latter possibility is not interesting, because if $T(t) = 0$ for all $t$, then $u(x,t) = X(x)T(t) = 0$, the trivial solution. Hence, the boundary condition $u(0,t) = 0$ implies that $X(0) = 0$. A similar argument shows that the boundary condition $u(L,t) = 0$ forces $X(L) = 0$.

**Step 3: Solve the equation subject to these boundary conditions.** We wish to seek non-trivial solutions of the two-point boundary value problem

$$X'' + \lambda X = 0, \quad \text{and} \quad X(0) = 0 = X(L). \tag{11.23}$$

The form of the solutions will depend upon the sign of $\lambda$, and there are three possibilities.

**Case 1:** $\lambda < 0$. We claim that if $\lambda < 0$, then the boundary value problem (11.23) has no non-trivial solutions. To see why, suppose that $\lambda = -\beta^2$ where $\beta > 0$. (Introducing $\beta$ is solely for the purpose of making the solutions look cleaner.) The ODE for $X$ becomes

$$X'' - \beta^2 X = 0.$$

The associated characteristic equation is $m^2 - \beta^2 = 0$, which has distinct, real roots $m = \pm\beta$. The general solution of this ODE is therefore

$$X(x) = Ce^{-\beta x} + De^{\beta x}, \tag{11.24}$$

where $C$ and $D$ are arbitrary constants. To solve for $C$ and $D$, we must incorporate the boundary conditions. Using $X(0) = 0$ yields $0 = C + D$, and using $X(L) = 0$ yields $0 = Ce^{-\beta L} + De^{\beta L}$. Since $D = -C$, the latter equation can be written as

$$C\left[e^{-\beta L} - e^{\beta L}\right] = 0.$$

We may exclude the possibility $C = 0$, because that would force $D = 0$, implying that $X(x) = 0$ and ultimately leading us to the trivial solution. If $C \neq 0$, then $e^{-\beta L} - e^{\beta L} = 0$. Multiplying both sides by $e^{\beta L}$ and rearranging terms yields $e^{2\beta L} = 1$. However, this equality is impossible because $2\beta L$ is positive, which would force $e^{2\beta L} > 1$.

**Case 2:** $\lambda = 0$. The boundary value problem (11.23) also has no non-trivial solutions if $\lambda = 0$. In this case, the $X$ equation reduces to $X'' = 0$. It is easy to solve this ODE, integrating twice to obtain the general solution $X(x) = Cx + D$. The only way this linear function can possibly satisfy the boundary conditions $X(0) = 0$ and $X(L) = 0$ is if $C = D = 0$. This means that $X(x) = 0$ and, consequently, that $u(x,t) = X(x)T(t) = 0$ as well. Unfortunately, we have yet to produce a single interesting solution of our PDE and boundary conditions, but our luck is about to change.

**Case 3:** $\lambda > 0$. Now assume that $\lambda = \beta^2$ where $\beta > 0$. (Again, introducing $\beta$ is merely for convenience.) The boundary value problem (11.23) becomes

$$X'' + \beta^2 X = 0, \quad \text{and} \quad X(0) = 0 = X(L).$$

The characteristic equation for the ODE is $m^2 + \beta^2 = 0$, which has pure imaginary roots $m = \pm\beta i$. Therefore, the general solution of the ODE is

$$X(x) = C\cos(\beta x) + D\sin(\beta x),$$

where $C$ and $D$ are constants. The boundary condition $X(0) = 0$ implies that $C = 0$, and we are left with $X(x) = D\sin(\beta x)$. The other boundary condition $X(L) = 0$ implies that $D\sin(\beta L) = 0$. One possibility is $D = 0$, but this would lead to the same trivial solution that we are trying to avoid. Fortunately, there is a much more interesting possibility: $\sin(\beta L) = 0$ if $\beta$ is chosen appropriately. We know that $\beta > 0$ by assumption, and $L > 0$ since it represents the length of our spatial domain. Therefore, if

$$\beta L = n\pi \qquad (n = 1, 2, 3, \ldots),$$

then we will have found non-zero solutions of the $X$ equation that satisfy both boundary conditions. In order to index the solutions, let us define

$$\beta_n = \frac{n\pi}{L}, \qquad \lambda_n = \beta_n^2 \qquad (n = 1, 2, 3, \ldots)$$

and

$$X_n(x) = D_n \sin(\beta_n x) = D_n \sin\left(\frac{n\pi x}{L}\right) \qquad (n = 1, 2, 3 \ldots), \tag{11.25}$$

where $D_n$ are arbitrary constants. The functions $X_n(x)$ form the set of all possible solutions of the boundary value problem (11.23).

Now that we have solved the $X$ equation in (11.22), let us turn our attention to

$$T'' + \lambda c^2 T = 0.$$

In light of the above calculations, we are only interested in the solution for special choices of $\lambda$, namely for $\lambda_n = \beta_n^2$. The equation

$$T'' + \frac{n^2\pi^2}{L^2} c^2 T = 0 \qquad (n = 1, 2, 3, \ldots)$$

has characteristic equation

$$m^2 + \left(\frac{n\pi c}{L}\right)^2 = 0.$$

The roots

$$m = \pm\left(\frac{n\pi c}{L}\right) \mathrm{i}$$

are pure imaginary, and therefore the solutions of the $T$ equation have the form

$$T_n(t) = E_n \cos\left(\frac{n\pi ct}{L}\right) + F_n \sin\left(\frac{n\pi ct}{L}\right) \qquad (n = 1, 2, 3, \ldots). \tag{11.26}$$

Here, $E_n$ and $F_n$ are arbitrary constants.

**Step 4: Building the general solution.** We will now attempt to build the set of all possible functions that simultaneously satisfy the PDE (11.17) and the boundary conditions (11.20)–(11.21). The initial conditions will be incorporated later. Now that we have found all non-trivial solutions $X_n(x)$ and $T_n(t)$ of the ODEs (11.22) and boundary conditions, recall that $u(x,t) = X(x)T(t)$. Thus, the functions

$$u_n(x,t) = X_n(x)T_n(t) \qquad (n = 1, 2, 3, \ldots) \tag{11.27}$$

are solutions of the original PDE (11.17) which also satisfy both boundary conditions (11.20)–(11.21). Since the wave equation (11.17) is linear and homogeneous, the Superposition Principle 8.1.12 states that any *finite* linear combination

$$\sum_{n=1}^{N} G_n u_n(x,t)$$

($G_n$ are constants) is also a solution of the PDE and its boundary conditions. Surprisingly, we can say much more: the general solution of the PDE and its boundary conditions is given by the *infinite* sum

$$\sum_{n=1}^{\infty} G_n u_n(x,t),$$

provided that this series converges in some sense. Understanding why this *infinite series* represents the general solution of the PDE with its boundary conditions requires some effort and is [part of] the subject of the next chapter. For the moment, we will proceed formally, assuming that this infinite sum really does represent the solution we seek. Justification will be provided later.

Recalling the definition of $u_n(x,t)$, we have claimed that the general solution of the wave equation (11.17) with its two boundary conditions (11.20)–(11.21) is given by

$$\begin{aligned}\sum_{n=1}^{\infty} G_n u_n(x,t) &= \sum_{n=1}^{\infty} G_n X_n(x) T_n(t) \\ &= \sum_{n=1}^{\infty} G_n \left[ E_n \cos\left(\frac{n\pi ct}{L}\right) + F_n \sin\left(\frac{n\pi ct}{L}\right)\right] D_n \sin\left(\frac{n\pi x}{L}\right).\end{aligned}$$

Products of arbitrary constants can be combined to make the solution a bit more concise: let $A_n = G_n E_n D_n$ and $B_n = G_n F_n D_n$ to obtain

$$u(x,t) = \sum_{n=1}^{\infty} \left[ A_n \cos\left(\frac{n\pi ct}{L}\right) + B_n \sin\left(\frac{n\pi ct}{L}\right)\right] \sin\left(\frac{n\pi x}{L}\right). \tag{11.28}$$

**Step 5: Use the initial conditions.** Armed with the general solution (11.28) of the wave equation and its two Dirichlet boundary conditions, the final step is to incorporate the initial conditions. Since $u(x,0) = \phi(x)$, setting $t = 0$ in (11.28) yields

$$\phi(x) = \sum_{n=1}^{\infty} A_n \sin\left(\frac{n\pi x}{L}\right). \tag{11.29}$$

Equation (11.29) may seem troubling and confusing at first: it says that our initial displacement $\phi$ *must* be a sum of sine waves with various frequencies and amplitudes. However, we never made *any* assumptions on $\phi(x)$ when we posed our original Dirichlet problem. Would it be a severe restriction to allow only those initial conditions $\phi(x)$ which have special sine series representations of the form (11.29)? How "big" is the class of functions that have sine series representations? If a function $\phi(x)$ has such a representation, is there a systematic way to determine constants $A_n$? Luckily, the answers to these three questions are "No", "Very Big", and "Yes", respectively. These issues will be studied in greater depth during the next chapter.

To use the other initial condition $u_t(x,0) = \psi(x)$, we must formally differentiate the solution (11.28) with respect to $t$:

$$u_t(x,t) = \sum_{n=1}^{\infty}\left[-\frac{n\pi c}{L}A_n \sin\left(\frac{n\pi ct}{L}\right) + B_n\frac{n\pi c}{L}\cos\left(\frac{n\pi ct}{L}\right)\right]\sin\left(\frac{n\pi x}{L}\right).$$

Now setting $t = 0$,

$$\psi(x) = u_t(x,0) = \sum_{n=1}^{\infty} B_n\frac{n\pi c}{L}\sin\left(\frac{n\pi x}{L}\right).$$

As with the other initial condition, we have found that $\psi(x)$ would need to have a special sine series representation in order for us to declare victory over the Dirichlet problem (11.17)–(11.21). To summarize: *IF* there exist constants $A_n$ and $B_n$ such that the initial conditions $\phi(x), \psi(x)$ can be represented as convergent sine series

$$\phi(x) = \sum_{n=1}^{\infty} A_n \sin\left(\frac{n\pi x}{L}\right) \quad \text{and} \quad \psi(x) = \sum_{n=1}^{\infty} B_n\frac{n\pi c}{L}\sin\left(\frac{n\pi x}{L}\right)$$

on $0 \le x \le L$, then the solution of the Dirichlet problem (11.17)–(11.21) is given by Equation (11.28) above. Series of this form are called *Fourier sine series*, and we will learn more about them in the next chapter.

**11.2.2 Heat Equation, Dirichlet Problem.** The homogeneous Dirichlet problem for the heat equation is given by

$$u_t = \kappa u_{xx} \qquad (0 < x < L) \tag{11.30}$$

$$u(x,0) = \phi(x) \qquad (0 < x < L) \tag{11.31}$$

$$u(0,t) = 0 \qquad (t \ge 0) \tag{11.32}$$

$$u(L,t) = 0 \qquad (t \ge 0). \tag{11.33}$$

Physically, this models the transfer of heat within a wire of length $L$, with initial temperature distribution $\phi(x)$ and with both ends of the wire in contact with zero-degree blocks of ice. We will follow the same steps used to solve the Dirichlet problem for the wave equation.

**Step 1: Separate the variables.** Substituting $u(x,t) = X(x)T(t)$ into the PDE in equation (11.30), we have $XT' = \kappa X''T$. By algebra,

$$\frac{X''}{X} = \frac{T'}{\kappa T}.$$

As before, the only way a function of $x$ could be identically equal to a function of $t$ is if both expressions are equal to a common constant:

$$\frac{X''}{X} = \frac{T'}{\kappa T} = -\lambda.$$

Equivalently, we have a system of two ODEs:

$$X'' + \lambda X = 0 \quad \text{and} \quad T' + \lambda\kappa T = 0.$$

**Step 2: Boundary conditions.** Using the fact that $u(x,t) = X(x)T(t)$, the homogeneous Dirichlet boundary conditions (11.32)–(11.33) yield $0 = X(0)T(t)$ and $0 = X(L)T(t)$. One possibility is $T(t) = 0$, but this would lead us to the trivial solution $u(x,t) = 0$. Hence, we may focus on the more interesting possibility that both $X(0) = 0$ and $X(L) = 0$.

**Step 3: Solve the equation subject to these boundary conditions.** Notice that $X$ satisfies the same two-point boundary value problem (11.23) that we encountered while solving the Dirichlet problem for the wave equation. As before, if $\lambda \leq 0$ there are no solutions. For $\lambda > 0$, it is convenient to let $\lambda = \beta^2$ where $\beta > 0$. The general solution of $X'' + \beta^2 X = 0$ is $X(x) = C\cos(\beta x) + D\sin(\beta x)$, where $C$ and $D$ are constants. Using the boundary condition $X(0) = 0$ yields $C = 0$, and the boundary condition $X(L) = 0$ yields $D\sin(\beta L) = 0$. We may assume $D \neq 0$, because otherwise $X(x)$ is identically zero and we are led to the trivial solution of the PDE. Thus, $\sin(\beta L) = 0$, and since both $\beta > 0$ and $L > 0$, it must be the case that $\beta L$ is a positive integer multiple of $\pi$. Define

$$\beta_n = \frac{n\pi}{L} \qquad \lambda_n = \beta_n^2 = \left(\frac{n\pi}{L}\right)^2 \qquad (n = 1, 2, 3, \ldots),$$

and

$$X_n(x) = D_n \sin(\beta_n x) = D_n \sin\left(\frac{n\pi x}{L}\right) \qquad (n = 1, 2, 3, \ldots).$$

The functions $X_n(x)$ satisfy the boundary value problem (11.23). With these special choices of $\lambda$ in mind, we return to the $T$ equation

$$T' + \lambda\kappa T = 0.$$

Replacing $\lambda$ with $\lambda_n$,

$$T' + \kappa\left(\frac{n\pi}{L}\right)^2 T = 0 \qquad (n = 1, 2, 3, \ldots),$$

a first-order constant-coefficient ODE. By separation of variables, the solutions are

$$T_n(t) = E_n e^{-\kappa(n\pi/L)^2 t} \qquad (n = 1, 2, 3, \ldots),$$

where $E_n$ are constants.

**Step 4: Building the general solution.** Since $u(x,t) = X(x)T(t)$, let us define

$$u_n(x,t) = X_n(x)T_n(t) = A_n \sin\left(\frac{n\pi x}{L}\right) e^{-\kappa(n\pi/L)^2 t},$$

where $A_n = D_n E_n$ are constants. Each function $u_n(x,t)$ satisfies the PDE (11.30) and both of its Dirichlet boundary conditions (11.32)–(11.33). The general solution of the PDE with its boundary conditions is given by

$$u(x,t) = \sum_{n=1}^{\infty} A_n \sin\left(\frac{n\pi x}{L}\right) e^{-\kappa(n\pi/L)^2 t}. \qquad (11.34)$$

**Step 5: Use the initial condition.** Finally, it remains to use the initial condition (11.31). Setting $t = 0$ in our formula for $u(x,t)$, we have

$$\phi(x) = \sum_{n=1}^{\infty} A_n \sin\left(\frac{n\pi x}{L}\right).$$

Thus, formula (11.34) is the solution of the Dirichlet problem (11.30)–(11.33) provided that the initial condition $\phi(x)$ has a Fourier sine series representation on the interval $0 < x < L$. Given a specific choice of initial condition $\phi(x)$, we will soon learn how to determine the values of the Fourier coefficients $A_n$.

In solving the homogeneous Dirichlet problems for the heat and wave equations, we encountered the two-point boundary value problem

$$X'' = -\lambda X \qquad X(0) = 0 = X(L).$$

Note the similarity between the differential equation

$$-\frac{d^2}{dx^2}X = \lambda X$$

and the matrix equation $A\mathbf{v} = \lambda\mathbf{v}$. The operator $-\frac{d^2}{dx^2}$ takes the place of the matrix $A$, and the function $X(x)$ takes the place of the vector $\mathbf{v}$. For certain special choices of $\lambda$, namely $\lambda_n = (n\pi/L)^2$, the boundary value problem has non-zero solutions $X$. By analogy with linear algebra jargon, the numbers $\lambda_n$ are called *eigenvalues* and the functions $X_n(x) = \sin(n\pi x/L)$ are called *eigenvectors* or *eigenfunctions* for the operator $-\frac{d^2}{dx^2}$. The problem $-\frac{d^2X}{dx^2} = \lambda X$ is called an *eigenvalue problem*. Notice that in this case, there are infinitely many eigenvalues. Unlike a square matrix $A$, the operator $-\frac{d^2}{dx^2}$ is an infinite-dimensional linear operator. Much of the branch of mathematics known as *functional analysis* involves extending various notions from linear algebra to infinite-dimensional settings.

**Example 11.2.1.** Suppose $r \neq 0$ is a constant. Solve

$$u_t = \kappa u_{xx} - ru \qquad (0 < x < L) \qquad (11.35)$$

$$u(x,0) = \phi(x) \qquad (0 < x < L) \qquad (11.36)$$

$$u(0,t) = 0 \qquad (t \geq 0) \qquad (11.37)$$

$$u(L,t) = 0 \qquad (t \geq 0). \qquad (11.38)$$

**Solution:** This is a homogeneous Dirichlet problem. The $-ru$ term in the PDE is known as a reaction term, and the PDE itself is an example of a reaction-diffusion equation. We will solve the PDE by separation of variables: substituting $u(x,t) = X(x)T(t)$ into (11.35) yields $XT' = \kappa X''T - rXT$. Some algebra will help us isolate the quantity $X''/X$ as in the previous examples. This time, we find that

$$\frac{X''}{X} = \frac{T' + rT}{\kappa T} = -\lambda,$$

where, as usual, $-\lambda$ is a constant. Solving the equation $X'' + \lambda X = 0$ subject to the boundary conditions (11.37)–(11.38) follows precisely the same steps as before. The only values of $\lambda$ for which the eigenvalue problem

$$X'' + \lambda X = 0, \qquad X(0) = 0 = X(L)$$

has non-zero solutions are the eigenvalues

$$\lambda_n = \left(\frac{n\pi}{L}\right)^2 \qquad (n = 1, 2, 3, \ldots).$$

The corresponding eigenfunctions are

$$X_n(x) = \sin\left(\frac{n\pi x}{L}\right) \qquad (n = 1, 2, 3, \ldots).$$

That is, the only non-trivial solutions of the eigenvalue problem are scalar multiples of these functions $X_n(x)$.

It remains to solve the $T$ equation $T' + rT = -\lambda\kappa T$ using the special $\lambda$ values (the eigenvalues) we found above. The first-order constant-coefficient ordinary differential equation $T' = (-\lambda\kappa - r)T$ has solution

$$T(t) = Ce^{(-\lambda\kappa - r)t},$$

where $C$ is a constant. Using the eigenvalues $\lambda = \lambda_n$, we are led to define

$$T_n(t) = C_n e^{-rt} e^{-\lambda_n \kappa t} = C_n e^{-rt} e^{-\kappa(n\pi/L)^2 t} \qquad (n = 1, 2, 3, \ldots).$$

Finally, we build the general solution by defining $u_n(x,t) = X_n(x)T_n(t)$ and summing:

$$u(x,t) = \sum_{n=1}^{\infty} B_n u_n(x,t),$$

where $B_n$ are constants. Inserting our expressions for $X_n$ and $T_n$ into this summation and abbreviating $A_n = B_n C_n$,

$$u(x,t) = e^{-rt} \sum_{n=1}^{\infty} A_n \sin\left(\frac{n\pi x}{L}\right) e^{-\kappa(n\pi/L)^2 t} \qquad (11.39)$$

is the general solution of the PDE (11.35) with boundary conditions (11.37)–(11.38).

The final step is to use the initial condition (11.36) by setting $t = 0$ in (11.39). The result is

$$\phi(x) = \sum_{n=1}^{\infty} A_n \sin\left(\frac{n\pi x}{L}\right),$$

implying once again that our initial condition $\phi(x)$ must have a Fourier sine series representation.

Take a moment to compare the solutions of the Dirichlet problem for the heat equation (11.30)–(11.33) and the Dirichlet problem for the reaction-diffusion system (11.35)–(11.38). The only difference between the PDEs (11.30) and (11.35) is the presence of the reaction term $-ru$ in the latter equation. By comparison, the only difference between the solutions (11.34) and (11.39) is the presence of the $e^{-rt}$ factor in the latter formula. The solution of $u_t = \kappa u_{xx} - ru$ blends the solutions of the heat equation $u_t = \kappa u_{xx}$ with the [exponential] solutions of the ODE $u_t = -ru$. Roughly speaking, the two terms on the right hand side of (11.35) seem to act independently of one another, and the overall solution is a hybrid of the behaviors we would see if either term were absent.

**11.2.3 Wave Equation, Neumann Problem.** The homogeneous Neumann problem for the wave equation is given by

$$u_{tt} = c^2 u_{xx} \qquad (0 < x < L) \qquad (11.40)$$

$$u(x,0) = \phi(x) \qquad (0 < x < L) \qquad (11.41)$$

$$u_t(x,0) = \psi(x) \qquad (0 < x < L) \qquad (11.42)$$

$$u_x(0,t) = 0 \qquad (t \geq 0) \qquad (11.43)$$

$$u_x(L,t) = 0 \qquad (t \geq 0). \qquad (11.44)$$

Until we incorporate the boundary conditions, the procedure used to solve this Neumann problem is identical to how we solved the Dirichlet problem (11.17)–(11.21).

**Step 1: Separate the variables.** Seeking separated solutions $u(x,t) = X(x)T(t)$ of the PDE (11.40) leads to

$$\frac{X''}{X} = \frac{T''}{c^2 T} = -\lambda,$$

where $\lambda$ is a constant. The pair of ODEs

$$X'' + \lambda X = 0 \quad \text{and} \quad T'' + \lambda c^2 T = 0 \tag{11.45}$$

is the same pair that we encountered when solving the Dirichlet problem for the wave equation.

**Step 2: Boundary Conditions.** Although the ODE for $X$ is quite familiar, the boundary conditions will be different this time. Since $u(x,t) = X(x)T(t)$, note that $u_x(x,t) = X'(x)T(t)$. Thus, the Neumann conditions (11.43)–(11.44) give

$$0 = u_x(0,t) = X'(0)T(t) \quad \text{and} \quad 0 = u_x(L,t) = X'(L)T(t).$$

These conditions are certainly satisfied if $T(t) = 0$, but this would result in the trivial solution $u(x,t) = X(x)T(t) = 0$. Instead, it must be the case that $X'(0) = 0 = X'(L)$.

**Step 3: Solve the equation subject to these boundary conditions.** We must seek non-trivial solutions of the eigenvalue problem

$$X'' = -\lambda X \quad \text{and} \quad X'(0) = 0 = X'(L). \tag{11.46}$$

The form of the solutions will depend upon the sign of $\lambda$, and there are three possibilities.

**Case 1:** $\lambda < 0$. We claim that there are no negative eigenvalues—i.e., if $\lambda < 0$ then the boundary value problem (11.46) has no non-trivial solutions. Suppose that $\lambda = -\beta^2$ where $\beta > 0$. The ODE

$$X'' - \beta^2 X = 0.$$

has associated characteristic equation $m^2 - \beta^2 = 0$, which has distinct, real roots $m = \pm\beta$. The general solution of this ODE is therefore

$$X(x) = Ce^{-\beta x} + De^{\beta x},$$

where $C$ and $D$ are arbitrary constants. Differentiating this expression yields

$$X'(x) = -\beta Ce^{-\beta x} + \beta De^{\beta x}.$$

The boundary condition $X'(0) = 0$ tells us that

$$0 = -\beta C + \beta D$$

and, since $\beta > 0$, we conclude that $C = D$. The other boundary condition states that

$$0 = X'(L) = -\beta Ce^{-\beta L} + \beta De^{\beta L}.$$

Since $\beta > 0$ and $C = D$, this equation reduces to $Ce^{-\beta L} = Ce^{\beta L}$. We may assume that $C \neq 0$, because otherwise we would have $D = 0$ as well, implying that $X(x) = 0$ and ultimately leading to the trivial solution $u(x,t) = 0$. Therefore, $e^{-\beta L} = e^{\beta L}$, or equivalently $e^{2\beta L} = 1$. However, since $\beta$ and $L$ are positive, it is impossible for $e^{2\beta L} = 1$. Thus, there are no negative eigenvalues.

**Case 2:** $\lambda = 0$. Unlike the homogeneous Dirichlet problems we solved in previous subsections, $\lambda = 0$ actually *is* an eigenvalue—i.e., the boundary value problem (11.46) *does* have non-trivial solutions if $\lambda = 0$. Integrating the equation $X'' = 0$ twice, the general solution is $X(x) = C + Dx$ and its derivative is $X'(x) = D$. The boundary conditions $X'(0) = 0$ and $X'(L) = 0$ imply that $D = 0$, but there are no restrictions on $C$. *Any* constant function $X(x) =$ constant will satisfy (11.46) if $\lambda = 0$. In particular, $X_0(x) = C_0$ is an eigenfunction corresponding to the eigenvalue $\lambda = 0$. The corresponding solution $T_0(t)$ of the $T$ equation in (11.45) will be obtained later.

**Case 3:** $\lambda > 0$. Now assume that $\lambda = \beta^2$ where $\beta > 0$. (Again, introducing $\beta$ is merely for convenience.) The boundary value problem (11.23) becomes

$$X'' + \beta^2 X = 0, \quad \text{and} \quad X'(0) = 0 = X'(L).$$

The characteristic equation for the ODE is $m^2 + \beta^2 = 0$, which has pure imaginary roots $m = \pm\beta i$. Therefore, the general solution of the ODE is

$$X(x) = C\cos(\beta x) + D\sin(\beta x),$$

where $C$ and $D$ are constants, and its derivative is

$$X'(x) = -\beta C\sin(\beta x) + \beta D\cos(\beta x).$$

The boundary condition $X'(0) = 0$ implies that $D = 0$, from which our expressions for $X$ and $X'$ reduce to

$$X(x) = C\cos(\beta x) \quad \text{and} \quad X'(x) = -\beta C\sin(\beta x).$$

The other boundary condition $X'(L) = 0$ implies that $-\beta C\sin(\beta L) = 0$. As usual, we avoid $C = 0$ and consider the more interesting possibility that $\sin(\beta L) = 0$. Since $\beta L > 0$, the only way to satisfy this equation is if

$$\beta L = n\pi \qquad (n = 1, 2, 3, \ldots).$$

Defining

$$\beta_n = \frac{n\pi}{L} \quad \text{and} \quad \lambda_n = \beta_n^2 \qquad (n = 1, 2, 3, \ldots),$$

we have the same positive eigenvalues $\lambda_n$ that we encountered when solving Dirichlet problems. However, the corresponding eigenfunctions $X_n(x)$ are *cosine* functions instead of sine functions:

$$X_n(x) = C_n\cos(\beta_n x) = C_n\cos\left(\frac{n\pi x}{L}\right) \qquad (n = 1, 2, 3\ldots), \qquad (11.47)$$

where $C_n$ are arbitrary constants. The functions $X_n(x)$ form the set of all possible solutions of the boundary value problem (11.46).

With these positive eigenvalues in mind, consider the $T$ equation in (11.45). The equation

$$T'' + \frac{n^2\pi^2}{L^2}c^2T = 0 \qquad (n = 1, 2, 3, \ldots)$$

has characteristic equation

$$m^2 + \left(\frac{n\pi c}{L}\right)^2 = 0.$$

The roots

$$m = \pm\left(\frac{n\pi c}{L}\right)\mathrm{i}$$

are pure imaginary, and therefore the solutions of the $T$ equation have the form

$$T_n(t) = E_n \cos\left(\frac{n\pi ct}{L}\right) + F_n \sin\left(\frac{n\pi ct}{L}\right) \qquad (n = 1, 2, 3, \ldots), \qquad (11.48)$$

where $E_n$ and $F_n$ are arbitrary constants.

Now, recall that $\lambda = 0$ is also an eigenvalue. In that case, the $T$ equation reduces to $T'' = 0$, and this ODE can be solved by integrating twice with respect to $t$. The solution has the form

$$T_0(t) = E_0 + F_0 t,$$

where $E_0$ and $F_0$ are arbitrary constants.

**Step 4: Building the general solution.** The general solution of the wave equation (11.40) with the two Neumann boundary conditions (11.43)–(11.44) is constructed in the usual way, by taking an infinite linear combination of the separated solutions that we found above. Define

$$u_n(x,t) = X_n(x)T_n(t) \qquad (n = 0, 1, 2, \ldots),$$

noting that $n = 0$ must be included this time because $\lambda = 0$ is an eigenvalue. The general solution is

$$u(x,t) = \sum_{n=0}^{\infty} G_n u_n(x,t),$$

where $G_n$ are constants. Because the form of $u_0(x,t)$ is different than the other terms in this summation, it is useful to present the general solution in the following way:

$$\begin{aligned} u(x,t) &= G_0 u_0(x,t) + \sum_{n=1}^{\infty} G_n u_n(x,t) = G_0 C_0 (E_0 + F_0 t) \\ &+ \sum_{n=1}^{\infty} G_n \left[ E_n \cos\left(\frac{n\pi ct}{L}\right) + F_n \sin\left(\frac{n\pi ct}{L}\right)\right] C_n \cos\left(\frac{n\pi x}{L}\right). \end{aligned}$$

This expression simplifies if we abbreviate various combinations of constants. For $n \geq 1$, let $A_n = G_n E_n C_n$ and $B_n = G_n F_n C_n$. For $n = 0$, we do something slightly different: let $G_0 C_0 E_0 = \frac{1}{2} A_0$ and $G_0 C_0 F_0 = \frac{1}{2} B_0$. The reason for including the factors of $\frac{1}{2}$ will be explained below. Overall, we have shown that the general solution of the PDE and the two Neumann conditions is

$$\begin{aligned} u(x,t) &= \frac{A_0}{2} + \frac{B_0}{2} t \\ &+ \sum_{n=1}^{\infty} \left[ A_n \cos\left(\frac{n\pi ct}{L}\right) + B_n \sin\left(\frac{n\pi ct}{L}\right)\right] \cos\left(\frac{n\pi x}{L}\right). \end{aligned} \tag{11.49}$$

**Step 5: Use the initial conditions.** Since $u(x,0) = \phi(x)$, setting $t = 0$ in (11.49) implies that

$$\phi(x) = \frac{A_0}{2} + \sum_{n=1}^{\infty} A_n \cos\left(\frac{n\pi x}{L}\right). \tag{11.50}$$

Equation (11.50) is in the standard form of a *Fourier cosine series*. The factor of $\frac{1}{2}$ in front of the leading term will be explained when we study Fourier series in the next chapter. As we shall see, insisting that $\phi(x)$ have a Fourier cosine series representation is not a severe restriction at all. Moreover, we will soon develop a systematic procedure for determining the Fourier coefficients $A_n$ for a function $\phi(x)$ defined on an interval $0 \leq x \leq L$. To use the other initial condition $u_t(x,0) = \psi(x)$, we must formally[2] differentiate (11.49) with respect to $t$:

$$u_t(x,t) = \frac{B_0}{2} + \sum_{n=1}^{\infty} \frac{n\pi c}{L} \left[ -A_n \sin\left(\frac{n\pi ct}{L}\right) + B_n \cos\left(\frac{n\pi ct}{L}\right)\right] \cos\left(\frac{n\pi x}{L}\right).$$

Setting $t = 0$, we have

$$\psi(x) = u_t(x,0) = \frac{B_0}{2} + \sum_{n=1}^{\infty} \frac{n\pi c}{L} B_n \cos\left(\frac{n\pi x}{L}\right). \tag{11.51}$$

[2]Strictly speaking, term-by-term differentiation of the series (11.49) is justified only under certain minor assumptions regarding the convergence of the series.

Thus, we must require that the initial velocity $\psi(x)$ also have a Fourier cosine series representation. In summary, the solution of the homogeneous Neumann problem (11.40)–(11.44) is given by formula (11.49), provided that the initial conditions $\phi(x)$ and $\psi(x)$ have Fourier cosine series representations (11.50) and (11.51).

**11.2.4 Heat Equation, Neumann Problem.** The homogeneous Neumann problem for the heat equation is given by

$$u_t = \kappa u_{xx} \qquad (0 < x < L) \tag{11.52}$$

$$u(x,0) = \phi(x) \qquad (0 < x < L) \tag{11.53}$$

$$u_x(0,t) = 0 \qquad (t \geq 0) \tag{11.54}$$

$$u_x(L,t) = 0 \qquad (t \geq 0). \tag{11.55}$$

Physically, this problem models the diffusion of heat within a one-dimensional wire whose ends are insulated to prevent heat flux across the boundaries. The general solution of the PDE with its two Neumann conditions can be presented as

$$u(x,t) = \frac{A_0}{2} + \sum_{n=1}^{\infty} A_n e^{-(n\pi/L)^2 \kappa t} \cos\left(\frac{n\pi x}{L}\right). \tag{11.56}$$

The reader is encouraged to use the separation of variables technique to derive formula (11.56). As in the Neumann problem for the wave equation, note that the initial condition $\phi(x)$ is required to have a Fourier cosine series representation of the form (11.50). For any reasonably well-behaved choice of initial condition $\phi(x)$ (e.g., if $\phi$ is continuous on $0 \leq x \leq L$), then there is a straightforward procedure for calculating the Fourier coefficients $A_n$ (see next chapter).

**11.2.5 Mixed Boundary Conditions: An Example.** As a final illustration of the separation of variables technique, we will solve the heat equation with mixed boundary conditions:

$$u_t = \kappa u_{xx} \qquad (0 < x < L) \tag{11.57}$$

$$u(x,0) = \phi(x) \qquad (0 < x < L) \tag{11.58}$$

$$u(0,t) = 0 \qquad (t \geq 0) \tag{11.59}$$

$$u_x(L,t) = 0 \qquad (t \geq 0). \tag{11.60}$$

Physically, the Dirichlet condition at $x = 0$ simulates placing the $x = 0$ end of the wire in contact with a block of ice, holding the temperature constant at zero degrees Celsius. The Neumann condition at $x = L$ simulates insulating that end of the wire, thereby preventing heat from entering/exitting the wire. If we seek separated solutions $u(x,t) = X(x)T(t)$, we find that $X$ and $T$ must satisfy ODEs of the form

$$X'' + \lambda X = 0 \quad \text{and} \quad T' + \kappa\lambda T = 0,$$

where $\lambda$ is a constant. The boundary conditions (11.59)–(11.60) imply that $X(0) = 0$ and $X'(L) = 0$. The reader should show that there are no negative eigenvalues—that is, if $\lambda < 0$, then it is impossible to simultaneously satisfy the ODE for $X$ as well as both of these boundary conditions. We also claim that $\lambda = 0$ is not an eigenvalue. If $\lambda = 0$, the ODE for $X$ reduces to $X'' = 0$ which has general solution $X(x) = Cx + D$, where $C$ and $D$ are constants. The condition $X(0) = 0$ implies that $D = 0$, and the condition $X'(L) = 0$ implies that $C = 0$ as well. Consequently, $X(x) = 0$ which leads to $u(x,t) = 0$, the trivial solution. Finally, let us seek positive eigenvalues by setting $\lambda = \beta^2$ where $\beta > 0$. The general solution of the $X$ equation is

$$X(x) = C\cos(\beta x) + D\sin(\beta x).$$

From the boundary condition $X(0) = 0$, we conclude that $C = 0$ and the expression for $X(x)$ reduces to $X(x) = D\sin(\beta x)$. To use the other boundary condition, we first compute the derivative $X'(x) = \beta D\cos(\beta x)$. Since $X'(L) = 0$, we obtain $0 = \beta D\cos(\beta L)$. We know that $\beta > 0$ by assumption and, as usual, we can rule out $D = 0$. However, if $\beta$ is chosen such that $\cos(\beta L) = 0$, then we will have produced non-trivial solutions of the boundary value problem for $X$. The only possible choices are

$$\beta L = -\frac{\pi}{2} + n\pi \qquad (n = 1, 2, 3, \ldots).$$

Defining

$$\beta_n = \frac{\left(n - \frac{1}{2}\right)\pi}{L} \qquad (n = 1, 2, 3, \ldots),$$

the eigenvalues are given by $\lambda_n = \beta_n^2$. The corresponding solutions of the $X$ equation are

$$X_n(x) = D_n \sin\left[\frac{\left(n-\frac{1}{2}\right)\pi x}{L}\right] \qquad (n = 1, 2, 3, \ldots),$$

where $D_n$ are constants.

We now turn our attention to the $T$ equation. The general solution of the first-order ordinary differential equation $T' + \kappa\lambda T = 0$ is given by

$$T(t) = Be^{-\kappa\lambda t},$$

where $B$ is a constant. The only $\lambda$ values of interest are the eigenvalues $\lambda_n$, which motivates us to define

$$T_n(t) = B_n e^{-\kappa\lambda_n t} = B_n e^{-\kappa\left[\left(n-\frac{1}{2}\right)\pi/L\right]^2 t}.$$

Letting $u_n(x,t) = X_n(x)T_n(t)$ for $n \geq 1$, the general solution of heat equation with our mixed boundary conditions is

$$u(x,t) = \sum_{n=1}^{\infty} F_n u_n(x,t) = \sum_{n=1}^{\infty} F_n B_n e^{-\kappa\left[\left(n-\frac{1}{2}\right)\pi/L\right]^2 t} D_n \sin\left[\frac{\left(n-\frac{1}{2}\right)\pi x}{L}\right].$$

Combining the various constants by introducing $A_n = F_n B_n D_n$, we have

$$u(x,t) = \sum_{n=1}^{\infty} A_n e^{-\kappa\left[\left(n-\frac{1}{2}\right)\pi/L\right]^2 t} \sin\left[\frac{\left(n-\frac{1}{2}\right)\pi x}{L}\right].$$

Finally, the initial condition (11.58) implies that

$$\phi(x) = \sum_{n=1}^{\infty} A_n \sin\left[\frac{\left(n-\frac{1}{2}\right)\pi x}{L}\right].$$

This expression for $\phi(x)$ is not quite in the form of a Fourier sine series. As we shall see, Fourier sine series and Fourier cosine series are special cases of a more general class of series: *(full) Fourier series*. For some functions $\phi(x)$, we will want to use (full) Fourier series representations instead of Fourier sine or cosine series.

---

## Exercises

1. Solve the homogeneous Neumann problem for the heat equation on a half-line:

$$\begin{aligned} u_t &= \kappa u_{xx} && (0 < x < \infty) \\ u(x,0) &= \phi(x) && (0 < x < \infty) \\ u_x(0,t) &= 0. \end{aligned}$$

2. Solve the homogeneous Neumann problem for the wave equation on a half-line:

$$\begin{aligned} u_{tt} - c^2 u_{xx} &= 0 && (0 < x < \infty) \\ u(x,0) &= \phi(x) && (0 < x < \infty) \\ u_t(x,0) &= \psi(x) && (0 < x < \infty) \\ u_x(0,t) &= 0. \end{aligned}$$

In case it helps reduce the number of cases you must consider, just give the solution for $t \geq 0$.

3. Solve the heat equation on a half-line with an inhomogeneous Dirichlet condition:

$$\begin{aligned} u_t &= \kappa u_{xx} && (0 < x < \infty) \\ u(x,0) &= \phi(x) && (0 < x < \infty) \\ u(0,t) &= g(t). \end{aligned}$$

To do so, first let $w(x,t) = u(x,t) - g(t)$ and show that $w$ satisfies an inhomogeneous PDE with a homogeneous Dirichlet boundary condition:

$$\begin{aligned} w_t &= \kappa w_{xx} + f(x,t) && (0 < x < \infty) \\ w(x,0) &= \tilde{\phi}(x) && (0 < x < \infty) \\ w(0,t) &= 0, \end{aligned}$$

where $\tilde{\phi}(x) = \phi(x) - g(0)$ and $f(x,t) = -g'(t)$. Then, solve for $w$ by combining the odd-reflection method with Duhamel's Principle (see (10.31)). Finally, obtain the overall solution by recalling that $u(x,t) = w(x,t) + g(t)$.

4. Solve the heat equation on a half-line with an inhomogeneous Neumann condition:

$$\begin{aligned} u_t &= \kappa u_{xx} \qquad & (0 < x < \infty) \\ u(x,0) &= \phi(x) \qquad & (0 < x < \infty) \\ u_x(0,t) &= g(t). \end{aligned}$$

To do so, first let $w(x,t) = u(x,t) - xg(t)$ and mimic the procedure outlined in the previous exercise.

5. Unfortunately, the separation of variables method does not work for all linear, constant-coefficient PDEs. For example, consider the homogeneous Dirichlet problem for the transport equation:

$$\begin{aligned} u_t - cu_x &= 0 \qquad & (0 < x < L), \\ u(0,t) &= 0 \\ u(L,t) &= 0, \end{aligned}$$

where $c$ is a positive constant. Show that there are no (non-zero) separated solutions of this problem. That is, there are no eigenvalues.

6. Use separation of variables to solve the homogeneous Neumann problem for the heat equation:

$$\begin{aligned} u_t &= \kappa u_{xx} \qquad & (0 < x < L) \\ u(x,0) &= \phi(x) \qquad & (0 < x < L) \\ u_x(0,t) &= 0 \\ u_x(L,t) &= 0. \end{aligned}$$

Your answer should be expressed in the form of a cosine series.

7. The wave equation can be used to model a vibrating string in a "vacuum" (i.e., we neglect friction/air resistance). The following Dirichlet problem models a

vibrating string in a *resistant* medium:

$$\begin{aligned} u_{tt} &= c^2 u_{xx} - r u_t && (0 < x < L) \\ u(x,0) &= \phi(x) && (0 < x < L) \\ u_t(x,0) &= \psi(x) && (0 < x < L) \\ u(0,t) &= 0 \\ u(L,t) &= 0. \end{aligned}$$

Assuming that $r$ is a constant and $0 < r < 2\pi c/L$, use separation of variables to find a series solution to this Dirichlet problem.

8. The purpose of this problem is to solve the heat equation on a one-dimensional ring-shaped domain of total length $2L$. The idea is to solve the heat equation on the domain $-L \leq x \leq L$, where $x = -L$ and $x = L$ correspond to the *same* physical location. To simulate this, we set up *periodic boundary conditions* as follows:

$$\begin{aligned} u_t &= \kappa u_{xx} && (-L \leq x \leq L) \\ u(-L,t) &= u(L,t) \\ u_x(-L,t) &= u_x(L,t). \end{aligned}$$

First, show that the eigenvalues are $\lambda_n = (n\pi/L)^2$ for $n \geq 0$. (Note that 0 is an eigenvalue in this case.) Then, show that the solution of this problem can be written in the form

$$u(x,t) = \frac{A_0}{2} + \sum_{n=1}^{\infty} \left[ A_n \cos\left(\frac{n\pi x}{L}\right) + B_n \sin\left(\frac{n\pi x}{L}\right) \right] e^{-(n\pi/L)^2 \kappa t}.$$

9. Solve the wave equation with *periodic* boundary conditions:

$$\begin{aligned} u_{tt} &= c^2 u_{xx} && (-L < x < L) \\ u(-L,t) &= u(L,t) \\ u_x(-L,t) &= u_x(L,t), \end{aligned}$$

where $c \neq 0$ is a constant.

# CHAPTER 12

## Introduction to Fourier Series

In the previous chapter, we developed the separation of variables technique to construct the solutions of homogeneous Dirichlet and Neumann problems. In each example, we were able to construct series representations of the solutions *provided* that the initial conditions themselves had special series representations (i.e., Fourier sine and cosine series). In this chapter, we will study Fourier series in greater depth, addressing three principal questions:

☞ Which functions $\phi(x)$ have Fourier series representations?

☞ Given a function $\phi(x)$ that does have such a representation, how can we calculate the coefficients that appear within the series?

☞ Can we *really* be sure that these series converge to $\phi(x)$?

Before beginning our study of Fourier series, let us briefly recall another type of series representation that you likely studied in calculus: Taylor series. Suppose that a function $f(x)$ has infinitely many derivatives in some open interval containing the point $x = a$. In calculus, you learned that the *Taylor series for $f(x)$ centered at $a$* is given by

$$\sum_{n=0}^{\infty} \frac{f^{(n)}(a)}{n!}(x-a)^n.$$

Computing the constants $f^{(n)}(a)$ requires that we first calculate the $n$th derivative of $f$ and then evaluate at $x = a$. For example, suppose we wish to compute the Taylor series for $f(x) = \mathrm{e}^x$ centered at 0. Since $f^{(n)}(x) = \mathrm{e}^x$ for all $n$, it follows that $f^{(n)}(0) = 1$ for all $n$. Thus, the Taylor series for $\mathrm{e}^x$ centered at 0 is given by

$$\sum_{n=0}^{\infty} \frac{1}{n!}x^n.$$

The ratio test shows that this series converges for all real $x$. Moreover, the series really *does* converge to $e^x$ for all real $x$.

Recall that an infinite series is defined as a limit of partial sums; e.g.,

$$e^x = \lim_{N\to\infty} \sum_{n=1}^{N} \frac{1}{n!} x^n.$$

Notice that partial sums of a Taylor series are nothing more than *polynomials*, as are the factors $(x-a)^n$ that appear in each individual term. In this sense, a Taylor series essentially represents a function $f(x)$ as a sum of polynomials.

Fourier series offer another way of representing functions as infinite series. Unlike Taylor series, which use polynomials as "building blocks", Fourier series are sums of sine and cosine functions. More specifically, a Fourier series effectively decomposes a function $\phi(x)$ into a sum of sine and cosine functions all of which have frequencies that are integer multiples of some "fundamental frequency".

**Definition 12.0.2.** Let $L > 0$. A *Fourier sine series* is a series of the form

$$\sum_{n=1}^{\infty} A_n \sin\left(\frac{n\pi x}{L}\right) \qquad (0 < x < L). \tag{12.1}$$

A *Fourier cosine series* is a series of the form

$$\frac{A_0}{2} + \sum_{n=1}^{\infty} A_n \cos\left(\frac{n\pi x}{L}\right) \qquad (0 < x < L). \tag{12.2}$$

A *(full) Fourier series* is a series of the form

$$\frac{A_0}{2} + \sum_{n=1}^{\infty} \left[A_n \cos\left(\frac{n\pi x}{L}\right) + B_n \sin\left(\frac{n\pi x}{L}\right)\right] \qquad (-L < x < L). \tag{12.3}$$

Notice that the interval over which the full Fourier series is defined is symmetric about $x = 0$, whereas the sine and cosine series are defined for $(0 < x < L)$.

Before we tackle the theoretical questions regarding which functions $\phi(x)$ have convergent Fourier series representations, we will explain how to calculate the coefficients assuming we are given an appropriate $\phi(x)$.

## 12.1. Fourier series

Computing the coefficients in a Taylor series for $\phi(x)$ requires that we calculate all of the *derivatives* of $\phi(x)$. By contrast, computing Fourier coefficients will require that we calculate *integrals* instead.

**12.1.1 Fourier sine series.** Suppose that $\phi(x)$ is defined and integrable on the interval $0 \leq x \leq L$, and assume that $\phi(x)$ has a convergent Fourier sine series representation

$$\phi(x) = \sum_{n=1}^{\infty} A_n \sin\left(\frac{n\pi x}{L}\right) \qquad (0 < x < L). \tag{12.4}$$

Our goal is to find the values of the Fourier coefficients $A_n$. First, we need a Lemma:

**Lemma 12.1.1.** Suppose $m$ and $n$ are positive integers. Then

$$\int_0^L \sin\left(\frac{m\pi x}{L}\right) \sin\left(\frac{n\pi x}{L}\right) dx = \begin{cases} 0 & \text{if } m \neq n \\ L/2 & \text{if } m = n. \end{cases}$$

*Proof.* The idea is to use some trigonometric identities to simplify the integrand. First, suppose $m = n$. The integral becomes

$$\int_0^L \sin^2\left(\frac{n\pi x}{L}\right) dx,$$

and the identity $\sin^2\theta = \frac{1}{2}[1 - \cos(2\theta)]$ leads us to

$$\frac{1}{2}\int_0^L 1 - \cos\left(\frac{2n\pi x}{L}\right) dx = \frac{1}{2}\left[x - \frac{L}{2n\pi}\sin\left(\frac{2n\pi x}{L}\right)\right]\Bigg|_0^L = \frac{L}{2}.$$

The case $m \neq n$ requires a more creative use of trigonometric identities. Specifically, we will use

$$\sin\alpha\sin\beta = \frac{1}{2}\cos(\alpha-\beta) - \frac{1}{2}\cos(\alpha+\beta). \tag{12.5}$$

As an exercise, you may wish to verify (12.5) using the more familiar "double-angle identity"

$$\cos(\alpha + \beta) = \cos\alpha\cos\beta - \sin\alpha\sin\beta.$$

Using (12.5) with $\alpha = m\pi x/L$ and $\beta = n\pi x/L$, we can re-write our integral as

$$\int_0^L \sin\left(\frac{m\pi x}{L}\right)\sin\left(\frac{n\pi x}{L}\right)\,dx = \frac{1}{2}\int_0^L \cos\left[\frac{(m-n)\pi x}{L}\right] - \cos\left[\frac{(m+n)\pi x}{L}\right]\,dx.$$

With the integrand expressed in this more convenient form, we evaluate

$$\frac{1}{2}\frac{L}{(m-n)\pi}\sin\left[\frac{(m-n)\pi x}{L}\right] - \frac{1}{2}\frac{L}{(m+n)\pi}\sin\left[\frac{(m+n)\pi x}{L}\right]\Bigg|_0^L$$
$$= \frac{1}{2}\frac{L}{(m-n)\pi}\sin\left[(m-n)\pi\right] - \frac{1}{2}\frac{L}{(m+n)\pi}\sin\left[(m+n)\pi\right] = 0,$$

where we have used the fact that the sine of an integer multiple of $\pi$ is always zero. The assumptions that $m$ and $n$ are positive integers and $m \neq n$ were important when writing down this antiderivative. Otherwise, we could not guarantee that both $m - n \neq 0$ and $m + n \neq 0$. □

Armed with Lemma 12.1.1, we can calculate the Fourier coefficients in the sine series representation of $\phi(x)$. Choose any positive integer $m \geq 1$ and multiply both sides (12.4) by $\sin(m\pi x/L)$:

$$\phi(x)\sin\left(\frac{m\pi x}{L}\right) = \sum_{n=1}^{\infty} A_n \sin\left(\frac{n\pi x}{L}\right)\sin\left(\frac{m\pi x}{L}\right).$$

Now integrate both sides over the interval $0 \leq x \leq L$:

$$\int_0^L \phi(x)\sin\left(\frac{m\pi x}{L}\right)\,dx = \int_0^L \sum_{n=1}^{\infty} A_n \sin\left(\frac{n\pi x}{L}\right)\sin\left(\frac{m\pi x}{L}\right)\,dx$$
$$= \sum_{n=1}^{\infty} A_n \int_0^L \sin\left(\frac{n\pi x}{L}\right)\sin\left(\frac{m\pi x}{L}\right)\,dx.$$

(When interchanging the sum and the integral, we have tacitly assumed that the Fourier sine series for $\phi(x)$ converges in an appropriate manner.) By Lemma 12.1.1, all of the integrals in the summand are zero *except* in the case the

$n = m$. Only one term in the summation survives:

$$\int_0^L \phi(x) \sin\left(\frac{m\pi x}{L}\right) \,\mathrm{d}x \;=\; \frac{L}{2} A_m.$$

Therefore, the Fourier sine series coefficients are given by

$$A_m \;=\; \frac{2}{L} \int_0^L \phi(x) \, \sin\left(\frac{m\pi x}{L}\right) \,\mathrm{d}x \qquad (m \geq 1). \tag{12.6}$$

**Example 12.1.2.** Calculate the Fourier sine series for $\phi(x) = \sin(2\pi x)$ on the interval $0 < x < 1$. *Solution:* In this case, the length of the interval is $L = 1$, so the Fourier sine series should have the form

$$\phi(x) \;=\; \sum_{n=1}^{\infty} A_n \sin(n\pi x).$$

The coefficients $A_n$ are given by (12.6):

$$A_n \;=\; 2\int_0^1 \phi(x) \, \sin(n\pi x) \,\mathrm{d}x \;=\; 2\int_0^1 \sin(n\pi x) \, \sin(2\pi x) \,\mathrm{d}x.$$

Applying Lemma 12.1.1 with $L = 1$, we have

$$2\int_0^1 \sin(n\pi x) \, \sin(2\pi x) \,\mathrm{d}x \;=\; \begin{cases} 1 & \text{if } n = 2 \\ 0 & \text{if } n \neq 2. \end{cases}$$

It follows that $A_2 = 1$ and $A_n = 0$ if $n \neq 2$. This is not at all surprising, as the function $\phi(x) = \sin(2\pi x)$ is *already* in the form of a Fourier sine series.

**Example 12.1.3.** Find the Fourier sine series for the constant function $\phi(x) = 1$ on the interval $0 < x < 1$. *Solution:* Again, $L = 1$ and the Fourier sine series has the form

$$\phi(x) \;=\; \sum_{n=1}^{\infty} A_n \sin(n\pi x).$$

The coefficients $A_n$ are given by (12.6):

$$\begin{aligned} A_n \;=\; 2\int_0^1 \phi(x) \, \sin(n\pi x) \,\mathrm{d}x \;=\; 2\int_0^1 \sin(n\pi x) \,\mathrm{d}x \;&=\; -\frac{2}{n\pi} \cos(n\pi x)\Big|_0^1 \\ &=\; -\frac{2}{n\pi}\left[\cos(n\pi) - 1\right]. \end{aligned}$$

Notice that $\cos(n\pi)$ alternates:

$$\cos(n\pi) = (-1)^n = \begin{cases} 1 & \text{if } n \text{ is even} \\ -1 & \text{if } n \text{ is odd.} \end{cases}$$

Therefore, the Fourier sine series coefficients are

$$A_n = -\frac{2}{n\pi}[(-1)^n - 1] = \frac{2}{n\pi}[1 - (-1)^n] = \begin{cases} 4/n\pi & \text{if } n \text{ is odd} \\ 0 & \text{if } n \text{ is even,} \end{cases}$$

and the Fourier series for $\phi(x)$ is

$$1 = \phi(x) = \frac{4}{\pi}\left[\sin(\pi x) + \frac{1}{3}\sin(3\pi x) + \frac{1}{5}\sin(5\pi x) + \cdots\right]$$

$$= \frac{4}{\pi}\sum_{n=1}^{\infty}\frac{1}{2n-1}\sin[(2n-1)\pi x]. \tag{12.7}$$

To visualize the convergence of the series (12.7), it is useful to plot the first few partial sums. The left panel of Figure 12.1 shows the first three partial sums, and the right panel shows the 20th partial sum. Notice that the sequence of partial sums does appear to converge to the function $\phi(x) = 1$ everywhere except at the endpoints. In fact, when $x = 0$ or $x = 1$, all of the terms in the sine series are equal to zero. Although the Fourier sine series cannot converge to $\phi(x)$ at the endpoints of the interval, we will soon see that this is not an issue for the Fourier cosine series of $\phi(x) = 1$.

As a side note, formula (12.7) actually provides some rather curious identities. For example, if we set $x = 1/2$, then the series becomes

$$1 = \frac{4}{\pi}\left(1 - \frac{1}{3} + \frac{1}{5} - \frac{1}{7} + \frac{1}{9} + \ldots\right).$$

Equivalently,

$$\pi = 4\sum_{n=0}^{\infty}(-1)^n\frac{1}{2n+1},$$

which is a series representation of an important mathematical constant.

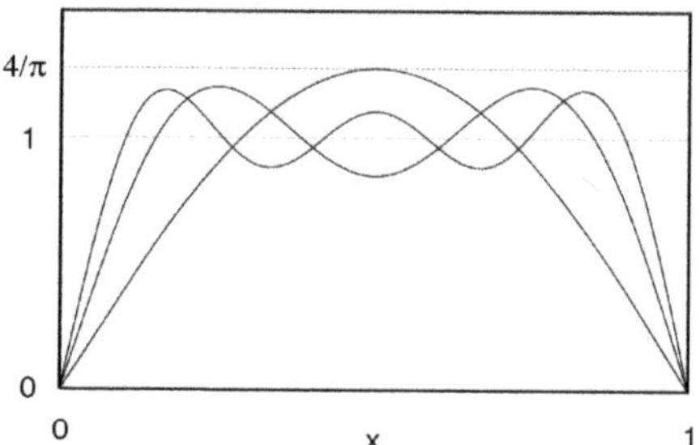


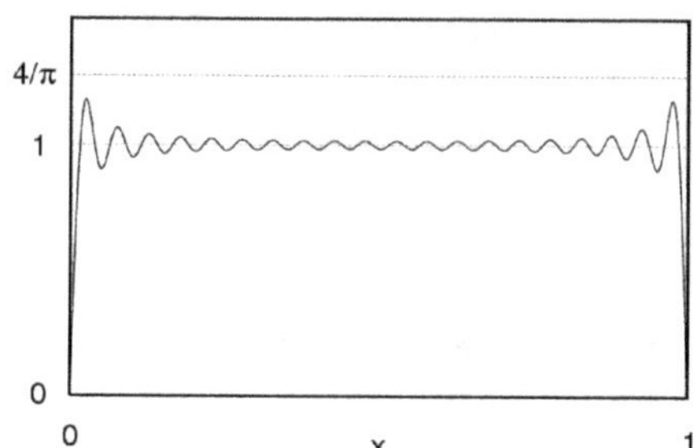


**Figure 12.1.** Left panel: The first, second, and third partial sums of the series (12.7). Right panel: The sum of the first 20 terms of the series (12.7). Horizontal lines corresponding to values of 1 and $4/\pi$ are included for reference.

**Example 12.1.4.** In the previous chapter, we showed that the solution of the homogeneous Dirichlet problem for the heat equation (11.30)–(11.33) is given by

$$u(x,t) = \sum_{n=1}^{\infty} A_n e^{-\kappa(n\pi/L)^2 t} \sin\left(\frac{n\pi x}{L}\right).$$

Setting $t = 0$, the initial condition (11.31) tells us that

$$\phi(x) = \sum_{n=1}^{\infty} A_n \sin\left(\frac{n\pi x}{L}\right).$$

As this is a Fourier sine series, the coefficients are given by formula (12.6). Assuming that $\phi(x)$ has such a representation, we are able to express the solution of this Dirichlet problem in terms of $\phi$ itself:

$$u(x,t) = \frac{2}{L}\sum_{n=1}^{\infty} e^{-\kappa(n\pi/L)^2 t} \sin\left(\frac{n\pi x}{L}\right) \int_0^L \phi(x) \sin\left(\frac{n\pi x}{L}\right)\, dx.$$

Let us consider a specific initial condition $\phi(x) = \sin(2\pi x)$ with $L = 1$ (that is, the spatial domain is $0 \le x \le 1$). In an example above, we showed that the Fourier coefficients for this function are $A_2 = 1$ and $A_n = 0$ for all $n \neq 2$. Only one of the terms in the formula for $u(x,t)$ is non-zero and, for this specific choice of initial condition, we have

$$u(x,t) = e^{-\kappa(2\pi)^2 t} \sin(2\pi x).$$

Visualizing the behavior of this solution is straightforward since the time-dependent and space-dependent parts are separated. As $t$ increases, the exponential factor decays to 0, effectively reducing the amplitude of the sinusoidal heat distribution. More specifically,

$$\lim_{t\to\infty} u(x,t) = 0$$

for each $x$ in the domain $0 \le x \le 1$. As an exercise, you may wish to choose some [positive] value for the diffusion constant $\kappa$ and use a calculator or computer to plot $u(x,t)$.

**Example 12.1.5.** In the previous chapter, we showed that the general solution of the wave equation (11.17) with homogeneous Dirichlet conditions (11.20)–(11.21) is given by

$$u(x,t) = \sum_{n=1}^{\infty} \left[ A_n \cos\left(\frac{n\pi ct}{L}\right) + B_n \sin\left(\frac{n\pi ct}{L}\right) \right] \sin\left(\frac{n\pi x}{L}\right).$$

Using the initial conditions (11.18)–(11.19), we found that

$$\phi(x) = \sum_{n=1}^{\infty} A_n \sin\left(\frac{n\pi x}{L}\right) \quad \text{and} \quad \psi(x) = \sum_{n=1}^{\infty} \frac{n\pi c}{L} B_n \sin\left(\frac{n\pi x}{L}\right).$$

Since $\phi(x)$ is in the form of a Fourier sine series, we know that the coefficients $A_n$ are

$$A_n = \frac{2}{L}\int_0^L \phi(x) \sin\left(\frac{n\pi x}{L}\right) dx \qquad (n = 1,2,3\ldots).$$

If we introduce $\tilde{B}_n = (n\pi c/L)B_n$, then the series for $\psi(x)$ also takes the standard form of a Fourier sine series. We conclude that

$$\tilde{B}_n = \frac{2}{L}\int_0^L \psi(x) \sin\left(\frac{n\pi x}{L}\right) dx \qquad (n = 1,2,3\ldots),$$

from which it follows that

$$B_n = \frac{2}{n\pi c}\int_0^L \psi(x) \sin\left(\frac{n\pi x}{L}\right) dx \qquad (n = 1,2,3\ldots).$$

Now that we have expressed the coefficients $A_n$ and $B_n$ in terms of the given initial data, we have completely solved the Dirichlet problem (11.17)–(11.21).

**12.1.2 Fourier cosine series.** When we solved the homogeneous Neumann problems for the heat and wave equations, we found that the initial conditions must have convergent Fourier cosine series representations. The process of determining the Fourier coefficients in the expansion

$$\phi(x) = \frac{A_0}{2} + \sum_{n=1}^{\infty} A_n \cos\left(\frac{n\pi x}{L}\right) \qquad (12.8)$$

is very similar to what we did for sine series expansions. First, a technical Lemma:

**Lemma 12.1.6.** Suppose $m$ and $n$ are non-negative integers and $L > 0$. Then

$$\int_0^L \cos\left(\frac{m\pi x}{L}\right) \cos\left(\frac{n\pi x}{L}\right) dx = \begin{cases} 0 & \text{if } m \neq n \\ L/2 & \text{if } m = n \neq 0 \\ L & \text{if } m = n = 0. \end{cases}$$

*Proof.* The proof is essentially the same as that of Lemma 12.1.1. To handle the first case, use the trigonometric identity

$$\cos\alpha \cos\beta = \frac{1}{2}\cos(\alpha - \beta) + \frac{1}{2}\cos(\alpha + \beta).$$

For the second case, use the identity

$$\cos^2\alpha = \frac{1}{2}\left[1 + \cos(2\alpha)\right],$$

which is actually a special case of the other identity in which $\alpha = \beta$. For the final case, notice that the integrand reduces to 1 if $m = n = 0$. □

To determine the Fourier cosine series coefficients in (12.8), we use a similar trick as with Fourier sine series. Choose any non-negative integer $m$ and multiply both sides of (12.8) by $\cos(m\pi x/L)$ to obtain

$$\phi(x)\cos\left(\frac{m\pi x}{L}\right) = \frac{A_0}{2}\cos\left(\frac{m\pi x}{L}\right) + \sum_{n=1}^{\infty} A_n \cos\left(\frac{n\pi x}{L}\right)\cos\left(\frac{m\pi x}{L}\right).$$

Now integrate both sides over the interval $0 \le x \le L$:

$$\int_0^L \phi(x) \cos\left(\frac{m\pi x}{L}\right) \, dx = \int_0^L \frac{A_0}{2} \cos\left(\frac{m\pi x}{L}\right) \, dx + \sum_{n=1}^{\infty} A_n \int_0^L \cos\left(\frac{n\pi x}{L}\right) \cos\left(\frac{m\pi x}{L}\right) \, dx.$$

First, suppose $m > 0$. Then according to Lemma (12.1.6), the only non-zero term in this series occurs when $n = m$. Thus, the equation reduces to

$$\int_0^L \phi(x) \cos\left(\frac{m\pi x}{L}\right) \, dx = \frac{L}{2} A_m$$

if $m > 0$. If $m = 0$, then

$$\int_0^L \phi(x) \cos\left(\frac{m\pi x}{L}\right) \, dx = \int_0^L \frac{A_0}{2} \cos\left(\frac{0\pi x}{L}\right) \, dx = \int_0^L \frac{A_0}{2} \, dx = \frac{L}{2} A_0.$$

Multiplying both sides of these equations by $2/L$, we have shown that the Fourier cosine series coefficients are given by

$$A_n = \frac{2}{L} \int_0^L \phi(x) \cos\left(\frac{n\pi x}{L}\right) \, dx \qquad (n = 0, 1, 2, \ldots). \qquad (12.9)$$

The reason for including the factor of $1/2$ in front of the coefficient $A_0$ in the Fourier cosine series is now evident. Namely, because the integral in Lemma 12.1.6 is twice as large if $m = n = 0$ than if $m = n \neq 0$, including the $1/2$ in front of $A_0$ lets us avoid listing multiple cases in formula (12.9).

**Example 12.1.7.** Find the Fourier cosine series representation for $\phi(x) = 1$ on the interval $0 \leq x \leq 1$. *Solution:* Since $L = 1$, the Fourier cosine coefficients are given by

$$A_n = 2 \int_0^L \cos(n\pi x) \, dx = \begin{cases} 2 & \text{if } n = 0 \\ 0 & \text{if } n \neq 0. \end{cases}$$

In other words, the only non-zero term in the Fourier cosine series for $\phi(x) = 1$ is the leading term: $\phi(x) = A_0/2 = 1$. This makes sense, because $\phi(x)$ is *already* in the form of a cosine series: $\phi(x) = 1 = \cos(0\pi x)$.

Compare this example with the earlier example in which we calculated the Fourier *sine* series representation for $\phi(x) = 1$. For this particular choice of $\phi(x)$, why is the Fourier cosine series so much cleaner than the sine series? We will explore this question in a later section.

**Example 12.1.8.** Find the Fourier cosine series for $\phi(x) = x$ on the interval $0 < x < 2$. *Solution:* Using formula (12.9) with $L = 2$,

$$A_n = \int_0^2 x \cos\left(\frac{n\pi x}{2}\right) \, dx.$$

We need to handle the $n = 0$ case separately, because the cosine function in the integrand reduces to 1 (which affects the form of our antiderivative):

$$A_0 = \int_0^2 x \, dx = \left.\frac{x^2}{2}\right|_0^2 = 2.$$

For $n \geq 1$, we integrate by parts:

$$\begin{aligned} A_n &= \frac{2}{n\pi} x \sin\left(\frac{n\pi x}{2}\right)\Big|_0^2 - \frac{2}{n\pi} \int_0^2 \sin\left(\frac{n\pi x}{2}\right) \, dx \\ &= \frac{2}{n\pi} x \sin\left(\frac{n\pi x}{2}\right)\Big|_0^2 + \left(\frac{2}{n\pi}\right)^2 \cos\left(\frac{n\pi x}{2}\right)\Big|_0^2. \end{aligned}$$

Since $\sin(n\pi) = 0$ and $\cos(n\pi) = (-1)^n$ for all integers $n$, substituting in the limits of integration yields

$$A_n = \begin{cases} 0 & \text{if } n \geq 1 \text{ is even} \\ -8/(n\pi)^2 & \text{if } n \geq 1 \text{ is odd.} \end{cases}$$

The Fourier cosine series for $\phi(x) = x$ on the interval $0 < x < 2$ is

$$\begin{aligned} \phi(x) &= \frac{A_0}{2} + \sum_{n=1}^{\infty} A_n \cos\left(\frac{n\pi x}{2}\right) \\ &= 1 - \frac{8}{\pi^2} \cos\left(\frac{\pi x}{2}\right) - \frac{8}{(3\pi)^2} \cos\left(\frac{3\pi x}{2}\right) - \frac{8}{(5\pi)^2} \cos\left(\frac{5\pi x}{2}\right) - \cdots \\ &= 1 - 8 \sum_{n=1}^{\infty} \frac{1}{[(2n-1)\pi]^2} \cos\left[\frac{(2n-1)\pi x}{2}\right]. \end{aligned}$$

You may wish to use a computer or calculator to plot the first few partial sums of this cosine series in order to visualize the convergence.

**Example 12.1.9.** Solve the Neumann problem

$$\begin{aligned} u_t &= 2\kappa u_{xx} & (0 < x < 7) \\ u(x,0) &= 10 + 3\cos\left(\frac{\pi x}{7}\right) & (0 < x < 7) \\ u_x(0,t) &= 0 & (t \geq 0) \\ u_x(L,t) &= 0 & (t \geq 0). \end{aligned}$$

*Solution:* This Neumann problem models heat transfer in a one-dimensional wire of length $L = 7$ in which the ends are insulated. The diffusion coefficient is $\kappa = 2$. The initial condition is illustrated in Figure 12.2—notice that initially the highest temperature is 13 (when $x = 0$) and the lowest temperature is 7 (when $x = 7$). According to formula (11.56), the general solution of the PDE together with these two Neumann boundary conditions is

$$u(x,t) = \frac{A_0}{2} + \sum_{n=1}^{\infty} A_n e^{-2(n\pi/7)^2 t} \cos\left(\frac{n\pi x}{7}\right).$$

To incorporate the initial condition, we set $t = 0$ to find that

$$\phi(x) = 10 + 3\cos\left(\frac{\pi x}{7}\right) = \frac{A_0}{2} + \sum_{n=1}^{\infty} A_n \cos\left(\frac{n\pi x}{7}\right),$$

a Fourier cosine series. Luckily, our initial condition is *already* in the form of a Fourier cosine series with $A_0 = 20$, $A_1 = 3$, and $A_n = 0$ for $n \geq 2$. Inserting these coefficients into the formula for $u(x,t)$, we find that the solution of our Neumann problem is given by

$$u(x,t) = 10 + 3e^{-2(\pi/7)^2 t} \cos\left(\frac{\pi x}{7}\right).$$

Observe that for all $x$ in the domain $0 \leq x \leq 7$, the temperature approaches 10 as $t$ increases. More precisely,

$$\lim_{t\to\infty} u(x,t) = 10$$

for all $x \in [0,7]$, because the exponential factor in $u(x,t)$ decays to zero while the cosine factor remains bounded between $-1$ and 1. Physically, the heat equilibrates within the wire, with the temperature converging to the average of

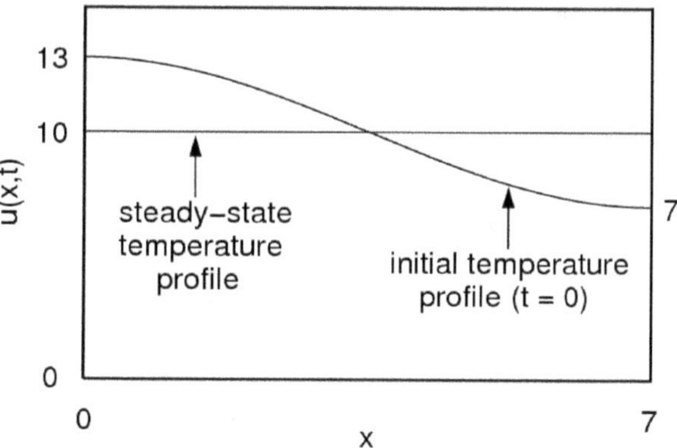


**Figure 12.2.** Initial ($t = 0$) and steady-state ($t = \infty$) heat distributions for the homogeneous Neumann problem example (see text). Because the boundaries are insulated, the temperature distribution eventually equilibrates to the average value of the initial temperature distribution.

the initial temperature profile (see Figure 12.2). This makes sense given that the ends of the wire are insulated to prevent heat from entering or exiting the wire at the boundaries.

Even if we had failed to notice that the initial condition $\phi(x)$ had the form of a Fourier cosine series, we could easily calculate the Fourier cosine coefficients by evaluating the integral (12.9) with the help of Lemma 12.1.6.

**12.1.3 Fourier series.** Recall that the Fourier series expansion for a function $\phi(x)$ on an interval $-L < x < L$ takes the form

$$\phi(x) = \frac{A_0}{2} + \sum_{n=1}^{\infty} \left[ A_n \cos\left(\frac{n\pi x}{L}\right) + B_n \sin\left(\frac{n\pi x}{L}\right) \right].$$

Again, notice that unlike the sine and cosine series we considered above, the Fourier series expansion is defined on an interval that is *symmetric* about $x = 0$.

Calculating the Fourier coefficients $A_n$ and $B_n$ is accomplished via the same sort of procedure that we used to compute the coefficients in sine and cosine series. Suppose that $m$ and $n$ are non-negative integers and that $L > 0$. As an exercise, you can show that, for all $m$ and $n$,

$$\int_{-L}^{L} \sin\left(\frac{m\pi x}{L}\right) \cos\left(\frac{n\pi x}{L}\right) dx = 0,$$

if $m \neq n$, then

$$\int_{-L}^{L} \sin\left(\frac{m\pi x}{L}\right) \sin\left(\frac{n\pi x}{L}\right) dx = \int_{-L}^{L} \cos\left(\frac{m\pi x}{L}\right) \cos\left(\frac{n\pi x}{L}\right) dx = 0,$$

and if $n \geq 1$, then

$$\int_{-L}^{L} \sin^2\left(\frac{n\pi x}{L}\right) dx = \int_{-L}^{L} \cos^2\left(\frac{n\pi x}{L}\right) dx = L.$$

With these integrals in mind, you may mimic the same procedure as before to show that the coefficients in a full Fourier series are given by

$$A_n = \frac{1}{L}\int_{-L}^{L} \phi(x) \cos\left(\frac{n\pi x}{L}\right) dx \qquad (n = 0, 1, 2, \ldots) \tag{12.10}$$

$$B_n = \frac{1}{L}\int_{-L}^{L} \phi(x) \sin\left(\frac{n\pi x}{L}\right) dx \qquad (n = 1, 2, 3, \ldots). \tag{12.11}$$

Observe that the scalar in front of these integrals is $\frac{1}{L}$, not $\frac{2}{L}$ as with Fourier sine and cosine coefficients. Moreover, the interval of integration is $[-L, L]$, not $[0, L]$.

**Example 12.1.10.** Find the Fourier series representation of the discontinuous function

$$\phi(x) = \begin{cases} -2 & \text{if } -3 < x < 0 \\ 6 & \text{if } 0 \leq x < 3. \end{cases}$$

on the interval $-3 < x < 3$. *Solution:* The fact that $\phi(x)$ has a jump discontinuity at $x = 0$ is not a problem—when computing the Fourier coefficients, we will simply split the region of integration at the point where the discontinuity occurs. Using (12.10)–(12.11) with $L = 3$, we compute

$$A_0 = \frac{1}{3}\int_{-3}^{3} \phi(x)\, dx = \frac{1}{3}\left[\int_{-3}^{0} -2\, dx + \int_{0}^{3} 6\, dx\right] = 4.$$

For $n \geq 1$,

$$\begin{aligned} A_n &= \frac{1}{3}\left[\int_{-3}^{0} -2\cos\left(\frac{n\pi x}{3}\right) dx + \int_{0}^{3} 6\cos\left(\frac{n\pi x}{3}\right) dx\right] \\ &= -\frac{2}{3}\cdot\frac{3}{n\pi}\sin\left(\frac{n\pi x}{3}\right)\Big|_{-3}^{0} + 2\cdot\frac{3}{n\pi}\sin\left(\frac{n\pi x}{3}\right)\Big|_{0}^{3}. \end{aligned}$$

Substituting in the limits of integration, we are left only with sines of integer multiples of $\pi$. Therefore, $A_n = 0$ for all $n \geq 1$. The other coefficients are given by

$$\begin{aligned} B_n &= \frac{1}{3}\left[\int_{-3}^{0} -2\sin\left(\frac{n\pi x}{3}\right)\,dx + \int_{0}^{3} 6\sin\left(\frac{n\pi x}{3}\right)\,dx\right] \\ &= \frac{2}{3}\cdot\frac{3}{n\pi}\cos\left(\frac{n\pi x}{3}\right)\Big|_{-3}^{0} - 2\cdot\frac{3}{n\pi}\cos\left(\frac{n\pi x}{3}\right)\Big|_{0}^{3} \\ &= \frac{2}{n\pi}\left[1-\cos(-n\pi)\right] - \frac{6}{n\pi}\left[\cos(n\pi)-1\right] \\ &= \frac{2}{n\pi}\left[1-(-1)^n\right] - \frac{6}{n\pi}\left[(-1)^n-1\right] = \frac{8}{n\pi}\left[1-(-1)^n\right]. \end{aligned}$$

Equivalently,

$$B_n = \begin{cases} 16/n\pi & \text{if } n \text{ is odd} \\ 0 & \text{if } n \text{ is even.} \end{cases}$$

In summary, the only non-zero Fourier coefficients are $A_0 = 4$ and $B_n$ for $n$ odd. The Fourier series for $\phi(x)$ on the interval $-3 < x < 3$ is given by

$$\begin{aligned} \phi(x) &= 2 + \frac{16}{\pi}\sin\left(\frac{\pi x}{3}\right) + \frac{16}{3\pi}\sin\left(\frac{3\pi x}{3}\right) + \frac{16}{5\pi}\sin\left(\frac{5\pi x}{3}\right) + \cdots \\ &= 2 + \frac{16}{\pi}\sum_{n=1}^{\infty}\frac{1}{2n-1}\sin\left[\frac{(2n-1)\pi x}{3}\right]. \end{aligned}$$

Notice that, apart from the leading constant term $A_0/2$, the only terms that are present in this Fourier series are *sine* functions. Why are none of the cosine terms present in this particular series? Why do some functions have series expansions that require only sine terms, others have expansions with only cosine terms, and some require a combination of both? We will address these questions in the next section.

---

## 12.2. Convergence of Fourier Series

When we used the separation of variables technique to solve initial-boundary value problems for linear, homogeneous PDEs, we presented the solutions as infinite series. In particular, this required the initial data to have Fourier series representations. Fortunately, this is not a severe restriction at all—any physically realistic choice of initial condition is guaranteed to have a convergent Fourier series representation. In this section, we will explore classes of functions that have convergent Fourier series representations. Virtually any "reasonable" function $\phi(x)$ can be represented as the sum of its Fourier sine *or* cosine series on the interval $(0, L)$, or as the sum of its Fourier series on the interval $(-L, L)$.

This last remark seems surprising, given that every term in a sine series is an odd, periodic function, and every term in a cosine series is an even, periodic function. Let us take a moment to review some properties of functions with these types of symmetry. Recall that a function $\phi(x)$ is *odd* if $\phi(-x) = -\phi(x)$ for all $x$ and $\phi(x)$ is *even* if $\phi(-x) = \phi(x)$ for all $x$. A non-constant function $\phi(x)$ is called *periodic* if there exists a constant $p > 0$ such that $\phi(x+p) = \phi(x)$ for all $x$. The least positive $p$ for which this statement holds is called the *period* of $\phi(x)$. For example, the period of $\phi(x) = \sin(3x)$ is $p = 2\pi/3$. Here are some properties of odd, even and periodic functions:

☞ If $\phi(x)$ is odd, then $\phi(0) = 0$. To see why, note that $\phi(-0) = -\phi(0)$.

☞ If $\phi(x)$ is even, then $\phi(x)$ is automatically continuous at $x = 0$. If $\phi(x)$ also differentiable at $x = 0$, then $\phi'(0) = 0$. This follows from the limit definition of the derivative, using centered differences:

$$\phi'(0) = \lim_{h \to 0} \frac{\phi(h) - \phi(-h)}{2h} = 0$$

since $\phi(h) = \phi(-h)$.

☞ If $f(x)$ and $g(x)$ are odd, then $f(x) + g(x)$ is odd and $f(x)g(x)$ is even. If $f(x)$ and $g(x)$ are even, then both $f(x) + g(x)$ and $f(x)g(x)$ are even.

☞ Suppose that $f(x)$ is differentiable and integrable. If $f(x)$ is odd then $f'(x)$ and $\int_0^x f(s)\mathrm{d}s$ are even. If $f(x)$ is even, then $f'(x)$ and $\int_0^x f(s)\mathrm{d}s$ are odd.

☞ Let $L > 0$. *Any* function $f(x)$ defined on an interval $[-L, L]$ can be written as a sum of an odd function and an even function. Namely, $f(x) = \mathcal{E}(x) + \mathcal{O}(x)$ where

$$\mathcal{E}(x) = \frac{f(x) + f(-x)}{2} \quad \text{and} \quad \mathcal{O}(x) = \frac{f(x) - f(-x)}{2}.$$

It is easy to check that $\mathcal{E}(x)$ and $\mathcal{O}(x)$ are even and odd, respectively.

☞ If $\phi(x)$ is odd on $[-L, L]$ then $\int_{-L}^{L} \phi(x)\mathrm{d}x = 0$.

The above list should convince you that odd and even *functions* obey considerably different properties than odd and even *numbers*. For example, although the sum of an odd number and an even number is always odd, the sum of an odd function and an even function need not be odd or even.

When we discussed reflection methods for solving PDEs with a single boundary (i.e., on the domain $[0, \infty)$), we introduced the notion of the odd/even extensions of functions. These concepts are defined analogously on finite domains—suppose that $\phi(x)$ is defined on the interval $[0, L]$. The *odd extension* of $\phi(x)$ is defined as

$$\phi_{\text{odd}}(x) = \begin{cases} \phi(x) & \text{if } 0 < x \leq L \\ 0 & \text{if } x = 0 \\ -\phi(-x) & \text{if } -L \leq x < 0. \end{cases}$$

Likewise, the *even extension* of $\phi(x)$ is defined as

$$\phi_{\text{even}}(x) = \begin{cases} \phi(x) & \text{if } 0 \leq x \leq L \\ \phi(-x) & \text{if } -L \leq x < 0. \end{cases}$$

We introduce one more notion that will be useful in our discussion of (full) Fourier series. Suppose $\phi(x)$ is defined on some interval $-L < x < L$ where $L > 0$. Then the *periodic extension* of $\phi(x)$ is defined as

$$\phi_p(x) = \phi(x - 2nL) \quad \text{if } -L + 2nL < x < L + 2nL,$$

for all integers $n$. Notice that when $n = 0$, the definition of $\phi_p$ reduces to that of $\phi(x)$.

What do these remarks have to do with Fourier series? In the Fourier sine series

$$\phi(x) = \sum_{n=1}^{\infty} A_n \sin\left(\frac{n\pi x}{L}\right) \qquad (0 < x < L),$$

all of the terms in the series are odd and periodic. Since the frequencies of the sine functions are positive integer multiples of frequency of the leading term, the period of the sum must equal the period of the lowest-frequency term, which is $2L$. What function does the series represent, assuming it actually converges? If $\phi(x)$ is defined on $0 < x < L$, first form its odd extension $\phi_{\text{odd}}$ to the interval $-L < x < L$, and then form the periodic extension of $\phi_{\text{odd}}$ to the entire real line. The sine series should represent that extended function on the entire real line.

Likewise, Fourier cosine series can be used to represent even, periodic functions with period $2L$.

When solving boundary value problems for PDEs, we rarely concern ourselves with the behavior of Fourier series representations of solutions outside the physical domain, typically $0 \leq x \leq L$. However, it is worth noting that the types of homogeneous boundary conditions that we most commonly encounter are related to the three types of extensions described above. Odd extensions are associated with homogeneous Dirichlet conditions $u(0,t) = 0 = u(L,t)$, whereas even extensions are associated with homogeneous Neumann conditions $u_x(0,t) = 0 = u_x(L,t)$. Finally, if our domain is $-L \leq x < L$, then the periodic boundary conditions $u(-L,t) = u(L,t)$ and $u_x(-L,t) = u_x(L,t)$ are associated with periodic extensions. Observe that the periodic extension of a function satisfying these periodic boundary conditions is differentiable at every odd integer multiple of $L$ (i.e., at every junction between consecutive subintervals in the periodic extension).

**12.2.1 Norms, distances, inner products, and convergence.** Our main goal for the remainder of this chapter is to classify the types of functions that have convergent Fourier series representations. In basic calculus, you learned what it means for a series of *numbers* to converge to a *number* $M$—namely, the sequence of partial sums must converge to $M$. It is less clear how to define what it means for a series of *functions* (e.g., a Fourier series) to converge to a limiting *function* over some interval. Indeed, there are many different notions of convergence that we could adopt, some more natural than others. There are three types of convergence that we shall introduce for sequences/series of functions. The main difference between these types of convergence lies in how we measure *distances* between functions.

**Norms.** Consider the class of continuous, real-valued functions defined on the interval $[a,b]$. There are many ways to quantify the "size" of a function $f(x)$. For example, if $f(x)$ is continuous then $|f(x)|$ is guaranteed to achieve some maximum value on the interval $[a,b]$. The quantity

$$\|f\|_\infty = \max_{a \leq x \leq b} |f(x)| \tag{12.12}$$

is an example of a *norm*. Norms are used to measure the sizes of various mathematical objects, such as vectors or functions. The norm defined in (12.12) above has several important features. First, it is clear that $\|f\|_\infty \geq 0$ for all continuous functions $f(x)$ on $[a,b]$ and that $\|f\|_\infty = 0$ if and only if $f(x) = 0$ everywhere on $[a,b]$. Second, if $\alpha$ is any real constant, then $\|\alpha f\|_\infty = |\alpha|\|f\|_\infty$. Third, if $f$ and $g$ are both continuous functions on $[a,b]$, then $\|f+g\|_\infty \leq \|f\|_\infty + \|g\|_\infty$, the triangle inequality. In general, *any* norm (not just the one defined in (12.12)) must satisfy the three properties listed here.

The norm (12.12) defined above is one of two important norms that we will use from now on. The other norm is called the $L^2$ *norm* (read "L-two norm"). For functions $f$ defined on $[a,b]$, the $L^2$ norm is defined as

$$\|f\|_{L^2} = \left(\int_a^b |f(x)|^2 \, dx\right)^{1/2}, \tag{12.13}$$

provided that the integral converges. Observe that every continuous function on $[a,b]$ certainly has a finite $L^2$ norm. However, the $L^2$ norm can be used to measure the "size" of functions that are not necessarily continuous. To check that (12.13) really is a norm, first observe[1] that $\|f\|_{L^2} \geq 0$ is clear from the definition. Moreover, if $\alpha$ is any real constant, then it is easy to check that $\|\alpha f\|_{L^2} = |\alpha|\|f\|_{L^2}$. Verifying the other properties requires considerably more effort. In fact, proving the inequality $\|f+g\|_{L^2} \leq \|f\|_{L^2} + \|g\|_{L^2}$, or equivalently

$$\left(\int_a^b |f(x)+g(x)|^2 \, dx\right)^{1/2} \leq \left(\int_a^b |f(x)|^2 \, dx\right)^{1/2} + \left(\int_a^b |g(x)|^2 \, dx\right)^{1/2},$$

is surprisingly non-trivial. This special instance of the triangle inequality is actually named in honor of the first person to prove it: *Minkowski's inequality* is proved in most textbooks on mathematical analysis.

**Distances.** Once a set of functions is equipped with a norm, we automatically inherit a natural notion of *distance* between functions. Namely, given two functions $f$ and $g$, we can define the distance between $f$ and $g$ as the norm of the difference $f - g$. Consider, for example, the two norms defined above. If $f(x)$ and $g(x)$ are continuous functions defined on $[a,b]$, one way of measuring their

---

[1] Since $f$ need not be continuous, it is actually *not* the case that $\|f\|_{L^2} = 0$ if and only if $f$ is zero everywhere. If $f$ is zero except at finitely many points, then certainly $\|f\|_{L^2} = 0$ as well. This seemingly subtle point is discussed in great detail in the mathematical subject of *measure theory*.

distance is to use

$$\|f - g\|_\infty = \max_{a \le x \le b} |f(x) - g(x)|.$$

Graphically, this crude notion of distance measures the maximum vertical gap between the graphs of $f$ and $g$ over the interval $[a, b]$. It is sometimes called the $L^\infty$ distance between $f$ and $g$. Alternatively, we could define the $L^2$ distance between $f$ and $g$ as

$$\|f - g\|_{L^2} = \left( \int_a^b [f(x) - g(x)]^2 \, dx \right)^{1/2}.$$

The $L^2$ distance gives a better sense of the "cumulative" discrepancy between the graphs of $f$ and $g$. Anytime $f(x)$ and $g(x)$ deviate from one another, the quantity $[f(x) - g(x)]^2$ makes a positive contribution to the value of the integral. The bigger the gap, the larger the contribution.

**Example 12.2.1.** Consider the functions $f(x) = x$ and $g(x) = x^2$ on the interval $[0, 1]$. Calculate the $L^\infty$ and $L^2$ distances between $f(x)$ and $g(x)$ over this interval.
*Solution:* The $L^\infty$ distance is given by

$$\|f - g\|_\infty = \max_{0 \le x \le 1} |x - x^2| = \max_{0 \le x \le 1} (x - x^2),$$

where we have dropped the absolute value bars because $x \ge x^2$ on the interval $[0, 1]$. Finding the maximum value amounts to a calculus problem: the extreme values of the continuous function $x - x^2$ must occur either (i) at one of the endpoints of the interval $[0, 1]$ or (ii) at an interior critical point. The derivative of $x - x^2$ is $1 - 2x$, which is zero when $x = \frac{1}{2}$. Therefore, we must compare the values of $x - x^2$ when $x = 0$, $x = \frac{1}{2}$, and $x = 1$. The maximum value of $\frac{1}{4}$ occurs at the interior critical point, and we conclude that

$$\|f - g\|_\infty = \frac{1}{4}.$$

As for the $L^2$ distance, we calculate

$$\|f - g\|_{L^2} = \left( \int_0^1 [x - x^2]^2 \, dx \right)^{1/2} = \left( \int_0^1 x^2 - 2x^3 + x^4 \, dx \right)^{1/2}$$

$$= \left( \frac{x^3}{3} - \frac{x^4}{2} + \frac{x^5}{5} \Big|_0^1 \right)^{1/2} = \sqrt{\frac{1}{30}}.$$

As is typically the case, the $L^\infty$ and $L^2$ distances are unequal, because they measure distance between functions in completely different ways. If two continuous functions $f$ and $g$ have finite $L^\infty$ distance on the interval $[a, b]$, then they automatically have finite $L^2$ distance (try to prove it). However, it is not hard to create an example of two functions $f$ and $g$ with infinite $L^\infty$ distance but finite $L^2$ distance!

**The $L^2$ inner product.** The set of functions on $[a, b]$ with finite $L^2$ norm, denoted by $L^2[a, b]$, has considerable structure. Algebraically, $L^2[a, b]$ is a vector space whose elements are functions, and the various vector space axioms are easy to check. Geometrically, $L^2[a, b]$ is actually endowed with an *inner product*: a generalized version of the familiar dot product from multivariable calculus. This enables us to define geometric notions such as orthogonality (perpendicularity) of functions.

**Definition 12.2.2.** Suppose $f(x)$ and $g(x)$ are real-valued functions belonging to the set $L^2[a, b]$. Then the $L^2$ *inner product* of $f$ and $g$ is defined as

$$\langle f, g \rangle = \int_a^b f(x) g(x) \, dx. \tag{12.14}$$

If $\langle f, g \rangle = 0$, we say that $f$ and $g$ are *orthogonal.*

Notice that the inner product of two functions is a *scalar.* Unlike the norm of a function, it is certainly possible for an inner product of two functions to be negative.

**Example 12.2.3.** Consider the functions $f(x) = x^2$, $g(x) = -1$ and $h(x) = x$ on the interval $-2 \leq x \leq 2$. Certainly all three of these functions belong to $L^2[-2, 2]$ because continuous functions defined on closed intervals are always square integrable. The $L^2$ norm of $f$ is given by

$$\|f\|_{L^2[-2,2]} = \left( \int_{-2}^2 |x^2|^2 \, dx \right)^{1/2} = \left( \int_{-2}^2 x^4 \, dx \right)^{1/2}$$

$$= \left( \frac{x^5}{5} \Big|_{-2}^2 \, dx \right)^{1/2} = \sqrt{\frac{64}{5}}.$$

Similarly, you can check that the $L^2$ norm of $g(x)$ is 2. The inner product of $f$ and $g$ is

$$\langle f, g \rangle = \int_{-2}^{2} -x^2 \, dx = -\left.\frac{x^3}{3}\right|_{-2}^{2} = -\frac{16}{3}$$

and the inner product of $g$ and $h$ is

$$\langle g, h \rangle = \int_{-2}^{2} -x \, dx = 0.$$

We conclude that $g$ and $h$ are orthogonal in the space $L^2[-2, 2]$.

Many vector spaces other than $L^2$ are equipped with their own inner products. For example, the familiar dot product is an inner product on three-dimensional Euclidean space $\mathbb{R}^3$. However, not every vector space can be endowed with an inner product—only those with a certain level of geometric structure. There are several properties that every inner product must satisfy. We list these properties for the $L^2$ inner product: suppose that $f$ and $g$ are real-valued functions in the space $L^2[a, b]$. Then

☞ $\langle f, g \rangle = \langle g, f \rangle$.

☞ $\langle f + g, h \rangle = \langle f, h \rangle + \langle g, h \rangle$.

☞ $\langle \alpha f, g \rangle = \alpha \langle f, g \rangle$ for all real constants $\alpha$.

☞ $\langle f, f \rangle \geq 0$.

Notice that $\langle f, f \rangle$ is equal to the square of the $L^2$ norm of $f$.

**Types of convergence.** Recall that the norms (12.12) and (12.13) give us different ways of measuring the *distance* between two functions $f(x)$ and $g(x)$ defined on an interval $I$. With notions of distance in mind, we are able to define what it means for a sequence of functions $\{f_n(x)\}_{n=1}^{\infty}$ to converge to a limiting function $f(x)$ on $I$. The first type of convergence that we shall introduce is called *uniform convergence*:

**Definition 12.2.4.** Suppose that $f_n(x)$ is a sequence of functions defined on an interval $I$. We say that the sequence converges *uniformly* to $f(x)$ on $I$ if

$$\|f_n(x) - f(x)\|_\infty = \max_{x \in I} |f_n(x) - f(x)| \to 0 \quad \text{as } n \to \infty.$$

Students who have completed courses in mathematical analysis will realize that the maximum appearing in Definition 12.2.4 should technically be a supremum.

In the examples below, we generally work with sequences of continuous functions over closed intervals $I$. That way, we are assured that the functions actually achieve maximum values.

**Example 12.2.5.** Consider the sequence of functions $f_n(x) = x^n$, where $n \geq 1$. We claim that this sequence converges uniformly to $f(x) = 0$ on the interval $I = [0, \frac{1}{2}]$. To see this, we calculate

$$\|f_n(x) - f(x)\|_\infty = \max_{0 \leq x \leq \frac{1}{2}} |x^n - 0| = \max_{0 \leq x \leq \frac{1}{2}} x^n = \left(\frac{1}{2}\right)^n.$$

Since $\left(\frac{1}{2}\right)^n \to 0$ as $n \to \infty$, we have $\|f_n(x) - f(x)\|_\infty \to 0$ as $n \to \infty$, as required.

**Example 12.2.6.** When testing the convergence of a sequence of functions, it is important to specify the domain. In the previous example, suppose that we had worked on the interval $[0, 1)$ instead of $[0, \frac{1}{2}]$. We claim that the sequence $f_n(x) = x^n$ does *not* converge uniformly to $f(x) = 0$ on the interval $[0, 1)$. To see why, notice that no matter how large $n$ is, there are always $x$ values within the interval $[0, 1)$ for which $f_n(x) \geq \frac{1}{2}$. Indeed, we have $x^n \geq \frac{1}{2}$ whenever

$$\sqrt[n]{\frac{1}{2}} \leq x < 1.$$

Regardless of $n$, it must be the case that $\|f_n(x) - f(x)\|_\infty \geq \frac{1}{2}$ on the interval $[0, 1)$. Hence, $f_n(x)$ does *not* converge uniformly to $f(x) = 0$ on that interval.

There is a subtle point here worth mentioning. Above, we noted that no matter how large $n$ is, there will always be $x$ values inside the interval $[0, 1)$ for which $f_n(x) \geq \frac{1}{2}$. By contrast, suppose we fix an $x$ value and examine what happens to $f_n(x)$ as $n \to \infty$. Since $0 \leq x < 1$, it must be the case that $x^n \to 0$ as $n \to \infty$. Therefore, for each *fixed* $x$ in the domain $[0, 1)$, the sequence $f_n(x)$ converges to 0 as $n \to \infty$.

The remarks in the preceding example may seem surprising. We demonstrated that it is possible to have a sequence of functions $f_n(x)$ that does *not* converge uniformly to a function $f(x)$ on an interval $I$ even though $f_n(x) \to f(x)$ for each *fixed* $x \in I$. In this case, the functions $f_n$ exhibit a somewhat "weaker" version of convergence that we now define.

**Definition 12.2.7.** Suppose that $f_n(x)$ is a sequence of functions defined on an interval $I$. We say that the sequence converges *pointwise* to $f(x)$ on $I$ if

$$|f_n(x) - f(x)| \to 0 \quad \text{as } n \to \infty$$

for each $x \in I$.

Take a moment to compare the definitions of pointwise and uniform convergence. At first glance, it may be difficult to distinguish these two notions of convergence. If a sequence of functions $f_n(x)$ converges uniformly to $f(x)$ on some interval $I$, then

$$\max_{x \in I} |f_n(x) - f(x)| \to 0 \quad \text{as } n \to \infty.$$

Since the maximum gap between the functions is approaching 0, then certainly

$$|f_n(x) - f(x)| \to 0 \quad \text{as } n \to \infty$$

for each *fixed* $x \in I$. This argument proves that

*Uniform convergence implies pointwise convergence.*

On the other hand, the converse is not true, as illustrated by the example $f_n(x) = x^n$ on the interval $I = [0, 1)$. That sequence of functions converges pointwise to $f(x) = 0$ on $I$, but the convergence is not uniform.

**Example 12.2.8.** Let $f_n(x) = \frac{1}{n}\sin(n\pi x/L)$ on the interval $[0, L]$. This sequence of functions converges uniformly to $f(x) = 0$ on $[0, L]$. To see why, we exploit the boundedness of the sine function to make the estimate

$$\|f_n(x) - f(x)\|_\infty = \max_{0 \le x \le L} \left| \frac{1}{n} \sin\left(\frac{n\pi x}{L}\right) \right| \le \frac{1}{n}.$$

Therefore, $\|f_n(x) - f(x)\|_\infty \to 0$ as $n \to \infty$, which means that the sequence $f_n(x)$ converges uniformly to the function $f(x) = 0$. Since the sequence converges uniformly, it automatically converges pointwise as well.

**Example 12.2.9.** Consider the sequence of functions

$$f_n(x) = \frac{x^n}{1 + x^n} \qquad n = 1, 2, 3, \ldots$$

on the interval $I = [0,2]$. To test for pointwise convergence, we will calculate $\lim_{n\to\infty} f_n(x)$ for each fixed $x \in I$. If $0 \le x < 1$, then $x^n \to 0$ as $n \to \infty$, and it follows that

$$\lim_{n\to\infty} f_n(x) = \lim_{n\to\infty} \frac{x^n}{1+x^n} = 0 \qquad \text{if } 0 \le x < 1.$$

If $x = 1$, then $f_n(x) = \frac{1}{2}$ for all $n$, and therefore $\lim_{n\to\infty} f_n(1) = \frac{1}{2}$. Finally, note that if $1 < x \le 2$, then $x^n \to \infty$ as $n \to \infty$. Using L'Hôpital's rule, we calculate

$$\lim_{n\to\infty} f_n(x) = \lim_{n\to\infty} \frac{x^n}{1+x^n} = \lim_{n\to\infty} \frac{nx^{n-1}}{nx^{n-1}} = 1 \qquad \text{if } 1 < x \le 2.$$

Putting everything together, we have shown that the sequence $f_n(x)$ converges pointwise to the discontinuous function

$$f(x) = \begin{cases} 0 & \text{if } 0 \le x < 1 \\ \frac{1}{2} & \text{if } x = 1 \\ 1 & \text{if } 1 < x \le 2. \end{cases}$$

The convergence is *not* uniform—can you explain why?

Now that we are more accustomed to pointwise and uniform convergence, we introduce a third type of convergence:

**Definition 12.2.10.** Suppose that $f_n(x)$ is a sequence of functions defined on an interval $I$. We say that the sequence converges to $f(x)$ *in the $L^2$ sense* on $I$ if

$$\|f_n(x) - f(x)\|_{L^2} \to 0 \quad \text{as } n \to \infty.$$

If $I = [a,b]$, stating that $f_n(x) \to f(x)$ in the $L^2$ sense is equivalent to saying that

$$\int_a^b [f_n(x) - f(x)]^2 \, dx \to 0 \quad \text{as } n \to \infty.$$

**Example 12.2.11.** In a previous example, we showed that the sequence $f_n(x) = x^n$ converges pointwise to $f(x) = 0$ on the interval $[0,1)$, but that the convergence is not uniform. Does the sequence converge to $f(x) = 0$ in the $L^2$ sense? We calculate

$$\|f_n(x) - f(x)\|_{L^2}^2 = \int_0^1 [x^n - 0]^2 \, dx = \int_0^1 x^{2n} \, dx = \left. \frac{x^{2n+1}}{2n+1} \right|_0^1 = \frac{1}{2n+1}.$$

Letting $n \to \infty$, we see that $\|f_n(x) - f(x)\|_{L^2}^2 \to 0$ as $n \to \infty$. It follows that the sequence $f_n(x)$ *does* converge to $f(x) = 0$ in the $L^2$ sense on $[0, 1)$.

Above, we noted that uniform convergence is "stronger" than pointwise convergence: if $f_n(x) \to f(x)$ uniformly on an interval $I$, then we automatically conclude that $f_n(x) \to f(x)$ pointwise on $I$ as well. A similar statement holds with respect to $L^2$ convergence, namely,

*Uniform convergence on a finite interval implies convergence in the $L^2$ sense.*

To see why, suppose that $f_n(x) \to f(x)$ uniformly on $[a, b]$, and let

$$M_n = \|f_n(x) - f(x)\|_\infty = \max_{a \le x \le b} |f_n(x) - f(x)|.$$

Our assumption of uniform convergence is equivalent to saying that $M_n \to 0$ as $n \to \infty$. To test for convergence in the $L^2$ sense, observe that

$$\begin{aligned} \|f_n(x) - f(x)\|_{L^2} &= \left( \int_a^b [f_n(x) - f(x)]^2 \, dx \right)^{1/2} \le \left( \int_a^b M_n^2 \, dx \right)^{1/2} \\ &= \sqrt{M_n^2 (b-a)} = M_n \sqrt{b-a}. \end{aligned}$$

Since $M_n \to 0$ as $n \to \infty$, it follows that $\|f_n(x) - f(x)\|_{L^2} \to 0$ as well. We have now proved that $f_n(x) \to f(x)$ in the $L^2$ sense.

**Example 12.2.12.** For sequences of functions defined over infinite domains, uniform convergence need not imply convergence in the $L^2$ sense. Consider the sequence of functions defined by

$$f_n(x) = \begin{cases} \frac{1}{\sqrt{n}} & \text{if } 0 \le x \le n \\ 0 & \text{if } x > n. \end{cases}$$

This sequence converges uniformly to $f(x) = 0$ on the infinite domain $0 \le x < \infty$ because

$$\|f_n(x) - f(x)\|_\infty = \max_{0 \le x < \infty} |f_n(x) - f(x)| = \frac{1}{\sqrt{n}}$$

approaches 0 as $n \to \infty$. However, the sequence $\{f_n(x)\}$ does *not* converge to $f(x)$ in the $L^2$ sense because

$$\|f_n(x) - f(x)\|_{L^2} = \left(\int_0^\infty [f_n(x) - f(x)]^2 \, dx\right)^{1/2} = \left(\int_0^n \frac{1}{n} \, dx\right)^{1/2} = 1$$

for all $n$. Since $\|f_n(x) - f(x)\|_{L^2} \to 1 \neq 0$ as $n \to \infty$, we do not have convergence in the $L^2$ sense.

In our study of Fourier series, we are interested in pointwise, uniform, and $L^2$ convergence of *series* of functions, not sequences. Fortunately, the definitions of these various modes of convergence can be easily extended. Given an infinite series of functions

$$\sum_{n=1}^{\infty} f_n(x), \tag{12.15}$$

the $N$th partial sum is defined as

$$S_N(x) = \sum_{n=1}^{N} f_n(x).$$

**Definition 12.2.13.** We say that the series (12.15) converges to a function $f(x)$

☞ *pointwise* on $(a,b)$ if $|f(x) - S_N(x)| \to 0$ as $N \to \infty$ for each $x \in (a,b)$,

☞ *uniformly* on $(a,b)$ if $\|f(x) - S_N(x)\|_\infty \to 0$ as $N \to \infty$,

☞ *in the $L^2$ sense* on $(a,b)$ if $\|f(x) - S_N(x)\|_{L^2} \to 0$ as $N \to \infty$.

Again, uniform convergence is the strongest of these three types of convergence: if a series converges uniformly on the finite interval $(a,b)$, then it automatically converges pointwise and in the $L^2$ sense.

**Example 12.2.14.** In calculus, you studied the geometric series

$$\sum_{n=0}^{\infty} x^n.$$

We claim that this series converges pointwise but *not* uniformly to $f(x) = 1/(1-x)$ on the interval $(-1,1)$. To verify pointwise convergence, consider the partial sum

$$S_N(x) = \sum_{n=0}^{N} x^n,$$

which is a polynomial of degree $N$. This summation simplifies nicely if we use an algebra trick: notice that

$$(1-x)S_N(x) \;=\; (1-x)(1+x+x^2+\cdots+x^N) \;=\; 1-x^{N+1}.$$

Therefore, if $x \neq 1$,

$$S_N(x) \;=\; \frac{1-x^{N+1}}{1-x}.$$

Testing for pointwise convergence, we measure

$$|f(x)-S_N(x)| \;=\; \left|\frac{1}{1-x}-\frac{1-x^{N+1}}{1-x}\right| \;=\; \left|\frac{x^{N+1}}{1-x}\right|.$$

Assuming that $-1 < x < 1$, notice that $x^{N+1} \to 0$ as $N \to \infty$. Therefore,

$$|f(x)-S_N(x)| \to 0 \quad \text{as } n \to \infty$$

for each fixed $x \in (-1,1)$. This is precisely what we needed in order to conclude that our series converges pointwise.

To show that the series does *not* converge uniformly, we must show[2] that

$$\|f(x)-S_N(x)\|_\infty \;=\; \max_{-1<x<1} |f(x)-S_N(x)|$$

does *not* converge to 0 as $N \to \infty$. It suffices to prove that for each choice of $N$, there are always $x$ values in the interval $(-1,1)$ for which $|f(x)-S_N(x)| \geq 1$. From our proof of pointwise convergence, we already know that

$$f(x)-S_N(x) \;=\; \frac{x^{N+1}}{1-x},$$

so let us try to show that the equation

$$\frac{x^{N+1}}{1-x} \;=\; 1$$

always has a solution $x \in (-1,1)$ regardless of $N$. The equation can be re-written as $x^{N+1}+x-1=0$. Letting $p(x) = x^{N+1}+x-1$ denote the left-hand side,

[2]A word of caution here. On open intervals such as $-1 < x < 1$, even continuous functions are not guaranteed to achieve a maximum/minimum value. Students who have taken a course in mathematical analysis will realize that we should really be taking a *supremum* instead of a maximum.

notice that $p(x)$ is continuous. Moreover, observe that $p(0) = -1 < 0$ and $p(1) = 1 > 0$ regardless of $N$. Since $p(x)$ is continuous and transitions from negative to positive as $x$ varies from 0 to 1, the Intermediate Value Theorem from calculus tells us that $p(x)$ must have a root on the interval $(0, 1)$. This is precisely what we needed to show, and we conclude that the series does *not* converge uniformly on the interval $(-1, 1)$.

Determining whether the series converges in the $L^2$ sense is more challenging. According to the definition of $L^2$ convergence, we would need to evaluate

$$\lim_{N\to\infty} \|f(x) - S_N(x)\|_{L^2}$$

and check whether the limit is zero. This requires that we integrate the square of $f(x) - S_n(x)$—i.e.,

$$\int_{-1}^{1} [f(x) - S_N(x)]^2 \, dx = \int_{-1}^{1} \frac{x^{2N+2}}{(1-x)^2} \, dx.$$

Notice that the integral is improper and should therefore be written as a limit

$$\lim_{b\to 1^-} \int_{-1}^{b} \frac{x^{2N+2}}{(1-x)^2} \, dx.$$

This integral seems a bit tricky—one way to evaluate it is to substitute $x = u + 1$, use the binomial theorem to expand the numerator, and integrate one term at a time. It turns out that the integral diverges regardless of $N$, from which it follows that the series we started with does *not* converge in the $L^2$ sense. Thankfully, we will soon state a much cleaner test for $L^2$ convergence of series.

**Example 12.2.15.** For non-negative integers $n$, define the functions $f_n(x) = x^n - x^{n+2}$ on the interval $0 < x < 1$. We claim that the infinite series

$$\sum_{n=0}^{\infty} f_n(x)$$

converges to $f(x) = 1 + x$ both pointwise and in the $L^2$ sense on $(0, 1)$, but *not* uniformly. The partial sums collapse nicely because this is a telescoping series:

$$S_N(x) = \sum_{n=0}^{N} f_n(x) = (1 - x^2) + (x - x^3) + (x^2 - x^4) + \cdots$$

$$+ (x^{N-1} - x^{N+1}) + (x^N - x^{N+2}) \;=\; 1 + x - x^{N+1} - x^{N+2}.$$

Testing for pointwise convergence, notice that

$$|f(x) - S_N(x)| \;=\; \left|x^{N+1} + x^{N+2}\right| \;=\; x^{N+1}(1+x) \to 0 \;\text{ as } N \to \infty,$$

for $0 < x < 1$. This establishes that $f_n(x) \to f(x)$ pointwise on $(0,1)$. To prove $L^2$ convergence, we calculate

$$\begin{aligned}
\|f(x) - S_N(x)\|_{L^2} &= \left(\int_0^1 [f(x) - S_N(x)]^2 \, dx\right)^{1/2} \\
&= \left(\int_0^1 [x^{N+1} + x^{N+2}]^2 \, dx\right)^{1/2} \\
&= \left(\int_0^1 x^{2N+2} + 2x^{2N+3} + x^{2N+4} \, dx\right)^{1/2} \\
&= \left(\frac{x^{2N+3}}{2N+3} + \frac{2x^{2N+4}}{2N+4} + \frac{x^{2N+5}}{2N+5}\Bigg|_0^1\right)^{1/2} \\
&= \left(\frac{1}{2N+3} + \frac{2}{2N+4} + \frac{1}{2N+5}\right)^{1/2} \qquad \to 0 \quad \text{as } N \to \infty.
\end{aligned}$$

Finally, to see that the convergence is not uniform, it suffices to note that the maximum value of

$$|f(x) - S_N(x)| \;=\; |x^{N+1} + x^{N+2}|$$

on the *closed* interval $[0,1]$ is 2, and is achieved at the right endpoint of that interval. Since $x^{N+1} - x^{N+2}$ is a continuous function, there must be $x$ values in the *open* interval $(0,1)$ for which $x^{N+1} - x^{N+2}$ is arbitrarily close to 2. Consequently, it is impossible for $\|f(x) - S_N(x)\|_\infty$ to converge to 0 as $N \to \infty$, and therefore the series does not converge uniformly to $f(x)$ on $(0,1)$.

**12.2.2 Convergence theorems.** Armed with our understanding of the three different notions of convergence that we have singled out, we now state theorems regarding convergence of Fourier series.

**Theorem 12.2.16. Convergence in the $L^2$ sense.** Suppose that $\phi(x)$ is defined on an interval $[a,b]$. If $\|\phi(x)\|_{L^2[a,b]}$ is finite, then Fourier series for $\phi(x)$ converges to $\phi(x)$ in the $L^2$ sense on $(a,b)$.

**Example 12.2.17.** In a previous example, we showed that the Fourier series for

$$\phi(x) = \begin{cases} -2 & \text{if } -3 < x < 0, \\ 6 & \text{if } 0 \le x < 3. \end{cases}$$

on the interval $-3 < x < 3$ is given by

$$2 + \frac{16}{\pi} \sum_{n=1}^{\infty} \frac{1}{2n-1} \sin\left[\frac{(2n-1)\pi x}{3}\right].$$

To verify that the series really *does* converge to the function $\phi(x)$ in the $L^2$ sense, we must check that $\|\phi(x)\|_{L^2} < \infty$. Indeed,

$$\|\phi(x)\|_{L^2} = \left(\int_{-3}^{3} \phi(x)^2 \, dx\right)^{1/2} = \left(\int_{-3}^{0} (-2)^2 \, dx + \int_{0}^{3} 6^2 \, dx\right)^{1/2} < \infty.$$

The exact value of the integral is irrelevant—the fact that it is finite assures that the Fourier series for $\phi(x)$ converges to $\phi(x)$.

Not surprisingly, since uniform convergence is stronger than convergence in the $L^2$ sense, the class of functions $\phi(x)$ whose Fourier series converge uniformly is considerably smaller:

**Theorem 12.2.18. Uniform convergence.** Suppose that $\phi(x)$ is defined and continuous on an interval $[-L, L]$ and $\phi(-L) = \phi(L)$. If $\phi'(x)$ is piecewise continuous (with only jump discontinuities) on $[-L, L]$, then the Fourier series for $\phi(x)$ converges to $\phi(x)$ uniformly.

There are variants of Theorem 12.2.18 that guarantee uniform convergence under slightly different conditions. Notice that this theorem would not guarantee uniform convergence of the Fourier series in the preceding example, because the function was discontinuous.

**Example 12.2.19.** Theorem 12.2.18 guarantees that the Fourier series for $\phi(x) = |x|$ on the interval $-10 \le x \le 10$ converges uniformly. To verify that the conditions of the theorem are satisfied, first observe that $\phi(x)$ is continuous on $[-10, 10]$ and $\phi(-10) = \phi(10)$. Also, $\phi'(x) = -1$ for $x < 0$ and $\phi'(x) = 1$ for

$x > 0$. Therefore, the only discontinuity of $\phi'(x)$ occurs at $x = 0$ and is a jump discontinuity.

Finally, we turn our attention to pointwise convergence of Fourier series. The criteria we shall state for pointwise convergence involve the notions of left and right-hand limits and derivatives. The left and right-hand limits of $f(x)$ at $x_0$ are defined as

$$f(x_0^-) = \lim_{x \to x_0^-} f(x) \quad \text{and} \quad f(x_0^+) = \lim_{x \to x_0^+} f(x),$$

respectively, provided that these one-sided limits exist. If $f(x_0^-)$ and $f(x_0^+)$ exist but are unequal, we say that $f(x)$ has a *jump discontinuity* at $x_0$. Furthermore, if $f(x_0^-) = f(x_0) = f(x_0^+)$, then $f(x)$ is continuous at $x_0$. The left and right-hand derivatives of $f(x)$ at $x_0$ are defined as

$$f'(x_0^-) = \lim_{h \to 0^+} \frac{f(x_0^-) - f(x_0 - h)}{h} \quad \text{and} \quad f'(x_0^+) = \lim_{h \to 0^+} \frac{f(x_0 + h) - f(x_0^+)}{h}$$

respectively, provided that these one-sided limits exist.

**Theorem 12.2.20** (Pointwise convergence.). Suppose that $\phi(x)$ is defined and piecewise continuous (with only jump discontinuities) on an interval $[-L, L]$. If the left and right-hand derivatives of $\phi(x)$ exist at each jump discontinuity, then the Fourier series for $\phi(x)$ converges pointwise to $\frac{1}{2}[\phi(x^+) + \phi(x^-)]$ for each $x \in (-L, L)$.

Here are some remarks that may help you interpret this theorem:

☞ Notice that $\frac{1}{2}[\phi(x^+) + \phi(x^-)]$ represents the average value of the left and right-hand limits of $\phi(x)$. If $\phi(x)$ is continuous at $x = x_0$, then $\phi(x_0^-) = \phi(x) = \phi(x_0^+)$, in which case the expression $\frac{1}{2}[\phi(x^+) + \phi(x^-)]$ reduces to $\phi(x)$. In other words, at points where $\phi(x)$ is continuous, its Fourier series is guaranteed to converge pointwise to $\phi(x)$ itself.

☞ To determine what happens at the endpoints $x = \pm L$, examine the periodic extension of $\phi(x)$. If the right-hand derivative $\phi'(-L^+)$ and the left-hand derivative $\phi'(L^-)$ exist, then the Fourier series for $\phi(x)$ converges pointwise to $\frac{1}{2}[\phi(-L) + \phi(L)]$ at both $x = \pm L$.

☞ Following up on the preceding two remarks, if $\phi(x)$ is continuous and $\phi(-L) = \phi(L)$, then the Fourier series for $\phi(x)$ automatically converges pointwise to $\phi(x)$ on the entire interval $[-L, L]$.

**Example 12.2.21.** Consider the piecewise continuous function

$$\phi(x) = \begin{cases} -2 & \text{if } -8 < x < -3 \\ 3 & \text{if } -3 \leq x \leq 2 \\ 7 & \text{if } 2 < x < 8. \end{cases}$$

This function satisfies the conditions of the pointwise convergence theorem (Theorem 12.2.20). Namely, $\phi(x)$ is continuous, and the only types of discontinuities are jump discontinuities. The Fourier series for $\phi(x)$ must converge pointwise to $\phi(x)$ itself everywhere except at the jumps. More exactly, the Fourier series for $\phi(x)$ would converge pointwise to

$$\tilde{\phi}(x) = \begin{cases} -2 & \text{if } -8 < x < -3 \\ \frac{1}{2} & \text{if } x = -3 \\ 3 & \text{if } -3 \leq x \leq 2 \\ 5 & \text{if } x = 2 \\ 7 & \text{if } 2 < x < 8. \end{cases}$$

Notice that at $x = -3$, the series converges to the average of the left and right-hand limits: $\frac{1}{2}[\phi(-3^+) + \phi(-3^-)] = \frac{1}{2}$. Similarly, at $x = 2$, the Fourier series converges to $\frac{1}{2}[3 + 7] = 5$.

At the endpoints $x = \pm 8$, the series would converge pointwise to the average of $f(-8^+) = -2$ and $f(8^-) = 7$. That is, the Fourier series would converge to $\frac{5}{2}$ when $x = \pm 8$.

We remark that the theorems regarding pointwise, uniform and $L^2$ convergence of Fourier series on $[-L, L]$ are easily adapted to handle Fourier sine and cosine series on an interval $[0, L]$. If (i) $\phi(x)$ is continuous on $[0, L]$ with $\phi(0) = \phi(L)$ and (ii) $\phi'(x)$ is piecewise continuous on $[0, L]$ (any discontinuities are jump discontinuities), then the Fourier sine & cosine series for $\phi(x)$ converge uniformly to $\phi(x)$ on $[0, L]$. On a similar note, if $\phi(x)$ has a finite $L^2$ norm over the interval $[0, L]$, then the Fourier sine & cosine series for $\phi(x)$ converge to $\phi(x)$ in the $L^2$ sense. Although the pointwise convergence theorem 12.2.20 has a natural counterpart for sine and cosine series, our earlier remarks concerning convergence at the endpoints $x = 0$ and $x = L$ are no longer valid. After all, a

Fourier sine series on $[0, L]$ must converge to zero at both endpoints, independent of how the function $\phi(x)$ is defined at those two points.

---

## Exercises

1. Suppose $m$ and $n$ are non-negative integers and let $L > 0$. Show that

$$\int_0^L \cos\left(\frac{m\pi x}{L}\right) \cos\left(\frac{n\pi x}{L}\right) \,\mathrm{d}x = \begin{cases} L & \text{if } m = n = 0 \\ L/2 & \text{if } m = n \neq 0 \\ 0 & \text{if } m \neq n. \end{cases}$$

You may wish to use the following trigonometric identities:

$$\cos\alpha\cos\beta = \frac{1}{2}\cos(\alpha-\beta) + \frac{1}{2}\cos(\alpha+\beta) \quad \text{and} \quad \cos^2\alpha = \frac{1}{2}\left[1+\cos(2\alpha)\right].$$

2. Solve the homogeneous Dirichlet problem

$$\begin{aligned} u_{tt} &= c^2 u_{xx} & (0 \leq x \leq 5) \\ u(x,0) &= \sin\left(\frac{17\pi x}{5}\right) \\ u_t(x,0) &= 0 \\ u(0,t) &= 0 \\ u(5,t) &= 0. \end{aligned}$$

Feel free to quote (11.28), the general solution of the wave equation on a finite interval with homogeneous Dirichlet boundary conditions. Describe how the solution behaves, as well as a physical interpretation.

3. Calculate the Fourier sine series for $\phi(x) = x$ on the interval $0 < x < 3$ and plot the first three partial sums of the series.

4. Calculate the Fourier cosine series for the discontinuous function

$$\phi(x) = \begin{cases} 1 & \text{if } 0 < x < \frac{5}{2} \\ 3 & \text{if } \frac{5}{2} \leq x < 5. \end{cases}$$

on the interval $0 < x < 5$.

5. Calculate the (full) Fourier series for $\phi(x) = x$ on the interval $-4 < x < 4$.
6. Show that for any choice of non-negative integers $m$ and $n$, the functions $f(x) = \sin(m\pi x/L)$ and $g(x) = \cos(n\pi x/L)$ on the interval $[-L, L]$ are orthogonal with respect to the $L^2$ inner product.
7. Consider the set of continuous functions on the interval $[a, b]$. The $L^1$ norm of such functions is defined as

$$\|f\|_{L^1} = \int_a^b |f(x)|\, dx.$$

Show that this really *is* a norm, by verifying that (i) $\|f\|_{L^1} \geq 0$, with equality if and only if $f = 0$; (ii) $\|\alpha f\|_{L^1} = |\alpha| \|f\|_{L^1}$ for any real constant $\alpha$; and (iii) $\|f + g\|_{L^1} \leq \|f\|_{L^1} + \|g\|_{L^1}$ for all functions $f$ and $g$ that are continuous on $[a, b]$.

8. By $L^1[a, b]$, we mean the set of integrable functions on $[a, b]$ which have finite $L^1$ norm (see previous problem). We say that a sequence of functions $\{f_n\}_{n=1}^{\infty}$ converges to $f$ in the $L^1$ sense if

$$\int_a^b |f(x) - f_n(x)|\, dx \to 0 \text{ as } n \to \infty.$$

Consider the functions

$$f_n(x) = \begin{cases} \frac{1}{n} & \text{if } 0 \leq x \leq n \\ 0 & \text{if } n < x \end{cases}$$

defined on the interval $[a, b] = [0, \infty)$. Show that $f_n(x) \to f(x) = 0$ pointwise as $n \to \infty$. Then show that the sequence $f_n(x)$ does *not* converge to $f(x) = 0$ in the $L^1$ sense.

9. Consider the functions $f_n(x) = x^{-n}$ on the interval $[1, 2]$. Does this sequence of functions converge pointwise as $n \to \infty$? If so, what function $f(x)$ does the sequence converge to? Does the sequence converge uniformly?
10. The purpose of this exercise is to show that pointwise convergence need not imply convergence in the $L^2$ sense. Consider the functions

$$f_n(x) = \begin{cases} n^2(1 - nx) & \text{if } 0 < x \leq \frac{1}{n} \\ 0 & \text{if } \frac{1}{n} < x < 1. \end{cases}$$

Show that this sequence converges pointwise to some function $f(x)$ (which you will need to find) on the interval $0 < x < 1$. Then, show that the sequence does *not* converge to $f(x)$ in the $L^2$ sense.

**11.** The purpose of this exercise is to show that convergence in the $L^2$ sense need not imply pointwise convergence. Consider the piecewise continuous functions

$$f_n(x) = \begin{cases} 1 & \text{if } \frac{1}{2} - \frac{1}{n} \le x \le \frac{1}{2} + \frac{1}{n} \\ 0 & \text{otherwise} \end{cases}$$

on the interval $0 \le x \le 1$. Show that this sequence converges in the $L^2$ sense to $f(x) = 0$, but that the sequence does *NOT* converge pointwise to $f(x)$.

**12.** Solve the Dirichlet problem

$$u_t + 2u = u_{xx} \qquad (0 < x < 1)$$

$$u(x,0) = \phi(x) = \begin{cases} x & \text{if } 0 \le x \le \frac{1}{2} \\ 1 - x & \text{if } \frac{1}{2} < x \le 1, \end{cases}$$

$$u(0,t) = u(1,t) = 0$$

expressing your solution in terms of a Fourier sine series.

**13.** Let $\phi(x)$ be defined as in the previous exercise. Explain why Fourier sine series expansion of $\phi(x)$ converges uniformly to $\phi(x)$ on $[0,1]$. Does it converge in the $L^2$ sense?

**14.** These questions concern the function

$$\phi(x) = \begin{cases} 0 & \text{if } -9 < x \le -3, \\ 2 & \text{if } -3 < x \le 3, \\ 0 & \text{if } \;\; 3 < x < 9. \end{cases}$$

(a) Without computing the Fourier series for $\phi(x)$, determine whether the series converges pointwise on the interval $(-9,9)$. If so, what does the series converge to?

(b) Without computing the Fourier series for $\phi(x)$, determine whether the series converges in the $L^2$ sense on the interval $(-9,9)$.

(c) Compute the (full) Fourier series representation of $\phi(x)$ on the interval $(-9,9)$.

(d) Let $S_3(x)$ denote the sum of the first three non-zero terms in the (full) Fourier series representation you calculated in Part (c). Carefully plot $S_3(x)$ and $\phi(x)$ on the same set of axes. Either use a computer to generate the plot for you, or give an accurate hand-sketch.

# CHAPTER 13

## The Laplace and Poisson Equations

Up to now, we have dealt almost exclusively with PDEs for which one independent variable corresponds to *time*. Now, we will analyze a PDE for which this is not the case: Laplace's equation. To motivate where Laplace's equation comes from and what it models, consider the heat equation with periodic boundary conditions:

$$\begin{aligned} u_t &= \kappa u_{xx} \qquad\qquad (-L \leq x \leq L) \\ u(-L,t) &= u(L,t) \\ u_x(-L,t) &= u_x(L,t). \end{aligned}$$

Physically, this system models diffusion of heat within a thin "one-dimensional" ring of wire. You can imagine "bending" the interval $-L \leq x \leq L$ into a ring. With the two ends $x = -L$ and $x = L$ in contact, they correspond to the same physical location (hence the two boundary conditions).

Using the separation of variables technique, you can show that the solution of the above system is given by

$$u(x,t) = \frac{A_0}{2} + \sum_{n=1}^{\infty} \left[ A_n \cos\left(\frac{n\pi x}{L}\right) + B_n \sin\left(\frac{n\pi x}{L}\right) \right] \mathrm{e}^{-n^2\pi^2\kappa t/L^2}.$$

Given an initial condition $u(x,0) = \phi(x)$ for $-L \leq x \leq L$, the constants appearing in the formula for $u(x,t)$ are given by the Fourier coefficients

$$A_n = \frac{1}{L}\int_{-L}^{L} \phi(x)\cos\left(\frac{n\pi x}{L}\right)\,\mathrm{d}x \quad \text{and} \quad B_n = \frac{1}{L}\int_{-L}^{L} \phi(x)\sin\left(\frac{n\pi x}{L}\right)\,\mathrm{d}x.$$

**Question:** How do solutions behave as $t \to \infty$? Intuitively, we would expect the exponential factor in the above series to decay rapidly to 0 as $t \to \infty$. This would suggest that only the leading term survives as $t \to \infty$:

$$\lim_{t \to \infty} u(x,t) = \frac{A_0}{2} = \frac{1}{2L} \int_{-L}^{L} \phi(x)\, dx$$

for all $x \in [-L, L]$. Notice that this integral represents the *average value* of the initial temperature distribution $\phi(x)$. Consequently, we expect the solution of the heat equation on a ring-shaped domain (periodic boundary conditions) to approach a *steady-state* in which heat is uniformly distributed around the ring.

Moving beyond the above example, let us ask a more general question: What are the steady-state solutions of the heat equation (or the wave equation)? In other words, are there solutions the heat equation that are time-independent? Physical intuition suggests that if we apply a fixed temperature distribution along the boundary of the spatial domain, then the heat profile should equilibrate to some steady configuration throughout the domain as $t \to \infty$.

**Example 13.0.22.** In one spatial dimension, finding steady-state solutions of the heat equation is rather easy. Suppose that we wish to find steady-state solutions of $u_t = \kappa u_{xx}$ on the domain $0 < x < L$, subject to homogeneous Dirichlet boundary conditions $u(0,t) = 0 = u(L,t)$. Since such solutions are time-independent, we should set $u_t = 0$, reducing the PDE to an ODE $\kappa u''(x) = 0$. Integrating twice, we find that $u(x) = C_1 x + C_2$, where $C_1$ and $C_2$ are constants. The boundary conditions imply that both of the constants are zero, which means that the only steady-state solution of this Dirichlet problem is the constant function $u = 0$. As an exercise, find the steady-state solution of the heat equation with the more general Dirichlet conditions $u(0,t) = \tau_1$ and $u(L,t) = \tau_2$. You should find that temperature varies linearly between $\tau_1$ and $\tau_2$ over the domain $0 \le x \le L$.

A similar argument shows that any constant function is a steady-state solution of the one-dimensional heat equation with homogeneous Neumann boundary conditions.

In higher dimensions, it is more challenging to find the steady-state solutions of the heat equation on a given spatial domain $\Omega$. In two or three dimensions, the heat equation takes the form $u_t = \kappa(u_{xx} + u_{yy})$ or $u_t = \kappa(u_{xx} + u_{yy} + u_{zz})$, respectively, where $\kappa > 0$ is the diffusion coefficient. These equations can be written more compactly if we introduce the *Laplace operator* (or *Laplacian*)

$$\begin{aligned} \Delta &= \left(\frac{\partial^2}{\partial x^2} + \frac{\partial^2}{\partial y^2}\right) && \text{in two space dimensions} \\ \Delta &= \left(\frac{\partial^2}{\partial x^2} + \frac{\partial^2}{\partial y^2} + \frac{\partial^2}{\partial z^2}\right) && \text{in three space dimensions.} \end{aligned} \tag{13.1}$$

The heat equation now takes the form $u_t = \kappa \Delta u$, and it should always be clear from the context as to whether we are working in two or three spatial dimensions. Likewise, the two and three-dimensional wave equations can be written as $u_{tt} = c^2 \Delta u$.

Steady-state solutions of the heat equation are time-independent, which implies that $u_t = 0$. Setting $u_t = 0$ in the heat equation leads to an important PDE that gets a special name:

**Definition 13.0.23.** The PDE

$$\Delta u(x, y) = \frac{\partial^2 u}{\partial x^2} + \frac{\partial^2 u}{\partial y^2} = 0$$

is called the two-dimensional *Laplace equation*. The PDE

$$\Delta u(x, y, z) = \frac{\partial^2 u}{\partial x^2} + \frac{\partial^2 u}{\partial y^2} + \frac{\partial^2 u}{\partial z^2} = 0$$

is called the three-dimensional *Laplace equation*. Solutions of the Laplace equation are called *harmonic functions*.

For emphasis, solutions of Laplace's equation can be thought of as solutions of the "steady-state heat equation", because Laplace's equation can be obtained by setting $u_t = 0$ in $u_t = \kappa \Delta u$. One way to interpret this physically is to imagine a pie that is placed in an oven of constant temperature. The oven temperature imposes a Dirichlet condition on the boundary (external surface) of the pie. As time passes, the temperature within the pie equilibrates to some steady-state distribution. This steady-state temperature profile will be a harmonic function—a solution to Laplace's equation on a pie-shaped domain.

Laplace's equation arises in a variety of physical contexts, not just steady-state heat distribution. For example, the velocity potential of an incompressible, irrotational fluid satisfies Laplace's equation. It is also worth noting that solutions of Laplace's equation can be thought of as solutions of the "steady-state *wave* equation" $u_{tt} = c^2 \Delta u = 0$.

The inhomogeneous version of Laplace's equation is also famous enough to receive a special name:

**Definition 13.0.24.** The PDE

$$\Delta u(x, y) = \frac{\partial^2 u}{\partial x^2} + \frac{\partial^2 u}{\partial y^2} = f(x, y)$$

is called the two-dimensional *Poisson equation*. The PDE

$$\Delta u(x, y, z) = \frac{\partial^2 u}{\partial x^2} + \frac{\partial^2 u}{\partial y^2} + \frac{\partial^2 u}{\partial z^2} = f(x, y, z)$$

is called the three-dimensional *Poisson equation*.

Perhaps the most well-known application of Poisson's equation arises in electrostatics. If $\varphi(x, y, z)$ denotes the electric potential at a point $(x, y, z)$ in a static (time-independent) electric field, then Gauss's Law states that $\Delta\varphi = f(x, y, z)$, where $f$ is proportional to the total charge density. In other words, electrostatic potential satisfies Poisson's equation.

---

## 13.1. Dirchlet and Neumann Problems

Our main goal for this chapter is to solve Laplace's equation on certain bounded domains $\Omega$. In two-dimensions, we will focus on domains $\Omega$ whose boundaries are formed by a simple, closed, piecewise smooth curve. By *simple*, we mean that the curve is not allowed to intersect itself, and by piecewise smooth, we mean that the curve can be subdivided into finitely many curves, each of which is infinitely differentiable. We denote the boundary of such domains by $\partial\Omega$—in this context, the use of the $\partial$ symbol is a notational standard, and has nothing to do with differentiation. Sketches of the types of domains we have in mind appear in Figure 13.1.

To further describe the domains $\Omega$ we have in mind, let us crudely define a few important topological notions. A set $\Omega$ is called *open* if it contains none of its boundary points. In two-dimensions, the set $S$ of points $(x, y)$ such that $x^2 + y^2 < 1$ forms an open set (a circular disc). The boundary $\partial S$ would be the

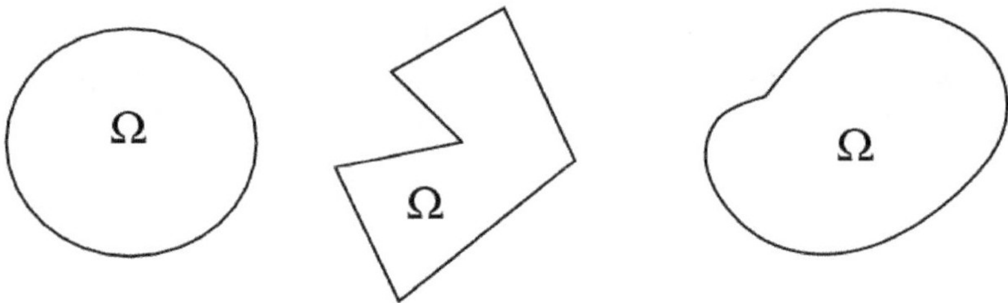


**Figure 13.1.** Three examples of two-dimensional domains $\Omega$ whose boundaries $\partial\Omega$ consist of simple, closed, piecewise smooth curves.

circle $x^2 + y^2 = 1$, none of which are contained in $S$. A set $\Omega$ in $\mathbb{R}^2$ is called *bounded* if there exists a positive number $R$ (possibly very large) such that $\Omega$ is contained within a disc or radius $R$. Finally, by a *connected* set $\Omega$, we have in mind a domain that consists of one contiguous "piece". For example, on a map of the United States, the state of Kansas is connected whereas the state of Hawaii is disconnected since it is composed of multiple islands. Readers interested in precise, technical definitions of openness, boundedness, and connectedness are encouraged to complete a course in topology. The Dirichlet problem for Laplace's equation is formulated as followed. Let $\Omega$ be a two or three-dimensional domain of the type described above. Then the Dirichlet problem is given by

$$\begin{aligned} \Delta u(x,y) &= 0 && \text{inside } \Omega \\ u(x,y) &= h(x,y) && \text{on the boundary, } \partial\Omega \end{aligned}$$

in two dimensions, or

$$\begin{aligned} \Delta u(x,y,z) &= 0 && \text{inside } \Omega \\ u(x,y,z) &= h(x,y,z) && \text{on the boundary, } \partial\Omega \end{aligned}$$

in three dimensions. As explained previously, the Dirichlet boundary condition prescribes the temperature distribution on $\partial\Omega$. For example, if $h(x,y) = x^2 + y^2$, then points on $\partial\Omega$ that are far from the origin will be hotter than points that are close to the origin.

The Neumann problem for Laplace's equation is a bit more difficult to state, because Neumann conditions describe heat *flux* across the boundary, not the actual temperature distribution. In two dimensions, suppose that $(x_0, y_0)$ is a point on the boundary $\partial\Omega$ and let $\mathbf{n}$ denote the outward unit normal vector at that point. Recall that the *normal derivative* of $u$ at $(x_0, y_0)$ is defined as

$$\frac{\partial u}{\partial \mathbf{n}} = \nabla u(x_0, y_0) \bullet \mathbf{n}.$$

(This definition is easily extended to three or more dimensions.) The Neumann problem for Laplace's equation is formulated by specifying the normal derivative of $u$ at all points of the boundary; i.e.,

$$\begin{aligned} \Delta u(x,y) &= 0 && \text{inside } \Omega \\ \frac{\partial u}{\partial \mathbf{n}} &= h(x,y) && \text{on the boundary, } \partial\Omega \end{aligned}$$

in two dimensions, or

$$\begin{aligned} \Delta u(x,y,z) &= 0 && \text{inside } \Omega \\ \frac{\partial u}{\partial \mathbf{n}} &= h(x,y,z) && \text{on the boundary, } \partial\Omega \end{aligned}$$

in three dimensions. In this case, the function $h$ describes the net outward heat flux at all points on the boundary. If the boundary is insulated to prevent entry/exit of heat across the boundary $\partial\Omega$, then $h = 0$.

---

## 13.2. Well-posedness and the Maximum Principle

Assuming that there actually exists a solution of the Dirichlet problem for Laplace's equation, we will prove that the solution is unique and is not sensitive to small changes in the boundary condition. The proof makes use of the Maximum Principle for Laplace's equation, which we now state:

**Theorem 13.2.1** (Maximum Principle for Laplace's Equation)**.** Let $\Omega$ be a set that is bounded, open, and connected in either two or three dimensions, and let $u$ be a function is harmonic inside $\Omega$ and continuous on $\Omega \cup \partial\Omega$. If $u$ attains its maximum or minimum value inside $\Omega$, then $u$ is a constant function.

In other words, if $u$ is a non-constant solution of Laplace's equation, then the only place $u$ can attain its maximum and minimum values is on $\partial\Omega$, the boundary of the domain. The proof is similar to that of the Maximum Principle for the heat equation. The idea is to consider what would happen if $u$ had a maximum (or minimum) inside $\Omega$. Then the second derivative test would require that both $u_{xx} \leq 0$ and $u_{yy} \leq 0$. If either of these were strictly negative, then we would have $u_{xx} + u_{yy} < 0$, which contradicts our assumption that $u$ is a solution of Laplace's equation $u_{xx} + u_{yy} = 0$. Therefore, we may assume without loss of generality that $u_{xx} = u_{yy} = 0$ at any maximum/minimum occurring inside $\Omega$. Explaining why this implies that $u$ would have to be constant throughout $\Omega$ requires care, and the interested reader is encouraged to consult a more comprehensive text on PDEs (such as Strauss [10]).

**Example 13.2.2.** The function $u(x,y) = x^2 - y^2 + 3$ is harmonic: $u_{xx} + u_{yy} = 0$. Suppose that we wish to find the maximum and minimum values of this function on the disc-shaped domain $x^2 + y^2 \leq 1$. By the Maximum Principle, the extreme values of $u(x,y)$ must occur on the boundary of the domain, which is the circle $x^2 + y^2 = 1$. One way to maximize $u(x,y) = x^2 - y^2 + 3$ is to use the method of Lagrange multipliers from multivariable calculus. However, in this example we can take a simpler approach: the constraint $x^2 + y^2 = 1$ tells us that $y^2 = 1 - x^2$. Substituting this into the expression for $u(x,y)$, we find that $u(x,y) = 2x^2 + 2$ for points $(x,y)$ on the boundary of our disc. Maximizing this function of a single variable is straightforward: the minimum of $2x^2 + 2$ occurs when $x = 0$. The corresponding $y$-values on the circular boundary are $y = \pm 1$. It follows that the minimum value of $u(x,y)$ is $u(0,1) = u(0,-1) = 2$. Similarly, the maximum value of $u(x,y)$ is $u(1,0) = u(-1,0) = 4$.

Now let us use the Maximum Principle to prove that the Dirichlet problem for Poisson's equation

$$\begin{aligned} \Delta u &= f \qquad \text{inside } \Omega \\ u &= h \qquad \text{on } \partial\Omega \end{aligned}$$

has a unique solution. (Note that Laplace's equation is a special case of Poisson's equation with $f = 0$.) Suppose that $u$ and $v$ are both solutions to this Dirichlet problem and define $w = u - v$. Then by linearity of the Laplace operator,

$$\Delta w = \Delta(u - v) = \Delta u - \Delta v = f - f = 0.$$

On the boundary, we find that $w = h - h = 0$. It follows that $w$ satisfies a homogeneous Dirichlet problem for *Laplace's* equation:

$$\begin{aligned} \Delta w &= 0 && \text{inside } \Omega \\ w &= 0 && \text{on } \partial\Omega. \end{aligned}$$

By the Maximum Principle, both the maximum and minimum values of $w$ are attained on the boundary of the domain. This implies that $w = 0$ throughout $\Omega \cup \partial\Omega$, from which we conclude that $u = v$. Hence, the solution of the Dirichlet problem for Poisson's equation (and therefore Laplace's equation) is unique.

To test for stability with respect to small changes in the boundary condition, let us compare the solutions of the Dirichlet problems

$$\begin{aligned} \Delta u &= f && \text{inside } \Omega \\ u &= g && \text{on } \partial\Omega \end{aligned}$$

and

$$\begin{aligned} \Delta v &= f && \text{inside } \Omega \\ v &= h && \text{on } \partial\Omega. \end{aligned}$$

In other words, $u$ and $v$ satisfy the same PDE but different boundary conditions. As before, we measure the gap between the solutions by defining $w = u - v$. Then $w$ satisfies a Dirichlet problem for Laplace's equation:

$$\begin{aligned} \Delta w &= 0 && \text{inside } \Omega \\ w &= g - h && \text{on } \partial\Omega. \end{aligned}$$

By the Maximum Principle,

$$\max_{\Omega \cup \partial\Omega} |u - v| \;=\; \max_{\Omega \cup \partial\Omega} |w| \;=\; \max_{\partial\Omega} |w| \;=\; \max_{\partial\Omega} |g - h|.$$

In words, this says that the maximum separation between $u$ and $v$ throughout the domain $\Omega \cup \partial\Omega$ is at most as large as the maximum separation between $g$ and $h$ on the boundary. This is precisely what we need in order to prove stability: small discrepancies in boundary conditions (i.e., $g$ and $h$ are "close") cannot cause major changes in how the solutions $u$ and $v$ behave.

Now we can rest assured that the Dirichlet problem for Poisson's equation is well-posed, provided that we can actually find a solution. We now turn our attention to this task.

---

### 13.3. Translation and Rotation Invariance

Understanding a few "geometric" properties of Laplace's equation will give us intuition regarding what sorts of solutions we should seek. Take a moment to compare what follows with the approach we took when solving the Cauchy problem for the heat equation, (10.16)–(10.17).

Like the heat equation, Laplace's equation is *translation invariant* as we shall illustrate in two dimensions. If we make the change of coordinates $\xi = x + \alpha$ and $\eta = y + \beta$, this has the effect of translating (shifting) every point in the plane in the direction of the vector $(\alpha, \beta)$. Laplace's equation $u_{xx} + u_{yy} = 0$ can be written in terms of the new coordinates $(\xi, \eta)$ by using the chain rule:

$$
\begin{aligned}
u_\xi &= u_x \frac{dx}{d\xi} + u_y \frac{dy}{d\xi} &&= u_x \\
u_\eta &= u_x \frac{dx}{d\eta} + u_y \frac{dy}{d\eta} &&= u_y,
\end{aligned}
$$

and calculating the second derivatives yields $u_{xx} = u_{\xi\xi}$ and $u_{yy} = u_{\eta\eta}$. Therefore $u_{\xi\xi} + u_{\eta\eta} = 0$, demonstrating that Laplace's equation is not affected by translating the coordinates.

More interestingly, Laplace's equation is also *rotation invariant*: making a change of coordinates that rotates the plane through an angle $\theta$ (see Figure 13.2) does not affect Laplace's equation. In two dimensions, such a change of coordinates takes the form

$$
\begin{bmatrix} \xi \\ \eta \end{bmatrix} = \begin{bmatrix} \cos\theta & -\sin\theta \\ \sin\theta & \cos\theta \end{bmatrix} \begin{bmatrix} x \\ y \end{bmatrix} = \begin{bmatrix} x\cos\theta - y\sin\theta \\ x\sin\theta + y\cos\theta \end{bmatrix}.
$$

Verifying that this coordinate transformation really *does* rotate the plane through an angle $\theta$ is not difficult. You should check that (i) the vectors $(x, y)$ and $(\xi, \eta)$ have the same length and (ii) the angle between these vectors is $\theta$ (the dot product will help you see this). As before, we use the chain rule to write $u_x$ and $u_y$ in

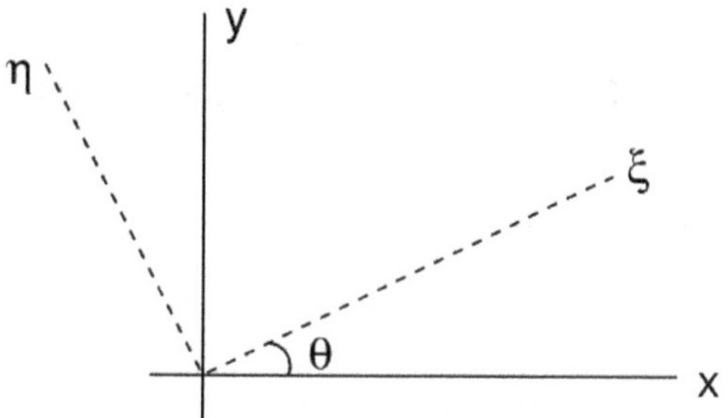


**Figure 13.2.** Laplace's equation is invariant under rotations of the coordinate system.

terms of the new coordinates:

$$\begin{aligned} u_x = u_\xi \frac{d\xi}{dx} + u_\eta \frac{d\eta}{dx} &= (\cos\theta)u_\xi + (\sin\theta)u_\eta \\ u_y = u_\xi \frac{d\xi}{dy} + u_\eta \frac{d\eta}{dy} &= (-\sin\theta)u_\xi + (\cos\theta)u_\eta. \end{aligned}$$

Calculating the second partial derivatives, we find that

$$u_{xx} + u_{yy} = (\sin^2\theta + \cos^2\theta)(u_{\xi\xi} + u_{\eta\eta}) = u_{\xi\xi} + u_{\eta\eta},$$

and it follows that rotating our coordinate system has no effect on Laplace's equation. We remark that Laplace's equation is also translation and rotation invariant in higher dimensions. Rotation invariance suggests that Laplace's equation may have solutions with radial symmetry. In two dimensions, let us seek solutions of $\Delta u(x,y) = 0$ of the form $u(x,y) = v(r)$, where $r = \sqrt{x^2+y^2}$ measures distance from the origin. Since $v$ is a function of one variable, we expect that substituting $u(x,y) = v(r)$ into Laplace's equation will reduce the PDE to an ODE. First, notice that

$$\frac{\partial r}{\partial x} = \frac{2x}{2\sqrt{x^2+y^2}} = \frac{x}{r} \quad \text{and} \quad \frac{\partial r}{\partial y} = \frac{2y}{2\sqrt{x^2+y^2}} = \frac{y}{r}.$$

Now, use the chain rule to convert $u_{xx} + u_{yy} = 0$ into an ODE for $v(r)$: the first partial derivatives of $u$ become

$$u_x = \frac{\partial}{\partial x} v(r) = v'(r)\frac{\partial r}{\partial x} = \frac{x}{r} v'(r)$$

$$u_y = \frac{\partial}{\partial y} v(r) \;=\; v'(r)\frac{\partial r}{\partial y} \;=\; \frac{y}{r} v'(r).$$

Since $r$ depends upon both $x$ and $y$, calculating the second partial derivatives requires careful use of the product/quotient and chain rules:

$$u_{xx} \;=\; \frac{\partial}{\partial x}\left[x \cdot \frac{1}{r} \cdot v'(r)\right] \;=\; \frac{1}{r}v'(r) \;+\; \left(-\frac{x}{r^2}\right)\left(\frac{x}{r}\right)v'(r) \;+\; \left(\frac{x}{r}\right)^2 v''(r),$$

or equivalently

$$u_{xx} \;=\; \left(\frac{1}{r} - \frac{x^2}{r^3}\right) v'(r) \;+\; \left(\frac{x}{r}\right)^2 v''(r).$$

Similarly,

$$u_{yy} \;=\; \left(\frac{1}{r} - \frac{y^2}{r^3}\right) v'(r) \;+\; \left(\frac{y}{r}\right)^2 v''(r).$$

Adding these two expressions,

$$0 \;=\; u_{xx} + u_{yy} \;=\; \left[\frac{2}{r} - \frac{x^2+y^2}{r^3}\right] v'(r) \;+\; \left[\frac{x^2+y^2}{r^2}\right] v''(r).$$

Finally, since $x^2 + y^2 = r^2$, we have shown that radially symmetric solutions of the two-dimensional Laplace equation would have to obey the ODE

$$v''(r) + \frac{1}{r}v'(r) \;=\; 0. \tag{13.2}$$

Equation (13.2) is a second-order, linear ODE, but does not have constant coefficients (which prevents us from solving it via the characteristic equation approach). There are several ways to solve this ODE—for example, we could reduce the order by defining $w = v'$, use separation of variables to solve the resulting first-order ODE for $w$, and then integrate $w$ to obtain $v$. Another approach is as follows: use algebra to rewrite (13.2) as

$$\frac{v''(r)}{v'(r)} \;=\; -\frac{1}{r},$$

and observe that

$$\frac{v''(r)}{v'(r)} \;=\; \frac{\mathrm{d}}{\mathrm{d}r}\ln[v'(r)].$$

The ODE becomes

$$\frac{\mathrm{d}}{\mathrm{d}r}\ln[v'(r)] \;=\; -\frac{1}{r},$$

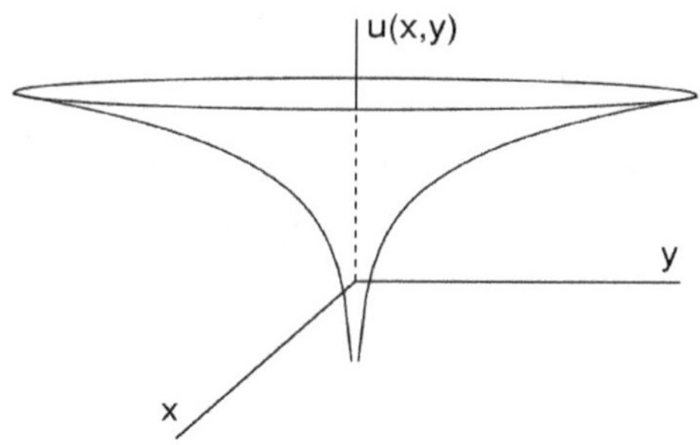


**Figure 13.3.** Graph of a radially symmetric solution (13.3) Laplace's equation $u_{xx} + u_{yy} = 0$.

and integrating both sides yields

$$\ln[v'(r)] = -\ln(r) + C = \ln\left(\frac{1}{r}\right) + C,$$

where $C$ is a constant of integration. Exponentiating both sides,

$$v'(r) = e^{\ln(1/r)+C} = \frac{A}{r},$$

where $A = e^C$ is a constant. Finally, integrating a second time reveals that the radially symmetric solutions of the two-dimensional Laplace equation are given by

$$v(r) = A\ln r + B, \tag{13.3}$$

where $A$ and $B$ are constants and $r > 0$. Equivalently, we have shown that

$$u(x,y) = A\ln\sqrt{x^2+y^2} + B$$

are harmonic functions in the plane. A sketch of one such function is given in Figure 13.3. The special case $A = -\frac{1}{2\pi}$ and $B = 0$ is sometimes called the *fundamental solution* of the Laplace equation in two dimensions. In three dimensions, the fundamental solution of Laplace's equation $\Delta u(x,y,z) = 0$ has a considerably different form. As before, let us seek solutions $u(x,y,z) = v(r)$, where $r = \sqrt{x^2+y^2+z^2}$. Recycling our previous calculations,

$$u_x = \frac{x}{r}v'(r) \quad u_y = \frac{y}{r}v'(r) \quad \text{and} \quad u_z = \frac{z}{r}v'(r),$$

and

$$u_{xx} = \left(\frac{1}{r} - \frac{x^2}{r^3}\right) v'(r) + \left(\frac{x}{r}\right)^2 v''(r)$$
$$u_{yy} = \left(\frac{1}{r} - \frac{y^2}{r^3}\right) v'(r) + \left(\frac{y}{r}\right)^2 v''(r)$$
$$u_{zz} = \left(\frac{1}{r} - \frac{z^2}{r^3}\right) v'(r) + \left(\frac{z}{r}\right)^2 v''(r).$$

Adding these yields

$$0 = u_{xx} + u_{yy} + u_{zz} = \left[\frac{3}{r} - \frac{x^2 + y^2 + z^2}{r^3}\right] v'(r) + \left[\frac{x^2 + y^2 + z^2}{r^2}\right] v''(r),$$

and since $x^2 + y^2 + z^2 = r^2$, we obtain the ODE

$$v''(r) + \frac{2}{r} v'(r) = 0. \tag{13.4}$$

Despite the apparent similarity between Equations (13.2) and (13.4), the solutions are quite different. Fortunately, the solution technique is identical: first, write (13.4) as

$$\frac{d}{dr} \ln[v'(r)] = -\frac{2}{r}.$$

Integrating both sides with respect to $r$,

$$\ln[v'(r)] = -2 \ln r + C = \ln\left(\frac{1}{r^2}\right) + C,$$

where $C$ is a constant of integration. Exponentiating both sides,

$$v'(r) = e^{\ln(1/r^2) + C} = \frac{A}{r^2},$$

where $A = e^C$ is a constant. Finally, integrating a second time reveals that the radially symmetric solutions of the three-dimensional Laplace equation are given by

$$v(r) = -\frac{A}{r} + B, \tag{13.5}$$

where $A$ and $B$ are constants and $r > 0$. Equivalently, we have shown that

$$u(x,y,z) = -\frac{A}{\sqrt{x^2+y^2+z^2}} + B \qquad (x,y,z) \neq (0,0,0)$$

are harmonic functions in three-dimensional space. The special case $A = -1$ and $B = 0$ is sometimes called the *fundamental solution* of the Laplace equation in three dimensions.

**Example 13.3.1.** Solve Laplace's equation on the annular (ring-shaped) domain $1 < r < e$, with Dirichlet boundary conditions $u = 8$ when $r = 1$ and $u = 12$ when $r = e$. *Solution:* Due to the symmetry of the domain, it is natural to work from (13.3) instead of using Cartesian coordinates. The two boundary conditions tell us that

$$8 = A\ln(1) + B = B \quad \text{and} \quad 12 = A\ln(e) + B = A + B.$$

It follows that $A = 4$ and $B = 8$, and the solution is given by $v(r) = 4\ln r + 8$. Equivalently,

$$u(x,y) = 4\ln\sqrt{x^2+y^2} + 8 = 2\ln(x^2+y^2) + 8.$$

By well-posedness of the Dirichlet problem, we know that this is the unique solution of Laplace's equation on the given domain.

In order to interpret the solution physically, you may wish to graph $u(x,y)$ over the annular domain. The boundary of the domain consists of two circles, and the Dirichlet boundary conditions tell us that the temperature at the "inner boundary" is held constant at 8 while the temperature at the outer boundary is held constant at 12. If we travel radially outward within the domain (from the inner boundary toward the outer boundary) the temperature increases *logarithmically*. This may seem counter-intuitive, since one might guess that temperature would increase linearly from 8 to 12 as we move radially outward.

**Example 13.3.2.** When we sought radial solutions $u(x,y) = v(r)$ of the Laplace equation in two dimensions, we calculated that

$$u_{xx} + u_{yy} = v''(r) + \frac{1}{r}v'(r) \qquad (r \neq 0).$$

Multiplying both sides by $r$, notice that

$$r(u_{xx} + u_{yy}) = rv''(r) + v'(r) = \frac{d}{dr}\left[rv'(r)\right].$$

This observation can be useful when solving the Laplace (or Poisson) equations on annular domains. For example, let us solve the Poisson equation $u_{xx} + u_{yy} = 1$ on the domain $3 < r < 5$, with Dirichlet boundary conditions $u = 10$ on the circle $r = 3$ and $u = 0$ on the circle $r = 5$. Multiplying both sides of Poisson's equation by $r$, we have

$$r(u_{xx} + u_{yy}) = r$$

and, from the above remarks, we conclude that

$$\frac{d}{dr}\left[rv'(r)\right] = r.$$

Integrating both sides with respect to $r$,

$$rv'(r) = \frac{r^2}{2} + A,$$

where $A$ is a constant of integration. Dividing both sides by $r$ and integrating a second time,

$$u(x,y) = v(r) = \frac{r^2}{4} + A\ln r + B.$$

**Remark.** Since Poisson's equation is nothing more than an inhomogeneous version of Laplace's equation, it is not surprising that this formula for $v(r)$ contains the terms $A\ln r + B$, which we recognize from formula (13.3).

Finally, we will use the two Dirichlet conditions to solve for the constants $A$ and $B$:

$$10 = v(3) = \frac{9}{4} + A\ln 3 + B$$
$$0 = v(5) = \frac{25}{4} + A\ln 5 + B.$$

This is a system of two equations in two unknowns:

$$A\ln 3 + B = \frac{31}{4}$$

$$A \ln 5 + B = -\frac{25}{4},$$

which has solution

$$A = \frac{14}{\ln(3/5)} \quad \text{and} \quad B = \frac{31}{4} - \frac{14 \ln 3}{\ln(3/5)}.$$

Inserting these into the above formula for $u(x, y)$, the solution of our Dirichlet problem is given by

$$u(x, y) = \frac{x^2 + y^2}{4} + \frac{14 \ln \sqrt{x^2 + y^2}}{\ln(3/5)} + \frac{31}{4} - \frac{14 \ln 3}{\ln(3/5)}.$$

**Example 13.3.3.** Solve the mixed Dirichlet-Neumann problem

$$\begin{aligned} \Delta u(x, y, z) &= \sqrt{x^2 + y^2 + z^2} && \text{inside } 1 < r < 2 \\ u &= 12 && \text{if } r = 1 \\ \frac{\partial u}{\partial \mathbf{n}} &= 3 && \text{if } r = 2. \end{aligned}$$

*Solution:* The domain is a spherical shell. Recall that when we sought solutions $u(x, y, z) = v(r)$ of the Laplace equation in three dimensions, we calculated that

$$\Delta u = u_{xx} + u_{yy} + u_{zz} = v''(r) + \frac{2}{r} v'(r) \qquad (r \neq 0).$$

If we multiply both sides of this equation by $r^2$, observe that

$$r^2 \Delta u = r^2 v''(r) + 2r v'(r) = \frac{\mathrm{d}}{\mathrm{d}r}[r^2 v'(r)].$$

In the present example, multiplying our Poisson equation $\Delta u = r$ by $r^2$ yields

$$\frac{\mathrm{d}}{\mathrm{d}r}[r^2 v'(r)] = r^3.$$

Integrating with respect to $r$ yields

$$r^2 v'(r) = \frac{r^4}{4} + A,$$

where $A$ is a constant of integration. Divide by $r^2$ and integrate a second time to get

$$v(r) = \frac{r^3}{12} - \frac{A}{r} + B.$$

We recognize the latter two terms from the formula (13.5) for radial solutions of Laplace's equation in three dimensions. (Again, this is not surprising since Poisson's equation is the inhomogeneous Laplace equation.)

The Dirichlet condition on the inner boundary ($r = 1$) tells us that $v(1) = 12$. The Neumann boundary condition specifies the net outward flux of heat at the outer boundary $r = 2$, and the vector $\mathbf{n}$ points radially outward in the direction of increasing $r$. In terms of $v$, the Neumann boundary condition becomes $v'(2) = 3$. Solving for $A$ and $B$, you will find that $A = 8$ and $B = 239/12$. In terms of the original coordinates, the overall solution is

$$u(x,y,z) = \frac{(x^2 + y^2 + z^2)^{3/2}}{12} - \frac{8}{\sqrt{x^2 + y^2 + z^2}} + \frac{239}{12}.$$

**Remark.** If a Neumann condition had been given at the *inner* boundary, we must exercise caution when writing the boundary condition in terms of our variable $v(r)$. Suppose, for example, that our boundary condition on the inner shell $r = 1$ had been the Neumann condition $\partial u/\partial \mathbf{n} = 12$. On the shell $r = 1$, the outward normal vector $\mathbf{n}$ is directed *towards the origin in the direction of decreasing* $r$. Hence, in terms of $v$, the boundary condition would be $v'(1) = -12$, not $v'(1) = 12$.

---

## 13.4. Laplace's Equation on Bounded Domains

In the previous section, we found all radial solutions of Laplace's equation on *unbounded* domains: two and three-dimensional Euclidean space. Then, we used our results to solve a few boundary value problems on domains with radial symmetry (annuli and spherical shells). Finding explicit solutions of Laplace's equation on general bounded domains is usually too much to expect. In this section, we will solve Dirichlet problems for Laplace's equation on two very special two-dimensional domains: rectangles and discs.

**13.4.1 Dirichlet problem on a rectangle.** Let $\Omega$ denote the open rectangular domain $0 < x < a$ and $0 < y < b$ in the plane. Our goal is to solve the Dirichlet

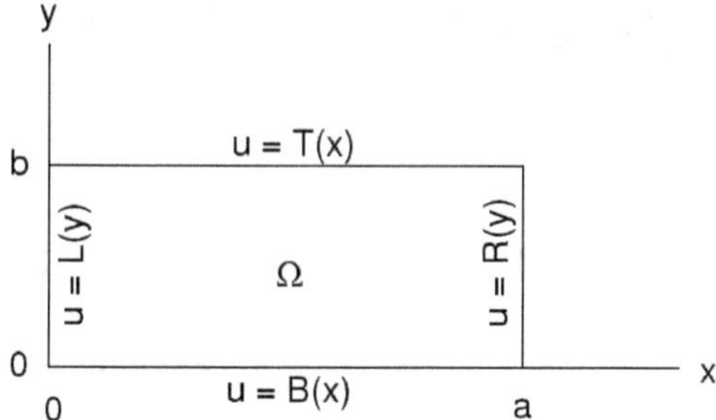


**Figure 13.4.** Illustration of the rectangular domain Ω with Dirichlet boundary conditions specified on each edge.

problem

$$\begin{aligned} \Delta u(x,y) &= 0 && \text{inside } \Omega \\ u(x,0) &= B(x) && 0 < x < a \\ u(x,b) &= T(x) && 0 < x < a \\ u(0,y) &= L(y) && 0 < y < b \\ u(a,y) &= R(y) && 0 < y < b. \end{aligned} \tag{13.6}$$

The functions $B(x)$, $T(x)$, $L(y)$, and $R(y)$ specify temperature distributions on the (B)ottom, (T)op, (L)eft, and (R)ight edges of the rectangle, respectively. Figure 13.4 illustrates this domain as well as the given boundary conditions. We can solve the system (13.6) by splitting this problem into four simpler Dirichlet problems in which three of the four edges are given homogeneous Dirichlet conditions. More precisely, suppose that $u(x,y)$ denotes the solution of the above Dirichlet problem with boundary conditions $(B,T,L,R)$. Let $u_B(x,y)$ denote the solution of the Dirichlet problem with boundary conditions $(B,0,0,0)$—i.e., let $T(x) = L(y) = R(y) = 0$ in (13.6) but leave the condition on the bottom edge of the rectangle unchanged. Similarly, let $u_T$, $u_L$ and $u_R$ denote the solution of the Dirichlet problem with boundary conditions $(0,T,0,0)$, $(0,0,L,0)$, and $(0,0,0,R)$, respectively. Then the solution of the general Dirichlet problem (13.6) is obtained by summing the solutions of the four special Dirichlet problems: $u = u_B + u_T + u_L + u_R$.

Let us solve one of these four simpler Dirichlet problems, in which only the top edge of the rectangle receives a non-homogeneous condition:

$$\begin{aligned} \Delta u(x,y) &= 0 && \text{inside } \Omega \\ u(x,0) &= 0 && 0 < x < a \\ u(x,b) &= T(x) && 0 < x < a \\ u(0,y) &= 0 && 0 < y < b \\ u(a,y) &= 0 && 0 < y < b. \end{aligned} \tag{13.7}$$

We will solve (13.7) via separation of variables—suppose that $u(x,y) = X(x)Y(y)$. Laplace's equation becomes $X''Y + XY'' = 0$ and, by algebra,

$$\frac{X''}{X} = -\frac{Y''}{Y}.$$

Since the left-hand side is a function of $x$ only whereas the right-hand side is a function of $y$ only, the only way this can happen is if both expressions are equal to a common constant, say $-\lambda$. The result is a system of two ODEs: $X'' + \lambda X = 0$ and $Y'' - \lambda Y = 0$. We will solve the $X$ equation first, because its boundary conditions are easy to incorporate. Since $u(x,y) = X(x)Y(y)$, the boundary conditions at the left and right edges of the rectangle tell us that

$$0 = u(0,y) = X(0)Y(y) \quad \text{and} \quad 0 = u(a,y) = X(a)Y(y).$$

We exclude the possibility that $Y(y) = 0$, since that would imply that $u(x,y) = 0$, which would not satisfy the boundary condition on the top edge of the rectangle. Therefore, the boundary conditions for our $X$ equation are $X(0) = 0 = X(a)$. We have encountered the two-point boundary value problem

$$X'' + \lambda X = 0 \qquad X(0) = 0 = X(a)$$

on numerous occasions. For $\lambda \leq 0$, this problem has no non-zero solutions, implying that all eigenvalues are positive. The eigenvalues are $\lambda_n = (n\pi/a)^2$ for $n \geq 1$, and corresponding eigenfunctions are

$$X_n(x) = \sin\left(\frac{n\pi x}{a}\right) \qquad (n \geq 1).$$

With these eigenvalues in mind, we turn our attention to the differential equation for $Y(y)$, namely $Y'' - \lambda Y = 0$. Let $Y_n(y)$ denote the solution of this equation corresponding to the eigenvalue $\lambda = \lambda_n$. The characteristic equation is $m^2 - (n\pi/a)^2 = 0$, which has distinct, real roots $\pm n\pi/a$. Therefore,

$$Y_n(y) = E_n e^{n\pi y/a} + F_n e^{-n\pi y/a},$$

where $E_n$ and $F_n$ are constants. The boundary condition on the lower edge of the rectangular domain tells us that $Y_n(0) = 0$, but it is less clear how to incorporate the boundary condition on the top edge of the rectangle. Ignoring the upper boundary for the moment, the condition $Y_n(0) = 0$ indicates that $E_n + F_n = 0$, from which

$$Y_n(y) = E_n \left[ e^{n\pi y/a} - e^{-n\pi y/a} \right] \qquad (n \geq 1).$$

Each of the functions

$$u_n(x,y) = X_n(x)Y_n(y) = E_n \left[ e^{n\pi y/a} - e^{-n\pi y/a} \right] \sin\left(\frac{n\pi x}{a}\right)$$

satisfies Laplace's equation as well as the Dirichlet boundary conditions on every edge of the rectangle except for the top one. Taking an infinite linear combination of these solutions, we find that

$$u(x,y) = \sum_{n=1}^{\infty} G_n u_n(x,y) = \sum_{n=1}^{\infty} E_n G_n \left[ e^{n\pi y/a} - e^{-n\pi y/a} \right] \sin\left(\frac{n\pi x}{a}\right)$$

is the general solution of the Laplace equation subject to the homogeneous Dirichlet conditions on the left, right, and bottom edges of the rectangular domain. The constants can be combined by introducing $A_n = E_n G_n$ so that our general solution takes the form

$$u(x,y) = \sum_{n=1}^{\infty} A_n \left[ e^{n\pi y/a} - e^{-n\pi y/a} \right] \sin\left(\frac{n\pi x}{a}\right).$$

It remains to incorporate the boundary condition on the upper edge of the rectangle, $u(x,b) = T(x)$. Setting $y = b$ in our series solution, we find that

$$T(x) = \sum_{n=1}^{\infty} A_n \left[ e^{n\pi b/a} - e^{-n\pi b/a} \right] \sin\left(\frac{n\pi x}{a}\right).$$

Because we have fixed $y = b$, notice that the two exponential terms are now independent of the two variables $x$ and $y$. Hence, we may absorb these terms into our coefficients $A_n$ by introducing

$$\tilde{A}_n = A_n \left[ e^{n\pi b/a} - e^{-n\pi b/a} \right].$$

In doing so, the function $T(x)$ takes the form of a Fourier sine series

$$T(x) = \sum_{n=1}^{\infty} \tilde{A}_n \sin\left(\frac{n\pi x}{a}\right).$$

The Fourier coefficients are given by

$$\tilde{A}_n = \frac{2}{a} \int_0^a T(x) \sin\left(\frac{n\pi x}{a}\right) \, dx$$

and, consequently,

$$A_n = \frac{2}{a} \left[ e^{n\pi b/a} - e^{-n\pi b/a} \right]^{-1} \int_0^a T(x) \sin\left(\frac{n\pi x}{a}\right) \, dx.$$

Inserting these Fourier coefficients into our general formula for $u(x,y)$, we have shown that the solution of (13.7) is given by

$$u(x,y) = \frac{2}{a} \sum_{n=1}^{\infty} \frac{e^{n\pi y/a} - e^{-n\pi y/a}}{e^{n\pi b/a} - e^{-n\pi b/a}} \left[ \int_0^a T(x) \sin\left(\frac{n\pi x}{a}\right) \, dx \right] \sin\left(\frac{n\pi x}{a}\right). \tag{13.8}$$

Formula (13.8) may seem complicated, but it is not too difficult to interpret if we make a few observations about the various terms/factors in the series. Notice that as $y$ increases from 0 to $b$, the factor

$$\frac{e^{n\pi y/a} - e^{-n\pi y/a}}{e^{n\pi b/a} - e^{-n\pi b/a}}$$

increases from 0 to 1. On the other hand, the other factors

$$\frac{2}{a} \left[ \int_0^a T(x) \sin\left(\frac{n\pi x}{a}\right) \, dx \right] \sin\left(\frac{n\pi x}{a}\right)$$

are nothing more than terms in a Fourier sine series representation for $T(x)$. Combining these two observations, the effect of the exponential factors should be to "damp" the function $T(x)$ as $y$ decreases from $b$ towards 0. This is illustrated

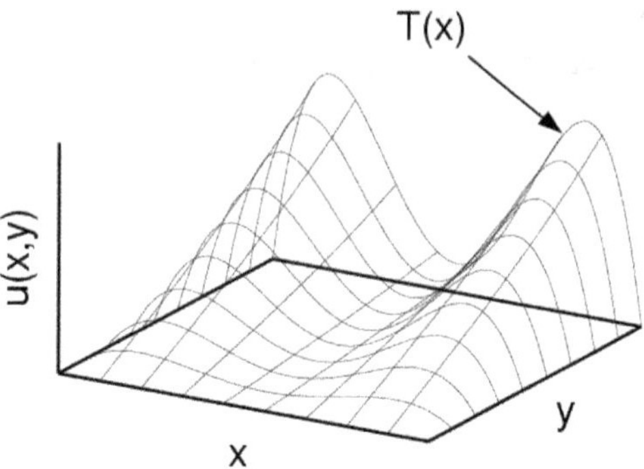


**Figure 13.5.** Solution of (13.7) for a particular choice of $T(x)$.

in Figure (13.5) for a particular choice of temperature distribution $T(x)$ applied at the top edge of the rectangle. Recall that (13.8) is merely the solution of *one out of four* Dirichlet problems that we must solve to build the solution of the general Dirichlet problem (13.6). Fortunately, we may recycle many of our calculations to solve the other Dirichlet problems. For example, suppose that we wish to solve the Dirichlet problem

$$\begin{aligned} \Delta u(x,y) &= 0 && \text{inside } \Omega \\ u(x,0) &= 0 && 0 < x < a \\ u(x,b) &= 0 && 0 < x < a \\ u(0,y) &= 0 && 0 < y < b \\ u(a,y) &= R(y) && 0 < y < b, \end{aligned} \tag{13.9}$$

in which only the right edge of the rectangle receives a non-homogeneous boundary condition. This is essentially identical to the problem (13.7), except that the roles of $x$ and $y$ are reversed and $R(y)$ now plays the role of $T(x)$. Retracing our steps, it is convenient to separate the variables as

$$-\frac{X''}{X} = \frac{Y''}{Y} = -\lambda,$$

and solve the $Y$ equation first because its boundary conditions will be homogeneous: $Y(0) = 0 = Y(b)$. Afterward, solving (13.9) is literally a matter of quoting formula (13.8) with (i) the variables $x$ and $y$ swapped; (ii) the constants $a$ and $b$ swapped; and (iii) the Fourier sine series for $T(x)$ replaced with the Fourier sine series for $R(y)$.

**Remark.** The same solution procedure works for other types of boundary conditions as well. We could impose Neumann boundary conditions along each edge of the rectangular domain, or we could mix and match Dirichlet and Neumann conditions along the different edges of the rectangle.

We can also use the separation of variables technique to solve Laplace's equation on a box-shaped domain in three dimensions. For example, let $\Omega$ denote the domain $0 < x < a$, $0 < y < b$, and $0 < z < c$, and consider the Dirichlet problem

$$\begin{aligned} \Delta u(x,y,z) &= 0 && \text{inside } \Omega \\ u(x,y,c) &= T(x,y) && \text{top face of } \partial\Omega \\ u &= 0 && \text{on the rest of } \partial\Omega. \end{aligned}$$

As usual, we seek separated solutions of the form $u(x,y,z) = X(x)Y(y)Z(z)$ so that Laplace's equation becomes $X''YZ + XY''Z + XYZ'' = 0$. Dividing by $XYZ$, we have

$$\frac{X''}{X} + \frac{Y''}{Y} + \frac{Z''}{Z} = 0.$$

The three ratios appearing in this sum must equal a common constant $-\lambda$ (explain). In particular, the $X$ equation satisfies the eigenvalue problem $X'' + \lambda X = 0$ with $X(0) = 0 = X(a)$. The eigenvalues and corresponding eigenfunctions are given by

$$\lambda_n = \left(\frac{n\pi}{a}\right)^2 \quad \text{and} \quad X_n(x) = \sin\left(\frac{n\pi x}{a}\right), \qquad (n \geq 1).$$

The $Y$ equation satisfies essentially the same eigenvalue problem, leading us to define

$$Y_m(y) = \sin\left(\frac{n\pi y}{b}\right), \qquad (m \geq 1).$$

Rewriting the $Z$ equation as

$$\frac{Z''}{Z} = -\frac{X''}{X} - \frac{Y''}{Y},$$

we seek solutions in which $X = X_n$ and $Y = Y_m$. Since

$$\frac{X_n''}{X_n} = -\frac{(n\pi/a)^2 \sin(n\pi x/a)}{\sin(n\pi x/a)} = -\left(\frac{n\pi}{a}\right)^2$$

and

$$\frac{Y_m''}{Y_m} = -\frac{(m\pi/a)^2 \sin(m\pi x/a)}{\sin(m\pi x/a)} = -\left(\frac{m\pi}{a}\right)^2,$$

we must solve

$$Z_{m,n}'' = \left[\left(\frac{n\pi}{a}\right)^2 + \left(\frac{m\pi}{a}\right)^2\right] Z_{m,n} \qquad (m, n \geq 1).$$

One of the boundary conditions on $Z$ is $Z(0) = 0$, which is easy to implement. However, the key observation regarding this ODE for $Z$ is that it is *doubly indexed*. When we build the general solution of this Dirichlet problem, we must sum over both $m$ and $n$, leading to a *double sum*. Ultimately, we would need to develop a theory of Fourier series in *several* variables and, although this is a straightforward extension of our efforts in the previous chapter, we will not deal with such series here.

**13.4.2 Dirichlet problem on a disc.** The fundamental solution of Laplace's equation in two dimensions has radial symmetry throughout the plane, so we may suspect that $\Delta u(x,y) = 0$ has "nice" solutions on (bounded) disc-shaped domains centered at the origin. In this section, we will solve a classic problem in PDEs: Find all bounded solutions of the Dirichlet problem

$$\begin{aligned} u_{xx} + u_{yy} &= 0 && \text{if } x^2 + y^2 < a^2 \\ u &= h(\theta) && \text{if } x^2 + y^2 = a^2. \end{aligned} \tag{13.10}$$

Here, $a > 0$ denotes the radius of the disc-shaped domain, and $\theta$ denotes an angle used to parametrize the boundary $(0 \leq \theta < 2\pi)$. The function $h(\theta)$ specifies a temperature distribution applied along the boundary of the disc. Problem (13.10) is famous enough in the study of PDEs that it is simply referred to as *the Dirichlet problem on a disc*. (When mathematicians mention this problem, it is understood that the underlying PDE is Laplace's equation.)

We will solve (13.10) by first writing Laplace's equation in polar coordinates $(r, \theta)$, and then applying the separation of variables technique. Let $x = r\cos\theta$

and $y = r\sin\theta$ and note that $x^2 + y^2 = r^2$ and $\theta = \arctan(y/x)$. By implicit differentiation, we calculate

$$\frac{\partial r}{\partial x} = \frac{x}{r}, \quad \frac{\partial r}{\partial y} = \frac{y}{r}, \quad \frac{\partial \theta}{\partial x} = -\frac{y}{r^2}, \quad \text{and} \quad \frac{\partial \theta}{\partial y} = \frac{x}{r^2}.$$

Now by the chain rule,

$$u_x = \frac{\partial u}{\partial r}\frac{\partial r}{\partial x} + \frac{\partial u}{\partial \theta}\frac{\partial \theta}{\partial x} = \frac{x}{r}u_r - \frac{y}{r^2}u_\theta$$

and

$$u_y = \frac{\partial u}{\partial r}\frac{\partial r}{\partial y} + \frac{\partial u}{\partial \theta}\frac{\partial \theta}{\partial y} = \frac{y}{r}u_r + \frac{x}{r^2}u_\theta.$$

Tedious use of the chain rule allows us to express the second partial derivatives $u_{xx}$ and $u_{yy}$ in terms of $u_{rr}$ and $u_{\theta\theta}$, and it turns out that

$$u_{xx} + u_{yy} = u_{rr} + \frac{1}{r}u_r + \frac{1}{r^2}u_{\theta\theta}.$$

Now that we know how to write the Laplacian in polar coordinates, the Dirichlet problem for a disc can be expressed as

$$\begin{aligned} u_{rr} + \frac{1}{r}u_r + \frac{1}{r^2}u_{\theta\theta} = 0 & \qquad \text{if } r < a \\ u = h(\theta) & \qquad \text{if } r = a. \end{aligned} \tag{13.11}$$

Our goal is to seek *bounded* solutions of (13.11). The boundedness will become important later when we attempt to solve a second-order ODE in the variable $r$ despite having only one boundary condition.

Equation (13.11) can be solved by separation of variables: let $u(r,\theta) = R(r)\Theta(\theta)$. Laplace's equation becomes

$$R''\Theta + \frac{1}{r}R'\Theta + \frac{1}{r^2}R\Theta'' = 0$$

which, after dividing by $R\Theta$ and multiplying by $r^2$, yields

$$\frac{r^2R'' + rR'}{R} + \frac{\Theta''}{\Theta} = 0.$$

With the variables thus separated, we conclude that

$$\frac{\Theta''}{\Theta} = -\frac{r^2R'' + rR'}{R} = -\lambda,$$

a constant. The result is a system of two second-order ODEs

$$\Theta'' + \lambda\Theta = 0$$
$$r^2R'' + rR' - \lambda R = 0.$$

At first, it seems unclear how to set up and solve an eigenvalue problem for $\Theta$. However, if we recall that $\Theta$ represents an angular variable, we may enforce a periodic boundary condition

$$\Theta(\theta + 2\pi) = \Theta(\theta) \qquad \text{for all choices of } \theta.$$

It is easy to check that if $\lambda < 0$, then $\Theta'' + \lambda\Theta = 0$ has no non-trivial solutions satisfying this periodic boundary condition. On the other hand, $\lambda = 0$ *is* an eigenvalue. In that case, the ODE reduces to $\Theta'' = 0$, implying that $\Theta(\theta) = C\theta + D$ where $C$ and $D$ are constants. The boundary condition $\Theta(\theta + 2\pi) = \Theta(\theta)$ implies that $C = 0$, but there are no restrictions on $D$. As for the positive eigenvalues, let $\lambda = \beta^2$ where $\beta > 0$. The general solution of the ODE for $\Theta$ is

$$\Theta(\theta) = C\cos(\beta\theta) + D\sin(\beta\theta),$$

where $C$ and $D$ are constants. The only way that the boundary condition $\Theta(\theta + 2\pi) = \Theta(\theta)$ can be satisfied for all choices of $\theta$ is if both

$$\cos(\beta\theta + 2\pi\beta) = \cos(\beta\theta) \quad \text{and} \quad \sin(\beta\theta + 2\pi\beta) = \sin(\beta\theta) \qquad \text{for all } \theta.$$

Thus, $2\pi\beta$ must be a positive[1] integer multiple of $2\pi$, implying that $\beta$ can be any positive integer. Defining $\beta_n = n$, the eigenvalues for the $\Theta$ equation are $\lambda_n = \beta_n^2 = n^2$ for $n \geq 0$. The eigenfunctions corresponding to these eigenvalues have the form

$$\Theta_0(\theta) = C_0$$

[1] Recall that $\beta$ was assumed positive.

and

$$\Theta_n(\theta) = C_n \cos(n\theta) + D_n \sin(n\theta) \qquad (n \geq 1)$$

where $C_n$ and $D_n$ are constants.

Now we turn our attention to the $R$ equation: corresponding to the eigenvalue $\lambda = 0$ we obtain the ODE $r^2R'' + rR' = 0$. By algebra,

$$\frac{R''}{R'} = -\frac{1}{r},$$

which we rewrite as

$$\frac{d}{dr}[\ln R'] = -\frac{1}{r}.$$

Integrating with respect to $r$,

$$\ln R' = -\ln r + C,$$

where $C$ is a constant. Exponentiate both sides and integrate a second time to obtain

$$R_0(r) = E_0 \ln r + F_0,$$

where $E_0 = e^C$ and $F_0$ are constants, and the subscript 0 was introduced to emphasize that this solution corresponds to the eigenvalue $\lambda = 0$.

For the positive eigenvalues $\lambda_n = n^2$, the ODE for $R$ takes the form

$$r^2R'' + rR' - n^2R = 0. \qquad (13.12)$$

Although this is a linear, second-order ODE, the variable coefficients prevent us from using a characteristic equation to solve it. Luckily, this is an example of an *Euler equation*: a special class of linear differential equations which can be solved analytically by seeking *power function* solutions $R(r) = r^m$ as opposed to exponential solutions $R(r) = e^{mr}$. For a more careful discussion of how to solve Euler-type ODEs, see the Appendix at the end of this section.

In light of these remarks, let us seek solutions of (13.12) of the form $R(r) = r^\alpha$, where $\alpha$ is a real number. The equation becomes

$$r^2\alpha(\alpha-1)r^{\alpha-2} + r\alpha r^{\alpha-1} - n^2r^\alpha = 0,$$

and dividing by $r^\alpha$ yields

$$\alpha(\alpha-1)+\alpha-n^2 = 0,$$

By algebra, $a^2-n^2=0$, which has solutions $\alpha=\pm n$. It follows that $r^{-n}$ and $r^n$ are solutions of (13.12). Moreover, since these are linearly independent functions of $r$ and are solutions to a linear, homogeneous ODE, the general solution of (13.12) is

$$R_n(r) = E_n r^{-n}+F_n r^n \qquad (n\geq 1),$$

where $E_n$ and $F_n$ are constants. Again, the reason for introducing the subscript $n$ is to emphasize the correspondence between the functions and the eigenvalues $\lambda_n$.

Now recall that we are seeking *bounded* solutions of the Dirichlet problem (13.11). Since the functions $\ln r$ and $r^{-n}$ blow up as $r\to 0$, we may exclude these terms in the solutions of the $R$ equation. In other words,

$$R_n(r) = F_n r^n \qquad (n\geq 0).$$

The next step is to form the general solution of (13.11) by assembling the solutions of the separated equations for $R$ and $\Theta$. Define

$$u_n(r,\theta) = R_n(r)\Theta_n(\theta) \qquad (n\geq 0)$$

and form

$$u(r,\theta) = \sum_{n=0}^{\infty} G_n u_n(r,\theta) = G_0F_0C_0 + \sum_{n=1}^{\infty} G_nF_nr^n\left[C_n\cos(n\theta)+D_n\sin(n\theta)\right].$$

Anticipating the use of a Fourier series representation, abbreviate the various combinations of constants as

$$\frac{A_0}{2} = G_0F_0C_0 \qquad A_n = G_nF_nC_n \quad \text{and} \quad B_n = G_nF_nD_n \quad (n\geq 1).$$

The general solution of (13.11) is

$$u(r,\theta) = \frac{A_0}{2} + \sum_{n=1}^{\infty} r^n\left[A_n\cos(n\theta)+B_n\sin(n\theta)\right]. \qquad (13.13)$$

The last step is to enforce the boundary condition $u(a, \theta) = h(\theta)$. Setting $r = a$ in (13.13),

$$h(\theta) = \frac{A_0}{2} + \sum_{n=1}^{\infty} a^n \left[A_n \cos(n\theta) + B_n \sin(n\theta)\right],$$

a (full) Fourier series representation of $h(\theta)$. Since $h(\theta)$ has period $2\pi$, we can obtain the Fourier coefficients $A_n$ and $B_n$ by integrating over *any* interval of length $2\pi$. It is convenient to choose the integral $[0, 2\pi]$ (as opposed to $[-\pi, \pi]$) and to define $\tilde{A}_n = a^n A_n$ and $\tilde{B}_n = a^n B_n$. Then

$$\tilde{A}_n = \frac{1}{\pi} \int_0^{2\pi} h(\psi) \cos(n\psi)\, \mathrm{d}\psi \qquad (n \geq 0)$$

and

$$\tilde{B}_n = \frac{1}{\pi} \int_0^{2\pi} h(\psi) \sin(n\psi)\, \mathrm{d}\psi \qquad (n \geq 1).$$

Inserting these Fourier coefficients into formula (13.13), we have shown that the bounded solution of the Dirichlet problem (13.11) is

$$\begin{aligned} u(r, \theta) &= \frac{1}{2\pi} \int_0^{2\pi} h(\psi)\, \mathrm{d}\psi \\ &\quad + \frac{1}{\pi} \sum_{n=1}^{\infty} \frac{r^n}{a^n} \int_0^{2\pi} h(\psi) \left[\cos(n\psi) \cos(n\theta) + \sin(n\psi) \sin(n\theta)\right]\, \mathrm{d}\psi. \end{aligned}$$

By the double-angle identities, the trigonometric terms in the integrand collapse into a simpler-looking expression:

$$\begin{aligned} u(r, \theta) &= \frac{1}{2\pi} \int_0^{2\pi} h(\psi)\, \mathrm{d}\psi + \frac{1}{\pi} \sum_{n=1}^{\infty} \frac{r^n}{a^n} \int_0^{2\pi} h(\psi) \cos[n(\theta - \psi)]\, \mathrm{d}\psi \\ &= \frac{1}{2\pi} \int_0^{2\pi} h(\psi) \left\{1 + 2 \sum_{n=1}^{\infty} \left(\frac{r}{a}\right)^n \cos[n(\theta - \psi)]\right\} \mathrm{d}\psi. \end{aligned} \tag{13.14}$$

**Observation.** The infinite series appearing in the integrand of (13.14) is actually a geometric series in disguise, and can be summed *explicitly*.

To see this, it helps to write the cosine function in terms of complex exponential functions via Euler's identity

$$\cos \theta = \frac{1}{2} \left[\mathrm{e}^{\mathrm{i}\theta} + \mathrm{e}^{-\mathrm{i}\theta}\right].$$

Focusing our attention on the series that appears in Equation (13.14), applying Euler's identity to the cosine term yields

$$1 + 2\sum_{n=1}^{\infty}\left(\frac{r}{a}\right)^n \cos[n(\theta-\psi)] = 1 + \sum_{n=1}^{\infty}\left(\frac{r}{a}\right)^n \left[\mathrm{e}^{\mathrm{i}n(\theta-\psi)} + \mathrm{e}^{-\mathrm{i}n(\theta-\psi)}\right].$$

$$= 1 + \underbrace{\sum_{n=1}^{\infty}\left(\frac{r}{a}\right)^n \mathrm{e}^{\mathrm{i}n(\theta-\psi)}}_{\text{Series 1}} + \underbrace{\sum_{n=1}^{\infty}\left(\frac{r}{a}\right)^n \mathrm{e}^{-\mathrm{i}n(\theta-\psi)}}_{\text{Series 2}}.$$

In the latter expression, Series 1 and Series 2 are both *geometric*: the ratio of consecutive terms is independent of $n$. In Series 1, the common ratio of consecutive terms is

$$\frac{r}{a}\mathrm{e}^{\mathrm{i}(\theta-\psi)},$$

a complex number with modulus *less* than one since $(r/a) < 1$ and the exponential factor has modulus exactly 1. Similar remarks hold for Series 2, and since the common ratios of these series have moduli less than 1, we can sum these geometric series explicitly to get

$$1 + \underbrace{\frac{(r/a)\mathrm{e}^{\mathrm{i}(\theta-\psi)}}{1-(r/a)\mathrm{e}^{\mathrm{i}(\theta-\psi)}}}_{\text{sum of Series 1}} + \underbrace{\frac{(r/a)\mathrm{e}^{-\mathrm{i}(\theta-\psi)}}{1-(r/a)\mathrm{e}^{-\mathrm{i}(\theta-\psi)}}}_{\text{sum of Series 2}} = 1 + \frac{r\mathrm{e}^{\mathrm{i}(\theta-\psi)}}{a - r\mathrm{e}^{\mathrm{i}(\theta-\psi)}} + \frac{r\mathrm{e}^{-\mathrm{i}(\theta-\psi)}}{a - r\mathrm{e}^{-\mathrm{i}(\theta-\psi)}}.$$

Introducing a common denominator,

$$1 + \frac{r\mathrm{e}^{\mathrm{i}(\theta-\psi)}\left[a - r\mathrm{e}^{-\mathrm{i}(\theta-\psi)}\right] + r\mathrm{e}^{-\mathrm{i}(\theta-\psi)}\left[a - r\mathrm{e}^{\mathrm{i}(\theta-\psi)}\right]}{\left[a - r\mathrm{e}^{\mathrm{i}(\theta-\psi)}\right]\left[a - r\mathrm{e}^{-\mathrm{i}(\theta-\psi)}\right]}$$

$$= 1 + \frac{-r^2 + ar\left[\mathrm{e}^{\mathrm{i}(\theta-\psi)} + \mathrm{e}^{-\mathrm{i}(\theta-\psi)}\right] - r^2}{a^2 - 2ar\left[\mathrm{e}^{\mathrm{i}(\theta-\psi)} + \mathrm{e}^{-\mathrm{i}(\theta-\psi)}\right] + r^2}.$$

Finally, Euler's identity allows us to express this using real-valued functions:

$$1 + \frac{-2r^2 + 2ar\cos(\theta-\psi)}{a^2 - 2ar\cos(\theta-\psi) + r^2} = \frac{a^2 - r^2}{a^2 - 2ar\cos(\theta-\psi) + r^2}.$$

Hence, we have *evaluated* the summation that appears in formula (13.14). It follows that the solution of the Dirichlet problem on a disc (13.11) is given by

$$u(r,\theta) \;=\; \frac{a^2-r^2}{2\pi}\int_0^{2\pi}\frac{h(\psi)}{a^2-2ar\cos(\theta-\psi)+r^2}\,\mathrm{d}\psi, \qquad (13.15)$$

for $0 \le r < a$ and $0 \le \theta < 2\pi$. Formula (13.15) is a famous result in PDEs, and is called *Poisson's formula*. One reason that Poisson's formula is so elegant is that it expresses the solution of our Dirichlet problem as a single integral, as opposed to the sort of infinite series representation that we have come to expect. Compare (13.15) to the solutions of Dirichlet problems for the wave and heat equations on the [one-dimensional] domain $0 < x < L$ (see Equations (11.28) and (11.34)). For those equations, we were forced to present the solutions as infinite series, because those series could not be summed explicitly like the one in (13.14). In that respect, Poisson's formula is one of the most remarkable (and lucky) results from basic PDEs.

Unfortunately, for most choices of boundary conditions $h$, the integral (13.15) is impossible to evaluate by hand. (We encountered similar difficulty when we solved the Cauchy problem for the heat equation.)

**Example 13.4.1.** Suppose that we wish to solve the Dirichlet problem on a disc (13.11) with the boundary condition $h(\theta) = 1$; i.e., a constant temperature distribution along the circular boundary. Poisson's formula (13.15) states that the solution is

$$u(r,\theta) \;=\; \frac{a^2-r^2}{2\pi}\int_0^{2\pi}\frac{1}{a^2-2ar\cos(\theta-\psi)+r^2}\,\mathrm{d}\psi.$$

Evaluating this integral directly is complicated, which is frustrating given the apparent simplicity of the boundary condition. Intuitively, we would expect the temperature within the disc to be a constant function, $u(r,\theta) = 1$ for $0 \le r \le a$ and $0 \le \theta < 2\pi$.

The problem is much simpler if we retrace our steps in the derivation of (13.15), instead referring to equation (13.14):

$$u(r,\theta) \;=\; \frac{1}{2\pi}\int_0^{2\pi} h(\psi)\,\mathrm{d}\psi + \frac{1}{\pi}\sum_{n=1}^{\infty}\frac{r^n}{a^n}\int_0^{2\pi} h(\psi)\cos[n(\theta-\psi)]\,\mathrm{d}\psi$$

for $0 \le r < a$ and $0 \le \theta < 2\pi$. With $h(\psi) = 1$, we find that

$$u(r,\theta) = \frac{1}{2\pi}\int_0^{2\pi} d\psi + \frac{1}{\pi}\sum_{n=1}^{\infty}\frac{r^n}{a^n}\int_0^{2\pi}\cos[n(\theta-\psi)]\,d\psi.$$

The integrals are easy to calculate:

$$\int_0^{2\pi} d\psi = 2\pi$$

and

$$\begin{aligned}\int_0^{2\pi}\cos[n(\theta-\psi)]\,d\psi &= -\left.\frac{\sin[n(\theta-\psi)]}{n}\right|_0^{2\pi}\\ &= -\frac{1}{n}\left[\sin(n\theta-2\pi n)-\sin(n\theta)\right] = 0 \qquad (n \ge 1).\end{aligned}$$

Therefore, all of the integrals appearing in the summation are zero, implying that the solution of the Dirichlet problem is $u(r,\theta) = 1$ throughout the disc-shaped domain (as expected).

**Example 13.4.2.** Solve the Dirichlet problem

$$\begin{aligned}\Delta u &= 0 & 0 &\le r < a\\ u(a,\theta) &= \cos(\theta) & r &= a.\end{aligned}$$

*Solution:* This time, we are given a non-constant heat distribution on the circular boundary of the domain. If we attempt to apply Poisson's formula directly, we obtain

$$u(r,\theta) = \frac{a^2-r^2}{2\pi}\int_0^{2\pi}\frac{\cos(\psi)}{a^2-2ar\cos(\theta-\psi)+r^2}\,d\psi$$

for $0 \le r < a$ and $0 \le \theta < 2\pi$. As in the preceding example, we encounter an integral that appears difficult to evaluate directly. Instead, we quote (13.14) where, in this case, $h(\psi) = \cos\psi$:

$$u(r,\theta) = \frac{1}{2\pi}\int_0^{2\pi}\cos\psi\,d\psi + \frac{1}{\pi}\sum_{n=1}^{\infty}\frac{r^n}{a^n}\int_0^{2\pi}\cos\psi\,\cos[n(\theta-\psi)]\,d\psi.$$

The leading term is easy to evaluate:

$$\frac{1}{2\pi}\int_0^{2\pi} \cos\psi \, d\psi \;=\; 0.$$

The integrals appearing in the summation are reminiscent of the ones we encountered during our study of Fourier series. It helps to apply the trigonometric identity

$$\cos\alpha\cos\beta \;=\; \frac{1}{2}\left[\cos(\alpha+\beta)+\cos(\alpha-\beta)\right]$$

to rewrite the integrals as

$$\begin{aligned}\int_0^{2\pi} \cos\psi\, \cos[n(\theta-\psi)]\, d\psi \;&=\; \int_0^{2\pi} \cos\psi\, \cos(n\theta - n\psi)\, d\psi \\ &= \frac{1}{2}\int_0^{2\pi} \cos[n\theta-(n-1)\psi)]\, d\psi \;+\; \frac{1}{2}\int_0^{2\pi} \cos[-n\theta+(n+1)\psi)]\, d\psi.\end{aligned}$$

When calculating antiderivatives of the integrands, we must distinguish between the case $n>1$ and $n=1$. If $n>1$, then these integrals evaluate to

$$\begin{aligned}&\frac{1}{2}\,\frac{\sin[n\theta-(n-1)\psi)]}{-(n-1)}\bigg|_0^{2\pi} \;+\; \frac{1}{2}\,\frac{\sin[-n\theta+(n+1)\psi)]}{n+1}\bigg|_0^{2\pi} \\ &\quad=\; \frac{\sin[n\theta-2\pi(n-1)]-\sin(n\theta)}{-2(n-1)} \;+\; \frac{\sin[-n\theta+2\pi(n+1)]-\sin(-n\theta)}{2(n+1)}.\end{aligned}$$

Since $2\pi(n-1)$ and $2\pi(n+1)$ are integer multiples of $2\pi$, this entire expression reduces to 0.

Finally, if $n=1$ we evaluate

$$\frac{1}{2}\int_0^{2\pi} \cos\theta\, d\psi \;+\; \frac{1}{2}\int_0^{2\pi} \cos(2\psi-\theta)\, d\psi \;=\; \pi\cos\theta,$$

the only non-zero integral in our infinite series representation of $u(r,\theta)$. The series collapses to a single term:

$$u(r,\theta) \;=\; \frac{1}{\pi}\left(\frac{r}{a}\right)\pi\cos\theta \;=\; \left(\frac{r}{a}\right)\cos\theta,$$

the solution of our Dirichlet problem for $0\le r\le a$ and $0\le\theta<2\pi$.

Take a moment to interpret this solution. The factor $\cos\theta$ comes from our boundary condition, while the factor $(r/a)$ varies from 0 to 1 as we move radially outward from the center of our disc toward the boundary. The temperature at the center of the disc ($r = 0$) is zero, which happens to be the average of the temperature along the circular boundary:

$$\frac{1}{2\pi}\int_0^{2\pi} h(\psi)\,\mathrm{d}\psi \;=\; \frac{1}{2\pi}\int_0^{2\pi} \cos\psi\,\mathrm{d}\psi \;=\; 0.$$

This is no accident, as we now show.

If we set $r = 0$ in Poisson's formula, we find that the temperature at the center of the disc is

$$u(0,\theta) \;=\; \frac{a^2}{2\pi}\int_0^{2\pi} \frac{h(\psi)}{a^2}\,\mathrm{d}\psi \;=\; \frac{1}{2\pi}\int_0^{2\pi} h(\psi)\,\mathrm{d}\psi,$$

the average value of the temperature along the boundary $r = a$. This observation is a special case of a remarkable property of solutions of Laplace's equation:

**Theorem 13.4.3. Mean value property for harmonic functions:** Suppose that $u$ is a harmonic function on an open disc $D$ and is continuous on the closed disc $D \cup \partial D$. Then the value of $u$ at the center of the disc is equal to the average value of $u$ on the boundary $\partial D$.

**Example 13.4.4.** Suppose that $u$ is a solution of the Dirichlet problem (13.11) with boundary condition $h(\theta) = 10 + \theta\sin\theta$. Then the value of $u$ at the origin is given by

$$\frac{1}{2\pi}\int_0^{2\pi} 10 + \theta\sin\theta\,\mathrm{d}\theta \;=\; 10 + \frac{1}{2\pi}\int_0^{2\pi} \theta\sin\theta\,\mathrm{d}\theta.$$

$$= 10 + \left.\frac{-\theta\cos\theta + \sin\theta}{2\pi}\right|_0^{2\pi} = 9.$$

Finding the maximum and minimum temperatures on this disc is tricky. By the Maximum Principle 13.2.1, we know that the maximum and minimum temperatures must be achieved on the boundary. Thus, we must find the extreme values of $h(\theta) = 10 + \theta\sin\theta$ on the interval $0 \le \theta \le 2\pi$. Since $h'(\theta) = \sin\theta + \theta\cos\theta$, critical points occur when $\theta = -\tan\theta$. Unfortunately, it is impossible to solve this transcendental equation by hand, and we would need to resort to computer assistance. It turns out that there are two critical points in the open interval $\theta \in (0, 2\pi)$, one of which corresponds to the global minimum temperature, and one of which corresponds to the global maximum temperature.

**Additional properties of harmonic functions.** Poisson's formula (13.15) and the mean value property (13.4.3) lead to a host of important theorems that characterize the behavior of harmonic functions. We state a few such results for reference—see the texts of McOwen [6] and Strauss [10] for technical proofs.

☞ **Smoothness Theorem:** *Suppose $u$ is a harmonic function on an open set $\Omega$. Then $u$ is smooth (i.e., has derivatives of all orders) on $\Omega$.* The Smoothness Theorem may seem surprising given that Laplace's equation $\Delta u = 0$ only requires existence of the second partial derivatives of $u$. In this sense, the class of harmonic functions is far more "selective" than we might expect.

☞ **Harnack's Theorem:** *Suppose that $u$ is a non-negative harmonic function on an open set $\Omega$, and let $\Omega_1$ be a closed subset of $\Omega$. Then there exists a constant $C$ depending only on $\Omega_1$ such that*

$$\max_{\Omega_1} u \;\le\; C \min_{\Omega_1} u.$$

The main implication of Harnack's Theorem is that *non-negative* harmonic functions cannot exhibit crazy oscillations on closed sets.

☞ **Liouville's Theorem:** *Suppose that $u$ is a harmonic function in the whole plane $\mathbb{R}^2$ (or in the whole space $\mathbb{R}^3$). If $u$ is bounded, then $u$ is a constant function.*

**Appendix: Solving Euler equations.** Consider the linear, homogeneous ODE

$$t^2 u''(t) + \beta t u'(t) + \gamma u(t) = 0,$$

where $\beta$ and $\gamma$ are constants. Notice that the exponents of the coefficient functions involve powers of the independent variable: $t^2$, $t^1$ and $t^0$. For this reason, it is natural to seek power function solutions of the form $u(t) = t^m$, because differentiating such functions reduces the exponent by one (unless $m = 0$). Inserting $u = t^m$ into the above equation yields

$$t^2 m(m-1)t^{m-2} + \beta t m t^{m-1} + \gamma t^m \;=\; 0,$$

which simplifies to

$$t^m \left[ m(m-1) + \beta m + \gamma \right] \;=\; 0.$$

Since the left hand side is identically equal to zero as a function of $t$, it must be the case that

$$m(m-1) + \beta m + \gamma \;=\; 0.$$

This quadratic equation for $m$ is called an *indicial equation* for our Euler equation, and is analogous to a characteristic equation for a constant-coefficient ODE. If the indicial equation has distinct, real roots $m = r$ and $m = s$, then the functions $t^r$ and $t^s$ are linearly independent solutions of the original ODE. By linearity, the general solution would then be given by

$$u(t) = C_1 t^r + C_2 t^s,$$

where $C_1$ and $C_2$ are arbitrary constants.

---

## Exercises

1. Consider the heat equation $u_t = \kappa u_{xx}$ on the interval $0 \leq x \leq L$ with Dirichlet boundary conditions $u(0,t) = \tau_1$ and $u(L,t) = \tau_2$. Find the steady-state solution of this Dirichlet problem, and give a mathematical/physical interpretation of the result.

2. Solve $u_{xx} + u_{yy} + u_{zz} = 0$ in the spherical shell $1 < r < 4$ with the Dirichlet boundary conditions $u = \tau_1$ when $r = 1$ and $u = \tau_2$ when $r = 4$. Here, $\tau_1$ and $\tau_2$ are constants, and $r = \sqrt{x^2 + y^2 + z^2}$ measures distance from the origin.

3. Find the *bounded* solution of $u_{xx} + u_{yy} = 3\sqrt{x^2 + y^2}$ in the disc $0 \leq r < 3$ with $u = 4$ on the boundary $r = 3$. As usual, $r = \sqrt{x^2 + y^2}$.

4. Solve $u_{xx} + u_{yy} + u_{zz} = 6$ in the spherical shell $2 < r < 3$ with $u(x, y, z) = 4$ on both the inner and outer shells of the boundary.

5. Solve $u_{xx} + u_{yy} + u_{zz} = 20(x^2 + y^2 + z^2)$ in the spherical shell $a < r < b$ with the Dirichlet condition $u = 0$ on the inner boundary $r = a$ and the Neumann condition $\partial u / \partial r = 0$ on the outer boundary $r = b$.

6. Solve $u_{xx} + u_{yy} = 0$ inside the rectangular domain $0 < x < a$ and $0 < y < b$ with Dirichlet boundary conditions $u = 0$ on the top, bottom, and left edges of the rectangle, but with $u(a, y) = R(y)$ on the right edge.

7. Solve $u_{xx} + u_{yy} = 0$ inside the square domain $0 < x < 1$ and $0 < y < 1$ with boundary conditions $u = 0$ on the bottom, left, and right edges of the rectangle, but with the *Neumann* condition $u_y(x, 1) = T(x)$ on the top edge.

8. Solve $u_{xx} + u_{yy} = 0$ inside the square domain $0 < x < 2$ and $0 < y < 2$ with homogeneous Dirichlet boundary conditions $u = 0$ on the bottom, left, and top edges of the rectangle, and with the inhomogeneous Dirichlet condition

$u(2, y) = 1$ on the right edge. Then, repeat the calculation with $u(2, y) = 1$ replaced by the inhomogeneous Neumann condition $u_x(2, y) = 1$ on the right edge.

9. Suppose $D$ is the open disc $r < 7$ (centered at the origin), $\bar{D}$ is the corresponding closed disc $r \leq 7$, and $\partial D$ denotes the boundary $r = 7$. Suppose that $u$ is a harmonic function inside $D$ and that $u = 50 + 10 \sin 2\theta + 10 \cos 2\theta$ on $\partial D$. *Without* solving for $u$, determine

   (a) the maximum value of $u$ in $\bar{D}$, and

   (b) the value of $u$ at the origin.

10. Solve $u_{xx} + u_{yy} = 0$ in the disc $r < a$ with the boundary condition $u = 8 - 5 \sin \theta$ when $r = a$. *Hint:* You may find it difficult to use Poisson's formula directly, so refer to (13.14) instead.

# Guide to Commonly Used Notation

| Symbol | Usual Meaning |
|---|---|
| $\mathbb{R}$ | the set of real numbers |
| $\mathbb{R}^n$ | $n$-dimensional Euclidean space |
| $t$ | independent variable (time) |
| $\mathbf{u}, \mathbf{v}$ | vectors consisting of real numbers |
| $\mathbf{x}, \mathbf{y}$ | vectors consisting of dependent variables |
| $\mathbf{x}_0, \mathbf{y}_0$ | vectors consisting of initial conditions $\mathbf{x}(0)$, $\mathbf{y}(0)$ |
| $\mathbf{x}^*, \mathbf{y}^*$ | equilibrium solution of a system of ODEs |
| $\mathbf{f}$ | a vector consisting of functions |
| $A, M, P$ | square $n \times n$ matrices |
| $D$ | a diagonal $n \times n$ matrix |
| $N$ | a nilpotent $n \times n$ matrix |
| $\text{tr}(A)$ | trace of a square matrix $A$ |
| $\det(A)$ | determinant of a square matrix $A$ |
| $\lambda$ | eigenvalue of a matrix, or a Lyapunov exponent |
| $\alpha, \beta$ | real, imaginary parts of an eigenvalue $\lambda = \alpha + \beta i$ |
| $E^s, E^u, E^c$ | stable, unstable, and center subspaces |
| $W^s, W^u, W^c$ | stable, unstable, and center manifolds |
| $A \oplus B$ | direct sum of subspaces $A$ and $B$ |
| $\text{span}\{\mathbf{v}_1, \mathbf{v}_2, \ldots \mathbf{v}_n\}$ | span of vectors $\mathbf{v}_1, \mathbf{v}_2, \ldots \mathbf{v}_n$ |
| $\phi_t(\mathbf{x}_0)$ | solution of $\mathbf{x}' = f(\mathbf{x})$ with $\mathbf{x}(0) = \mathbf{x}_0$ |
| $\phi_t$ | flow of a system of ODEs |
| $f : \mathbb{R}^n \to \mathbb{R}^m$ | a function from $\mathbb{R}^n$ into $\mathbb{R}^m$ |
| $\nabla f$ | gradient of a function from $\mathbb{R}^n$ into $\mathbb{R}$ |
| $Jf$ | Jacobian matrix of a function from $\mathbb{R}^n$ into $\mathbb{R}^m$ |
| $\mathbf{u} \bullet \mathbf{v}$ | dot product of vectors $\mathbf{u}$ and $\mathbf{v}$ |
| $\|\mathbf{v}\|_2$ | Euclidean norm (length) of a vector $\mathbf{v}$ in $\mathbb{R}^n$ |
| $\|\mathbf{u} - \mathbf{v}\|_2$ | Euclidean distance between points $\mathbf{u}$ and $\mathbf{v}$ in $\mathbb{R}^n$ |
| $B(\mathbf{x}, \epsilon)$ | open ball of radius $\epsilon$ centered at $\mathbf{x}$ |
| $V(\mathbf{x})$ | a Lyapunov function |
| $\Gamma(t)$ | a parametrized curve, $\Gamma : \mathbb{R} \to \mathbb{R}^n$ |

| **Symbol** | **Usual Meaning** |
|---|---|
| $\mathbf{n}$ | a normal vector to a parametrized curve |
| $r, \theta$ | radius and angle (polar coordinates) |
| $\mu$ | a bifurcation parameter |
| $\sigma$ | Lyapunov number (associated with Hopf bifurcations) |
| $\tau$ | time delay (for delay differential equations) |
| $x(a^+)$ | the right-hand limit $\lim_{t \to a^+} x(t)$ |
| $x(a^-)$ | the left-hand limit $\lim_{t \to a^-} x(t)$ |
| $x^*$ | a fixed-point of a first-order difference equation |
| $Sf(x)$ | Schwarzian derivative of a function $f : \mathbb{R} \to \mathbb{R}$ |
| $\gamma$ | feedback gain parameter in the TDAS algorithm |
| $L$ | linear operator, or length of the interval $[0, L]$ |
| $\phi(x)$ | initial condition for a PDE |
| $\partial\Omega$ | boundary of the domain $\Omega$ |
| $\kappa$ | positive-valued diffusion constant (heat equation) |
| $c$ | wave speed (wave/transport equations) |
| $\psi(x)$ | initial velocity (wave equation) |
| $S \star \phi$ | convolution of functions $S$ and $\phi$ |
| $S(x, t)$ | (one-dimensional) heat kernel |
| $\phi_{\text{odd}}, \phi_{\text{even}}$ | odd, even extensions of a function $\phi$ |
| $X(x), Y(y), T(t)$ | separated solutions of PDEs (separation of variables) |
| $R(r), \Theta(\theta)$ | separated solutions of PDEs (polar coordinates) |
| $\beta$ | abbreviation for square roots of eigenvalues $\lambda$ |
| $A_n, B_n$ | Fourier coefficients |
| $\\|f\\|_\infty, \\|f\\|_{L^2}$ | different types of norms of the function $f$ |
| $\langle f, g \rangle$ | inner product of functions $f$ and $g$ |
| $S_N(x)$ | partial sum of an infinite series of functions |
| $f(x_0^-), f(x_0^+)$ | left and right-hand limits of $f(x)$ as $x \to x_0$ |
| $f'(x_0^-), f'(x_0^+)$ | left and right-hand derivatives of $f(x)$ as $x \to x_0$ |
| $\Delta$ | Laplace operator |
| $v(r)$ | a radial solution of Laplace's or Poisson's equation |
| $h(\theta)$ | boundary condition for Laplace's equation on a disc |

# References


[1] G. Bachman, L. Narici, and E. Beckenstein, *Fourier and wavelet analysis*, Springer-Verlag, New York, 2000.

[2] R. Bellman and K. L. Cooke, *Differential-difference equations*, Academic Press, New York, 1963.

[3] S. Elaydi, *An introduction to difference equations*, Springer, New York, 1999.

[4] B.D. Hassard, N.D. Kazarinoff, and Y.-H. Wan, *Theory and applications of hopf bifurcation*, Cambridge University Press, New York, 1981.

[5] J. H. Hubbard and B. H. West, *Differential equations: A dynamical systems approach*, Springer-Verlag, New York, 1997.

[6] R. C. McOwen, *Partial differential equations: Methods and applications*, Prentice Hall, Upper Saddle River, 2003.

[7] J. D. Meiss, *Differential dynamical systems*, SIAM, Philadelphia, 2007.

[8] L. Perko, *Differential equations and dynamical systems*, Springer-Verlag, New York, 2001.

[9] J. E. S. Socolar, D. W. Sukow, and D. J. Gauthier, *Stabilizing unstable periodic orbits in fast dynamical systems*, Physical Review E **50** (1994), 3245–3248.

[10] W. A. Strauss, *Partial differential equations: An introduction, 2nd ed.*, Wiley, New York, 2007.

[11] S. Strogatz, *Nonlinear dynamics and chaos*, Addison-Wesley, 1994.

# Index

www.ingramcontent.com/pod-product-compliance
Lightning Source LLC
LaVergne TN
LVHW010940100826
845153LV00001B/99